Advances in Mapping Land Cover and Land Use Based on Remotely Sensed Data

Advances in Mapping Land Cover and Land Use Based on Remotely Sensed Data

Guest Editors

Yaqian He
Fang Fang
Christopher A. Ramezan

MDPI

Basel • Beijing • Wuhan • Barcelona • Belgrade • Novi Sad • Cluj • Manchester

Guest Editors

Yaqian He
Department of Geography
Indiana University Bloomington
Bloomington
USA

Fang Fang
School of Public Policy and Urban Affairs
Northeastern University
Boston
USA

Christopher A. Ramezan
Department of Management Information Systems
West Virginia University
Morgantown
USA

Editorial Office
MDPI AG
Grosspeteranlage 5
4052 Basel, Switzerland

This is a reprint of the Special Issue, published open access by the journal *Remote Sensing* (ISSN 2072-4292), freely accessible at: https://www.mdpi.com/journal/remotesensing/special_issues/97ZA9O7R4N.

For citation purposes, cite each article independently as indicated on the article page online and as indicated below:

Lastname, A.A.; Lastname, B.B. Article Title. *Journal Name* **Year**, *Volume Number*, Page Range.

ISBN 978-3-7258-7170-4 (Hbk)
ISBN 978-3-7258-7171-1 (PDF)
https://doi.org/10.3390/books978-3-7258-7171-1

Contents

remote sensing

MDPI

Article

WenSiM: A Relative Accuracy Assessment Method for Land Cover Products Based on Optimal Transportation Theory

Rui Zhu [1], Yumin Tan [1], Ziqing Luo [1], Yanzhe Shi [1], Jiale Wang [1], Guifei Jing [2] and Xiaolu Wang [1,*]

[1] School of Transportation Science and Engineering, Beihang University, Beijing 102206, China; buaa_zhurui@buaa.edu.cn (R.Z.); tanym@buaa.edu.cn (Y.T.); zqluo@buaa.edu.cn (Z.L.); syz_cannot@buaa.edu.cn (Y.S.); wangjiale13@buaa.edu.cn (J.W.)

[2] Research Institute for Frontier Science, Beihang University, Beijing 100191, China; guifeijing@buaa.edu.cn

* Correspondence: zy2113207@buaa.edu.cn

Abstract: Land cover (LC) products play a crucial role in various fields such as change detection, resource management, and urban planning. The diversity in methods and principles used to create different products poses a challenge for researchers in choosing the most suitable one for research needs. Mainstream evaluation methods typically consider only a portion of the accuracy information from the product and require a significant effort in creating validation datasets. Here, we propose a relative accuracy assessment method for LC products based on optimal transport theory, which provides a comprehensive evaluation by utilizing a broader range of accuracy information within the product. The method can directly compute the similarity between the target product and the reference truth at a global scale, addressing the issue of quantitatively assessing product accuracy in the absence of a validation dataset. To validate the effectiveness of the method, we select Beijing as the study area to assess the accuracy of four LC products. The results suggest that the method allows for precise quantification of product accuracy, aligning closely with validation outcomes, which can provide valuable guidance to researchers in both product creation and selection.

Keywords: accuracy assessment; land cover products; optimal transportation; relative accuracy

1. Introduction

Land cover (LC) products, as common remote sensing products, depict the interactions between human activities and the natural environment, which are widely applied in environmental change monitoring, ecosystem assessment, and sustainable development planning [1–3]. With the rapid advancement of satellite remote sensing technology and the continuous improvement in data accessibility, LC products continue to evolve and be updated. However, due to differences in the classification algorithms and schemes of products, researchers find it challenging to directly assess the quality of these products [4]. The accuracy assessment and rational selection of LC products have become key issues in research [5,6]. It is a widely held view that the absolute ground truth does not exist, making it impossible to directly compare remote sensing products with it [7].

Previous research has found that a validation dataset can be created for the study area using a full-sampling method and make the dataset I reference truth to evaluate the overall accuracy of LC products. But the high cost of creating the dataset has made the full-sampling method rarely used directly [8]. To strike a balance between cost and accuracy, a variety of studies have proposed two major categories of accuracy assessment methods: direct assessment methods and indirect assessment methods. The direct assessment method involves first obtaining samples through random sampling and then creating a validation dataset through field surveys or visual interpretation based on high-resolution satellite images such as Google Earth [9]. Finally, a confusion matrix is computed to determine the overall accuracy of the product [10]. Although the method saves costs, the number and methods of sampling can introduce errors and lead to unstable results. Furthermore, most validation datasets are not interoperable, making it difficult for researchers to select the highest-quality one. Moreover, the rationale of relying solely on one validation dataset to assess the accuracy of all LC products need to be questioned [11]. To avoid the creation of validation datasets, indirect evaluation methods that compare the similarity between products have been reported. Unlike direct assessment, the indirect assessment method evaluates relative accuracy by analyzing differences between products, primarily through regional consistency and spatial consistency analysis [12–14]. Specifically, regional consistency involves statistically assessing the composition of different land cover types within the study area [15]. Spatial consistency compares whether different products have the same land cover type at the same location through pixel-by-pixel spatial overlay analysis [16]. The number of matching pixels indicates spatial consistency between products, implying the degree of similarity [17]. However, it is important to note that the method primarily provides qualitative assessments and may not offer a definitive ranking of product quality. The methods for assessing the accuracy of LC products described above can be summarized in Figure 1 based on their distinctive characteristics and developmental trajectories. As can be seen from the figure, the full-sampling method is rarely used due to its high cost. Although direct assessment methods and indirect assessment methods are cost-effective, they result in the loss of a significant amount of information accuracy.

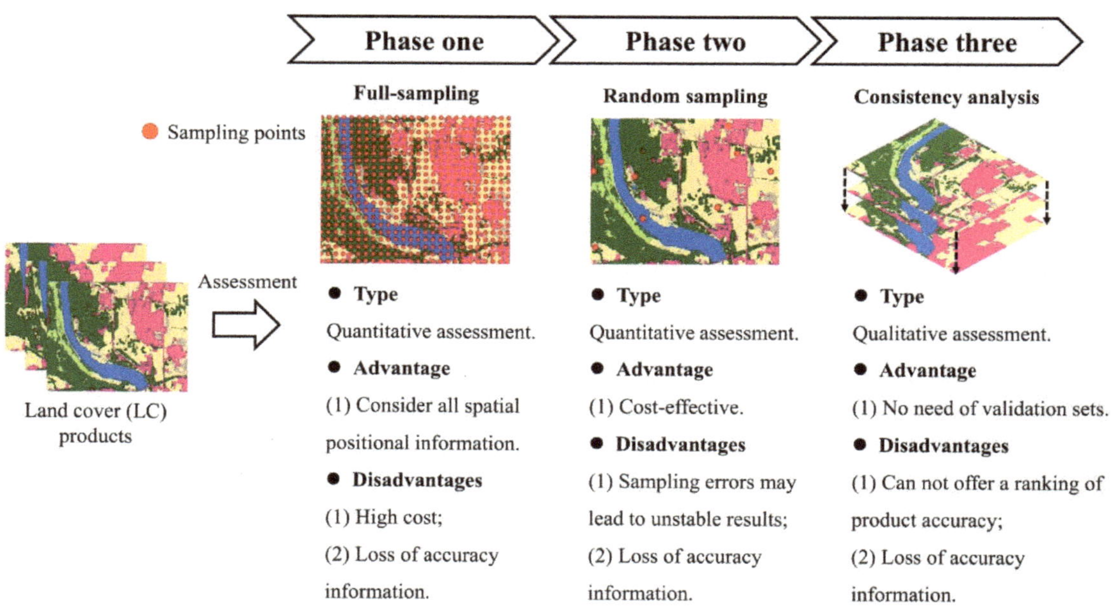

Figure 1. The characteristics and development history of accuracy assessment methods for LC products.

This paper argues that the accuracy information determined by LC products can be divided into two types: spatial position information and global feature information. Most studies that assess LC product accuracy have only focused on spatial position information, which means the classification correctness of each pixel, disregarding the distribution of features formed by all pixels [18,19]. Every land cover type exhibits distinctive morphologic characteristics, which are expressed through the arrangement of pixels. Therefore, even with the absolute ground truth, the methods depicted in Figure 1 are still unable to fully exploit the accuracy information of products and the ground truth for a comprehensive assessment. This paper extends the research of Tan et al. [20], which primarily took account of the global features of pixels to quantitatively compare the similarity of remote sensing products. However, the similarity index proposed in Tan's study also only considers a portion of the accuracy information in the products and may be challenging to explain in some application scenarios. In this paper, we attempt to utilize more accuracy information to assess LC products and ensure that the new similarity index maintains strict mathematical logic and interpretability, revealing both commonalities and differences among products. The new metric will assist researchers in making optimal selections quickly from LC products.

The primary innovations and contributions of this paper include:

- Proposal of a relative accuracy assessment method for LC products based on optimal transport theory. The method considers a broader range of accuracy information to compute the Wasserstein similarity (WenSiM) between the target product and the reference ground truth at a global level without the need for registration and validation datasets.
- Evaluation of the classification accuracy of four LC products in Beijing. Through a comparative analysis of WenSiM and the results obtained from mainstream approaches, the effectiveness of the relative accuracy assessment method is validated.

2. Data

2.1. Data Preprocessing

This study made use of Dynamic World [21], ESA WorldCover v100 [22], FROM_GLC10 [23], and SinoLC_1 [24] in Beijing to conduct accuracy assessments. ESRI World Cover [25] was chosen as the reference ground truth because of the highest overall official validation accuracy. Table 1 shows the basic information for every LC product.

Table 1. Basic information of the five LC products.

	ESRI_GLC10	DW_GLC10	ESA_GLC10	FROM_GLC10	SinoLC_1
Institution	Esri	WRI	ESA	THU	WHU
Resolution (m)	10	10	10	10	1
Coverage	Global	Global	Global	Global	National (China)
Classification scheme	10 classes	9 classes	11 classes	10 classes	11 classes
Version and timeline	2020	2020	2020	2017	2020
Overall accuracy (%)	85.96	77.80	74.40	72.76	73.61 55.55 (Beijing)

Notes: ESRI = Environmental Systems Research Institute; WRI = World Resources Institute; ESA = European Space Agency; THU = Tsinghua University; WHU = Wuhan University.

In contrast to common data preprocessing procedures, this study only dealt with resampling and reclassifying the products without requiring georeferencing. According to Table 1, only the SinoLC_1 product had a resolution of 1 m. To facilitate a more comprehensive comparison of each product's accuracy and ensure fairness in the process, we standardized the resolutions of the five products to 10 m. It is also important to emphasize that, although the FROM_GLC10 was produced in 2017, existing research has demonstrated that, compared to classification errors in LC products, land cover changes in large regions within a five-year timeframe can be negligible [26,27]. Additionally, for the DW_GLC10, as a continuously updated image collection, the study utilized the confusion matrix results obtained during the testing process as its overall accuracy. Finally, the land cover categories for products were reclassified into forestland, cropland, grassland, wetland, settlements, and other land according to the classification rules of the Intergovernmental Panel on Climate Change (IPCC), with specific classification criteria and details provided in Tables A1 and A2.

2.2. Study Area

The study area was Beijing (Lat 39°56′N, Long 116°20′E), which covers an area of 1,641,054 hectares and is renowned as both an ancient capital and a modern international city. The city is surrounded by mountains on its western, northern, and northeastern sides, whereas the southeastern part consists of a plain that slopes toward the Bohai Sea. Detailed geographic information about Beijing, as well as the visual comparative effects of LC products in Changping District, are depicted in Figure 2.

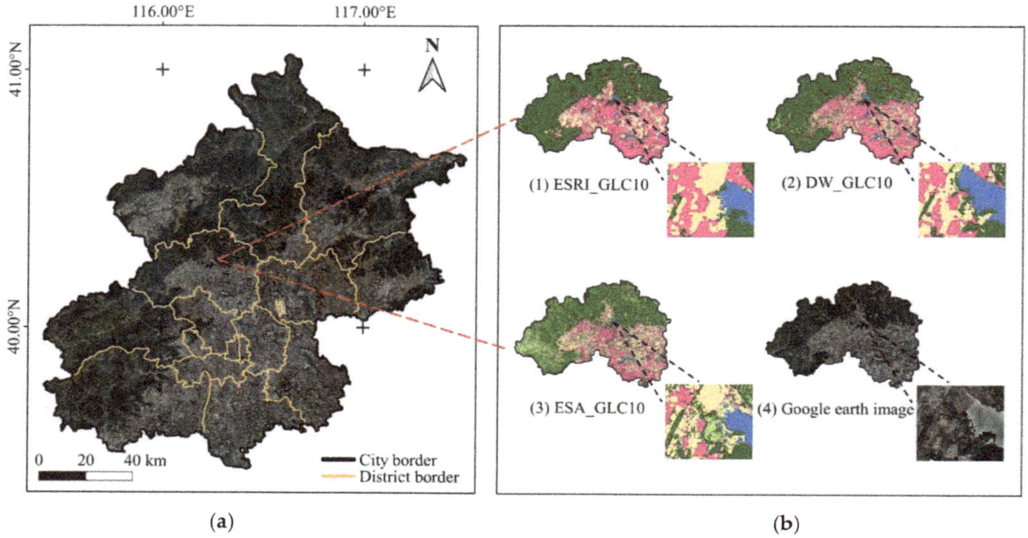

Figure 2. (**a**) Geographical location information of Beijing. (**b**) Depictions of the visual comparison of Changping District.

3. Methods

The section begins by highlighting the theoretical underpinnings of the relative accuracy assessment method. Following that, the section elaborates on the characteristics and computational process of the WenSiM index. Finally, the section introduces the creation of a validation dataset for the study area by using stratified sampling theory.

3.1. Wasserstein-p Distance

The Wasserstein distance (earth mover's distance, EMD) is a type of histogram similarity measure [28,29]. Compared to other similarity measures, like the Kullback–Leibler (KL) divergence and the Jensen–Shannon (JS) divergence, the Wasserstein distance has the advantage of being able to assess the similarity between two probability distributions that do not overlap at all. In addition, its value represents the minimum "cost" when two distributions are transformed into each other.

Let ϕ be a metric space. For this study, we considered ϕ to be compact d-dimensional Euclidean spaces, i.e., $\phi = [0,1]^d$. Let ϕ_μ and ϕ_v denote the set of Borel probability measures defined on ϕ. The Wasserstein-p distance for $p \in [1, \infty)$ between two distributions $P_\mu \in \phi_\mu$ and $P_v \in \phi_v$ can be defined as Equation (1) with the cost functions $c(x,y) = d^p(x,y)$.

$$W_p(P_\mu, P_v) \triangleq \left\{ \inf_{\gamma \in \Gamma(P_\mu, P_v)} \int_{\phi_\mu \times \phi_v} d^p(x,y) d\gamma(x,y) \right\}^{\frac{1}{p}} \quad (1)$$

Here, $\Gamma(P_\mu, P_v)$ is the set of all transportation plans $\gamma(x,y)$ whose marginals are P_μ and P_v, respectively, $\gamma \in \Gamma(P_\mu, P_v)$. Additionally, $d^p(x,y)$ represents the "cost" associated with transforming x in probability distribution P_μ into y in P_v, and p is the dimension size of the probability distribution. Thus, the Wasserstein p-distance reflects the "cost" of the optimal transportation plan.

According to Brenier's theorem [30], if P_μ and P_ν (with respect to the Lebesgue measure) are absolutely continuous probability measures, the Wasserstein p-distance can be equivalently calculated using Equation (2).

$$W_p(P_\mu, P_\nu) = \left\{ \inf_{f \in MP(P_\mu, P_\nu)} \int_{\phi_\mu} d^p(f(x), x) dP_\mu(x) \right\}^{\frac{1}{p}} \quad (2)$$

where $MP(P_\mu, P_\nu) = \left\{ f : \phi_\mu \to \phi_\nu \middle| f_{\#P_\mu} = P_\nu \right\}$ and $f_{\#P_\mu}$ is used to indicate the pushforward of measure P_μ.

Since LC products consist of pixels, they belong to a special two-dimensional discrete probability distribution and can be calculated numerically using a matrix. Thus, we focused on how to calculate the Wasserstein distance of absolute discrete probability measures in various dimensions.

The process of calculating higher-dimensional Wasserstein distance is difficult and complicated. But the development of the sliced Wasserstein distance provides a closed-form solution to the Wasserstein-p distance calculation problem [31]. By reducing the probability distribution to one dimension through random projection, computing the Wasserstein distance becomes fast and straightforward [32,33].

The idea behind the sliced Wasserstein distance is to obtain the marginal distribution family (i.e., one-dimensional distribution) of high-dimensional probability distribution through linear random projection and then calculate the Wasserstein distance of two marginal distributions. The method transforms the challenging high-dimensional optimal transport problem into several one-dimensional optimal transport problems with closed-form solutions. Then, the sliced Wasserstein-p distance between two probability distributions P_μ and P_ν can be defined as in Equation (3).

$$\widetilde{W}_p(P_\mu, P_\nu) = \left[\int_{\theta \in \Omega} W_p^p \left(P_\mu^\theta, P_\nu^\theta \right) d\theta \right]^{\frac{1}{p}} \quad (3)$$

where P_μ^θ and P_ν^θ denote all projections of marginal distributions P_μ and P_ν on the direction θ, respectively, and Ω is the set of all possible directions on the unit sphere. It has also been proven that $\widetilde{W}_p$ satisfies sub-additivity and coincidence axioms, making it a genuine metric [34].

Due to the various direction selections, calculating the corresponding projections during the actual process is still very complicated. To address the challenge of projection complexity while keeping the sample complexity constant, Deshpande et al. [35] proposed a method to first find the most meaningful projection direction. The result obtained by Equation (4) in the projection direction θ is known as the max sliced Wasserstein distance.

$$\widetilde{W}_{2-max}(P_\mu, P_\nu) = \left[\max_{\theta \in \Omega} W_2^2 \left(P_\mu^\theta, P_\nu^\theta \right) \right]^{\frac{1}{2}} \quad (4)$$

$\widetilde{W}_{2-max}$ is an effective measure that overcomes the limitations of projection complexity. It can directly compute the difference between two-dimensional distributions. Therefore, in this study, it was employed to calculate the Wasserstein distance between LC products and the reference ground truth within the study area.

3.2. Wasserstein Similarity Index

This study divided the accuracy information carried by the LC products into two types: global feature information and spatial position information. Although the Wasserstein distance is effective in quantifying the similarity between products and the reference truth from a global feature perspective, its interpretability is affected because the total weight is evenly distributed among all pixels. Therefore, we normalized $\widetilde{W}_{2-max}$ to ensure that the global feature similarity $\mathcal{F}$ had a strict mathematical significance and was independent of the size of the study area.

The existence of target product A and reference truth B was assumed, whose matrix spaces are denoted as A_μ and B_v, respectively. In the study area, there was a total of k land cover types, and the transformed numerical matrices satisfied $P_\mu \in A_\mu$ and $P_v \in B_v$. Then, $\mathcal{F}$ can be defined as Equation (5).

$$\mathcal{F}_i = 1 - \frac{\widetilde{W}_{2-max}^{(i)}}{D^{(i)}} \tag{5}$$

where $\mathcal{F}_i$ represents the global feature similarity for the type $i(i = 1, 2, \cdots, k)$ in the study area, with $0 < \mathcal{F}_i \leq 1$. $D^{(i)}$ denoting the maximum centroid distance for the pixels of the land cover type i in A and B, which can be defined with Equation (6).

$$D^{(i)} = max\left\{d\left(n_A^{(i)}, n_B^{(i)}\right)\right\} \tag{6}$$

where $n_A^{(i)}$ and $n_B^{(i)}$ represent the pixels of the land cover type i in A and B, respectively. d denotes the centroid distance for the pixels of land cover type i in A and B. When the pixels in the two products are distributed in the diagonal corners of the area, the maximum centroid distance can be computed.

After considering the global feature information of the products comprehensively, there is also a need for an evaluation metric to assess spatial positional information. This study used the correlation coefficient to quantify the spatial position similarity between the target product and the reference truth. The correlation coefficient K can be defined with Equation (7), primarily reflecting the commonalities and differences in category composition between the product and the truth [36].

$$K = \frac{\sum_{i=1}^{k}\left(P_\mu^{(i)} - \overline{P}_\mu\right)\left(P_v^{(i)} - \overline{P}_v\right)}{\sqrt{\sum_{i=1}^{k}\left(P_\mu^{(i)} - \overline{P}_\mu\right)^2 \sum_{i=1}^{k}\left(P_v^{(i)} - \overline{P}_v\right)^2}} \tag{7}$$

Here, $P_\mu^{(i)}$ and $P_v^{(i)}$ represent the areas of type $i(i = 1, 2, \cdots, k)$ for A and B, respectively. $\overline{P}_\mu$ and $\overline{P}_v$ are the average areas of each type in A and B, respectively within the study area.

With $\mathcal{F}$ and K serving as measures for different types of accuracy information, this study further introduced the Wasserstein similarity (WenSiM) between the target product A and reference truth B. The index comprehensively extracts accuracy information to reflect the overall similarity, and it is defined as follows:

$$WenSiM = \frac{1}{k}\sum_{i=1}^{k} K\mathcal{F}_i \tag{8}$$

where the $WenSiM$ satisfies the condition $0 < WenSiM \leq 1$ and $WenSiM = 1$ only when $K = \mathcal{F}_i = 1$, indicating that the target product and the reference truth are entirely identical in both spatial position and global features within the study area.

The WenSiM index fully takes use of more accuracy information to evaluate LC products. While maintaining properties such as symmetry, non-negativity, and identity, the WenSiM also possesses strict mathematical significance and interpretability. It is an excellent metric for accurately and rapidly assessing LC product accuracy. The detailed calculation process is summarized in Algorithm 1.

Algorithm 1. $WenSiM(P_\mu, P_v, k, \theta, \Omega)$

Initialize θ, Ω
Data preprocessing for (P_μ, P_v)
Repeat
$\left(A^{(i)}, B^{(i)}, P_\mu^{(i)}, P_v^{(i)}, n_A^{(i)}, n_B^{(i)}\right) \leftarrow (P_\mu, P_v)$
$D^{(i)}, K \leftarrow \left(A^{(i)}, B^{(i)}, n_A^{(i)}, n_B^{(i)}\right)$
$\tilde{W}_{2-max}^{(i)} \leftarrow \left(P_\mu^{(i)}, P_v^{(i)}, \theta, \Omega\right)$
$\mathcal{F}^{(i)} \leftarrow \left(\tilde{W}_{2-max}^{(i)}, D^{(i)}\right)$
$WenSiM^{(i)} \leftarrow \left(K, \mathcal{F}^{(i)}\right)$
$i \leftarrow i+1$
Until $i > k$
$WenSiM \leftarrow \left(WenSiM^{(i)}, k\right)$
Output $WenSiM$

3.3. Production of Validation Dataset

The evaluation method proposed in this paper can directly assess the accuracy of LC products without registration and validation datasets. In order to demonstrate the effectiveness and rationality of the WenSiM index, it is necessary to create a validation dataset to analyze the consistency between validation accuracy and assessment results.

Sampling is a crucial step in creating a validation dataset, as the sample size and sample variance determine the accuracy of the dataset. Although increasing the sample size is beneficial for improving accuracy, it also comes with the added cost of production. The optimal sampling scheme aims to obtain the most reliable validation results with the lowest cost [37]. Therefore, once the sample size is determined, the key to improving the accuracy lies in reducing sample variance [38].

Stratified sampling (SS) is a great method to reduce variance and improve sampling accuracy when the sample size is constant [39,40]. To implement SS, it is crucial to confirm that every object in the sampling range shares similar properties. Subsequently, researchers can use a particular characteristic or rule to divide the population into L non-repeating sub-groups, each referred to as a layer. Finally, samples for each layer can be acquired through random sampling.

LC products possess both geographical and administrative attributes, allowing them to be divided into multiple levels using administrative boundaries. Therefore, researchers can divide the study area into L layers based on the characteristic. After getting sampling points through SS, an unbiased estimate of the variance $V(y_{ss})$ of all samples can be calculated by using Equation (9).

$$(y_{ss}) = \sum_{l=1}^{L} W_l^2 \cdot V(y_l) = \sum_{l=1}^{L} W_l^2 \cdot \frac{1-f_l}{m_l} S_l^2 = \sum_{l=1}^{L} \frac{W_l^2 S_l^2}{m_l} - \sum_{l=1}^{L} \frac{W_l S_l^2}{M_{ss}} \quad (9)$$

where y_{ss} represents all samples, and M_{ss} is the total population size. $W_l = \frac{M_l}{M_{ss}}$ and $f_l = \frac{m_l}{M_l}$ are the weight parameter and the sampling ratio of layer l $(l = 1, 2, \ldots, L)$, respectively. m_l and M_l denote the number of samples and pixels in layer l, respectively; y_l is the sample mean of layer l; and the population variance of layer l can be written as S_l^2.

To minimize the objective function $V(y_{ss})$, it is essential to choose an appropriate sample size and allocation method. This study adopted the Neyman optimal allocation method, which maximizes sampling accuracy by achieving a proportional distribution between m_l and $W_l S_l$. $V_{min}(y_{ss})$ can be computed by Equation (10), where m_{ss} is used to indicate the total number of samples.

$$V_{min}(y_{ss}) = \frac{1}{m_{ss}} \left(\sum_{l=1}^{L} W_l S_l \right)^2 - \frac{1}{M_{ss}} \sum_{l=1}^{L} W_l S_l^2 \quad (10)$$

The production method is based on SS theory and considers factors such as economic cost and sample quality, which involves scientifically selecting sample points to ensure that the accuracy meets research requirements. However, it is important to emphasize that the sample points selected through SS only reflect the spatial position information of the pixels and do not effectively reveal the global features formed by the pixels' interactions.

4. Results

This section is divided into two parts: the computation results of the Wasserstein similarity and the validation results obtained through the validation dataset within the study area. Ultimately, the section computes the confusion matrix to obtain the validation accuracy of each product and compares the results with the WenSiM and official validation accuracy.

4.1. Wasserstein Similarity Results

On completing data preprocessing, the product's classification scheme was modified to the IPCC format, categorizing the types into forestland, cropland, grassland, wetlands, settlements, and other land. Since the absolute ground truth did not exist, the paper selected ESRI_GLC10, with the highest overall official validation accuracy from Table 1, as the reference truth. Accuracy assessment of DW_GLC10, ESA_GLC10, FROM_GLC10, and SinoLC_10 was performed by computing the WenSiM between the remaining products and the reference truth. The results for global feature similarity and correlation coefficients are shown in Tables 2 and 3, respectively.

Table 2. The global feature similarity between four LC products and ESRI_GLC10 in Beijing.

Types	DW_GLC10	ESA_GLC10	FROM_GLC10	SinoLC_10
FL (%)	**98.05**	97.57	97.17	93.00
CL (%)	96.37	**98.66**	96.51	96.15
GL (%)	85.86	85.06	**93.09**	75.55
WL (%)	98.86	95.45	**99.10**	88.26
SL (%)	98.45	**98.72**	96.40	97.95
OL (%)	**96.02**	93.52	84.06	77.89

Note: FL = forestland; CL = cropland; GL = grassland; WL = wetlands; SL = settlements; OL = other land. The bold results represent the maximum values of all products.

Table 3. The correlation coefficient between four LC products and ESRI_GLC10 in Beijing.

	DW_GLC10	ESA_GLC10	FROM_GLC10	SinoLC_10
ESRI_GLC10	**0.9985**	0.9596	0.9611	0.9925

The bold results represent the maximum values.

What stands out in the tables is that DW_GLC10 exhibited the strongest correlation with the reference truth, with the highest global feature similarity in forestland and other land of 98.05% and 96.02%, respectively. It also can be seen from the data in Table 2 that ESA_GLC10 had the most similar global features with the reference truth in cropland and settlements, with a value of 98.66%, and in wetlands, at 98.72%. The max similarity results for grassland of 93.09% and wetlands of 99.10% were from FROM_GLC10.

Table 4 displays the WenSiM results obtained from the preliminary analysis of global feature similarity and correlation coefficients. Following the results, researchers can evaluate the product accuracy at different levels, such as for each land cover type or the overall product, to meet the various requirements. From the data in Table 4, it is apparent that FROM_GLC10 exhibited the highest WenSiM to the reference ground truth in grassland, with a value of 89.47%. The single most striking observation to emerge from the results is that DW_GLC10 was the most similar to ESA_GLC10 in all aspects except grassland. In summary, for the overall WenSiM results in Table 4, the products can be ranked in order of the overall WenSiM results as follows: DW_GLC10 > ESA_GLC10 > FROM_GLC10 > SinoLC_10.

Table 4. The WenSiM results between four LC products and ESRI_GLC10 in Beijing.

Types	DW_GLC10	ESA_GLC10	FROM_GLC10	SinoLC_10
FL (%)	**97.90**	93.63	93.40	92.31
CL (%)	**96.23**	94.68	92.76	95.43
GL (%)	85.73	81.63	**89.47**	74.99
WL (%)	**98.71**	91.60	95.25	87.60
SL (%)	**98.30**	94.73	92.65	97.22
OL (%)	**95.88**	89.74	80.80	77.31
OS (%)	**95.46**	91.00	90.72	87.48

Note: FL = forestland; CL = cropland; GL = grassland; WL = wetlands; SL = settlements; OL = other land; OS = overall WenSiM. The bold results represent the maximum values of all products.

4.2. Validation Results

Following the SS theory outlined in Section 3.3, the study initially divided Beijing into 16 layers, corresponding to each administrative district. Subsequently, a total of 2001 samples were randomly selected within the study area. Table A3 shows the sampling results and computational parameters for each district. Table A4 displays the sample numbers for each land cover type in each layer. Figure 3 illustrates the visual distribution of districts in Beijing, as well as the distribution of sample points in Changping and Chaoyang districts.

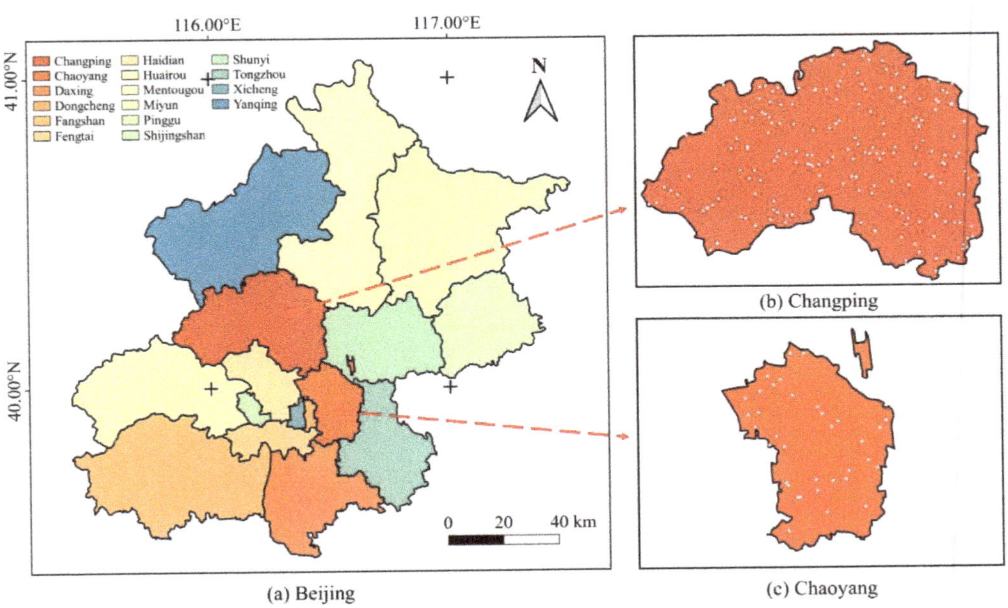

Figure 3. (**a**) Visual distribution of districts in Beijing. (**b**) Distribution of the samples in Changping District. (**c**) Distribution of the samples in Chaoyang District.

In the follow-up phase of the study, multiple experts were invited to determine the land cover types of the sample points based on Google Earth and Sentinel-2 high-resolution remote sensing images. For samples with disputed categorizations, experts engaged in collective discussions and voting to assign the type with the highest number of votes, ensuring that the accuracy of the validation dataset met the task requirements. Figure 4 below illustrates the number of samples for each land cover type in the validation dataset.

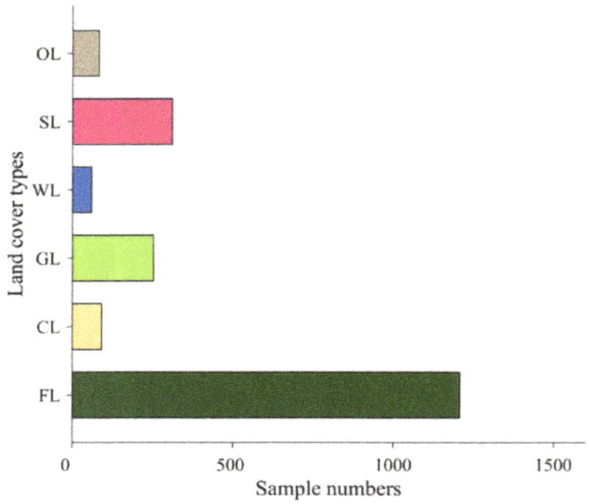

Figure 4. Sample numbers for each land cover type in the validation dataset.

After the validation dataset was created, the confusion matrices were computed and the quantitative metrics for evaluating the product performance were provided, including user accuracy (U.A.), producer accuracy (P.A.), overall accuracy (O.A.), and the kappa coefficient. The results of the confusion matrices for each product can be found in Tables A5–A9. For clarity, this paper refers to the accuracy computed from the validation dataset as the validation accuracy (V.A.), whereas the accuracy reported by the production agencies is referred to as the official validation accuracy (O.V.A.). Table 5 displays the V.A. and O.V.A. for every product in the study area.

Table 5. The V.A. and O.V.A. of four LC products in Beijing.

Accuracy (%)	DW_GLC10	ESA_GLC10	FROM_GLC10	SinoLC_10
V.A. (FL)	**77.96**	69.09	76.39	71.91
V.A. (CL)	27.78	**43.33**	34.44	28.89
V.A. (GL)	0.40	**26.59**	4.37	1.19
V.A. (WL)	**77.97**	61.01	54.23	15.25
V.A. (SL)	**95.18**	76.53	69.77	85.21
V.A. (OL)	6.10	**14.63**	1.22	0.85
V.A. (O.A.)	**65.67**	61.23	60.67	58.52
O.V.A. (O.A.)	**77.80**	75.00	72.76	55.55

Note: V.A. = validation accuracy; FL = forestland; CL = cropland; GL = grassland; WL = wetlands; SL = settlements; OL = other land; O.A. = overall accuracy; V.O.A. = official validation accuracy. The bold results represent the maximum values of all products.

The most surprising aspect of Table 5 is that DW_GLC10 and ESA_GLC10 achieved the highest validation accuracy across all land cover types. As shown in the table, the peak validation accuracy for forestland of 77.96%, wetlands of 77.97%, and settlements of 95.18% came from DW_GLC10. ESA_GLC10 had the highest validation accuracy for cropland of 43.33%, grassland of 26.59%, and other land of 14.63%. Additionally, DW_GLC10 had the maximum overall V.A. of 65.67% and O.V.A. of 77.80%. To further demonstrate the superiority of the WenSiM metric, Figures 5 and 6 present a comparison of the WenSiM, V.A., and O.V.A. results.

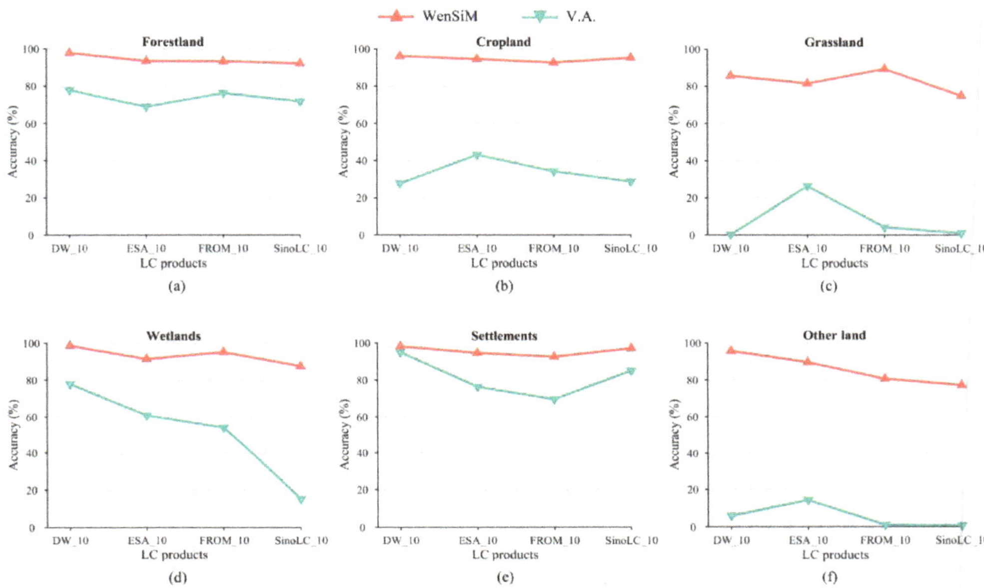

Figure 5. Comparison between the WenSiM and V.A. results for every land cover type. (**a**) Forestland. (**b**) Cropland. (**c**) Grassland. (**d**) Wetlands. (**e**) Settlements. (**f**) Other land.

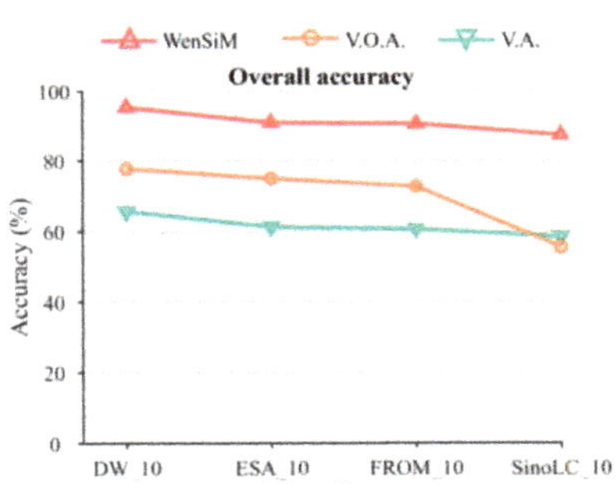

Figure 6. Comparison of the overall WenSiM, V.A., and O.V.A. results for LC products.

Looking at the figures, it is apparent that the WenSiM consistently had higher values compared to the V.A. and O.V.A. Figure 5 shows that the WenSiM exhibited similar trends to the V.A. for forestland, wetlands, and settlements, whereas it differed significantly for cropland, grassland, and other land categories. What is interesting about the trend in Figure 6 is that the overall WenSiM, V.A. and O.V.A. displayed the same accuracy assessment results: DW_GLC10 > ESA_GLC10 > FROM_GLC10 > SinoLC_10.

5. Discussion

This section initially delves into latent information within the results of the WenSiM and validation. Through a comparison with mainstream methods, the section further analyzes the strengths and weaknesses of the relative accuracy assessment method based on optimal transport theory. Finally, the section provides a reasonable accuracy assessment and ranking of LC products within the study area.

5.1. Analysis and Comparison of Assessment Results

Due to the fact that both the relative accuracy assessment method and the direct assessment method require the reference ground truth for accuracy evaluation, it is imperative to engage in a discussion about the scientific basis for the selection standards. Firstly, a number of recent studies have already demonstrated the rationality of using a validation dataset as the reference ground truth [41–43]. Although the V.A. and O.V.A. do not fully exploit all accuracy information, they are still considered the primary indicators for effectively assessing LC product accuracy [44]. Therefore, it is scientific to choose O.V.A. as the standard for selecting the reference ground truth. Moreover, the results of the V.A. and O.V.A. can also serve as important references for evaluating the WenSiM index.

As shown in Figure 5, nearly all quantitative results of the validation accuracy were lower than those of the WenSiM, which is because the WenSiM results encompass more accuracy information about the products. Figure 6 demonstrates that the overall WenSiM, V.A., and O.V.A. results had consistent trends, further confirming the effectiveness of the WenSiM index. The reason for the low V.A. of grassland and other land was attributed to the limitations of direct evaluation method.

The 10 m resolution LC product had hundreds of millions of pixels in Beijing. Considering cost constraints, it was not feasible to validate the true land cover type for each pixel. Despite the fact that the sample proportion chosen in this study surpassed that of other validation datasets, it remained insufficient for a region of this magnitude. The situation had the potential to create imbalanced sampling data ratios among different classes, ultimately leading to an unreasonable accuracy assessment [45].

On the one hand, when there is an imbalance in land cover composition within a study area, the direct evaluation method, with random sampling, will inevitably result in insufficient or even zero samples for certain land cover categories. This phenomenon may lead to unstable validation accuracy for those land cover types. On the other hand, the direct evaluation method only considers the spatial position information of sample points to quantify product accuracy, which can unavoidably yield lower results.

The indirect evaluation method achieves cost-effective utilization of spatial position information for all pixels by comparing the consistency between products [46]. The study categorized spatial consistency into five levels from high to low. Taking forestland as an example, level 5 indicates that the classification results of all LC products at that pixel were forest land, whereas level 1 means that only one LC product classified the pixel as forestland. Figures 7 and 8 present the regional consistency analysis results and the spatial consistency analysis results, respectively, for the five LC products.

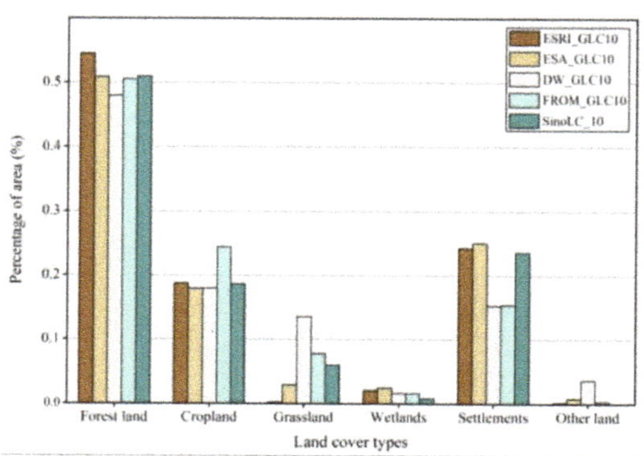

Figure 7. Regional consistency analysis of LC products.

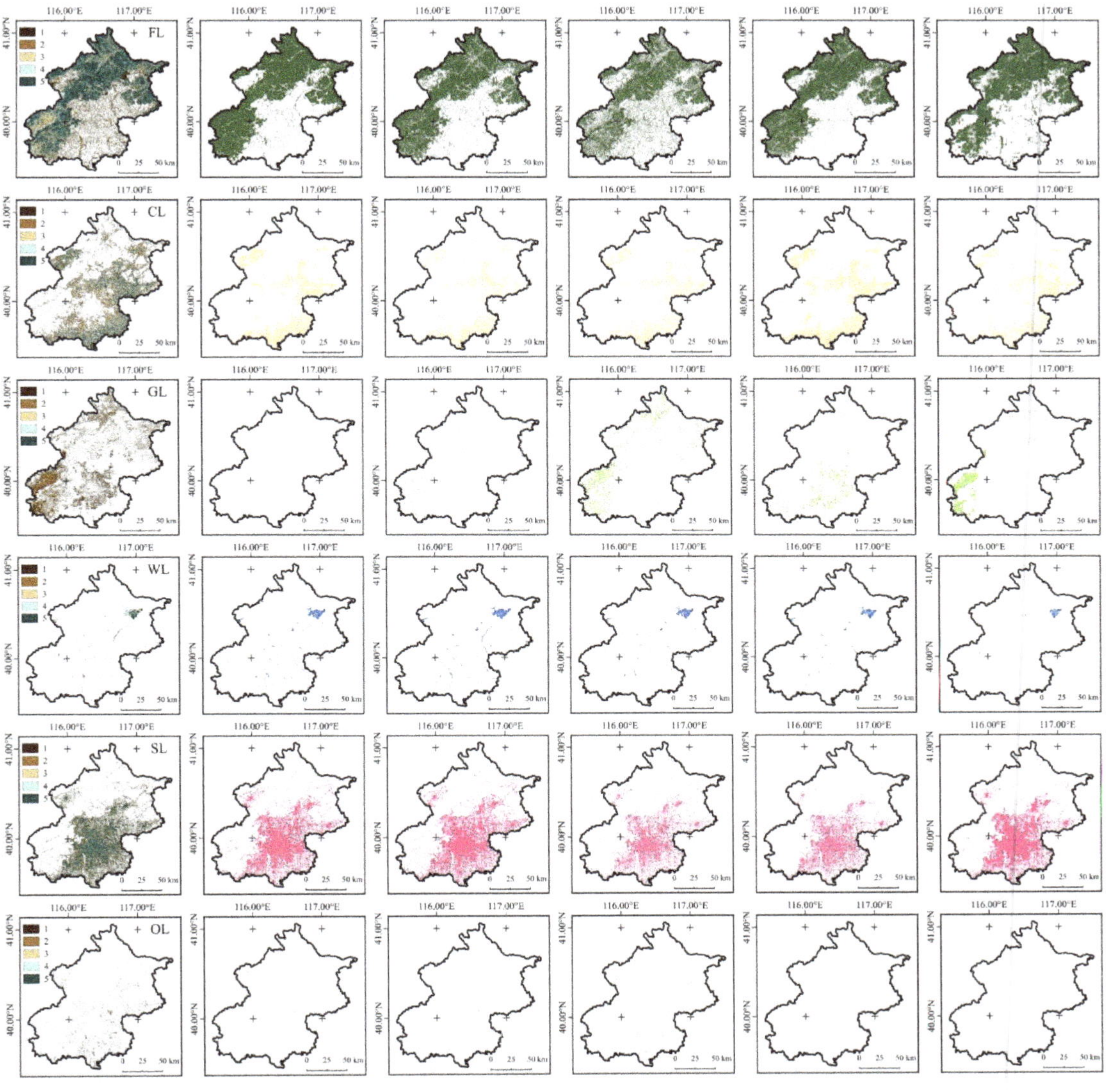

Figure 8. Spatial consistency analysis of LC products.

Looking at Figure 7, it is apparent that there was a significant difference for each product in grassland and other land, whereas the differences are relatively small in the forestland and wetlands. Interestingly, a noticeable grouping effect was observed in the regional consistency in cropland and settlements. Except for FROM_GLC10, the products exhibited stronger consistency in cropland. The consistency results of settlements in ESRI_GLC10, ESA_GLC10, and SinoLC_10 were similar, whereas DW_GLC10 was more consistent with FROM_GLC10.

Figure 8 shows that the spatial consistency analysis not only revealed the distribution of land cover types in the study area but also indirectly assessed the classification accuracy of the products. For instance, the analysis results indicate that forestland was mainly distributed in the northern and southwestern parts of Beijing. Through visual interpretation,

SinoLC_10 exhibited the highest spatial consistency in forestland, which suggests that SinoLC_10 had a higher classification accuracy in forestland.

The indirect evaluation method can effectively mitigate issues arising from low samples, preventing insufficient utilization of spatial location information in products. However, the method lacks on-site validation and the ability to quantify product accuracy, providing only qualitative suggestions for product selection [47].

5.2. Advantages and Limitations of Wasserstein Similarity

Land cover types often exhibit anomalous morphologies, manifesting as anomalous pixels. Anomalous pixels carry accuracy information that is markedly different from other pixels of the same type, representing unique features. The causes of anomalous pixels can be attributed to three main factors.

1. The random distribution of the actual class;
2. Errors introduced during development;
3. Misclassification resulting from insufficient accuracy in the classification algorithms.

Therefore, anomalous pixels primarily reflect the commonalities and differences between products. The direct evaluation method, due to the randomness of sampling, cannot effectively utilize the accuracy information of anomalous pixels. Although the indirect evaluation method considers the spatial position information of anomalous pixels, it falls short in quantifying the accuracy information. This paper provides the first method to take into account various types of accuracy information, with a specific emphasis on anomalous pixels, to quantify product accuracy globally.

The advantages of Wasserstein similarity are primarily evident in two application scenarios. Firstly, when the absolute ground truth is available, serving as a reference for LC products, the method reasonably evaluates the products by thoroughly exploring their accuracy information. This results in more comprehensive findings compared to other approaches. Secondly, when the absolute ground truth is unattainable, WenSiM enables a cost-effective evaluation of new LC products. Unlike direct evaluation methods, the method selects the product with the highest overall accuracy as the reference truth for computing WenSiM, eliminating the need for a validation dataset. In contrast to indirect evaluation methods, it allows for a quantitative assessment of product accuracy and provides a clear ranking of accuracy for each product at different levels.

However, the method also has some limitations. Firstly, the presence of anomalous pixels can make the evaluation results overly stringent. Secondly, when performing assessments on an extensive scale, such as at the global or super-regional level, the large number of pixels can lead to slower computation speeds. Furthermore, in the absence of the absolute ground truth, the evaluation results are influenced by the choice of reference truth. If the criteria are chosen arbitrarily, the reliability of the results may significantly decrease. Therefore, the criteria for selecting the reference truth should be flexibly determined in applications. In the future, the method could be expanded for the accuracy assessment of a broader range of remote sensing products, such as land surface temperature and digital elevation models, thereby aiding researchers in rapidly selecting the most suitable products based on task requirements [48,49].

After elucidating the characteristics of the accuracy evaluation method based on optimal transportation theory, the final evaluation ranking was obtained by combining the WenSiM with the validation results: DW_GLC10 > ESA_GLC10 > FROM_GLC10 > SinoLC_10.

6. Conclusions

The paper set out to find a relative accuracy assessment method based on optimal transport theory for LC products and demonstrates its effectiveness through a research case. Three key conclusions are drawn from the paper. First, the method enables a rapid quantification of LC product accuracy without the need for registration and validation datasets. Second, the WenSiM index derived from the method measures the similarity between the target product and the reference truth at a global scale, utilizing a broader range of accuracy information and resulting in more accurate assessment. Third, this study evaluated the accuracy of four LC products in Beijing, with results closely aligning with validation outcomes. Although the method is limited by the small number of cases, it possesses universality and can be extended to the accuracy assessment of various remote sensing products in global regions. This would be a fruitful area for further work, with the goal of providing researchers with a fast and reliable accuracy assessment tool for the selection and production of remote sensing products.

Author Contributions: Conceptualization, R.Z. and Y.T.; methodology, R.Z., Y.T., Y.S. and G.J.; software, R.Z. and X.W.; validation, R.Z., Z.L. and J.W.; formal analysis, R.Z.; investigation, R.Z.; resources, R.Z.; data curation, R.Z.; writing—original draft preparation, R.Z.; writing—review and editing, and Y.T.; visualization, R.Z.; supervision, Y.T. and G.J.; project administration, X.W.; funding acquisition, Y.T. and G.J. All authors have read and agreed to the published version of the manuscript.

Funding: This research was co-funded by Special Fund for the follow-up work of the Three Gorges project (grant number 102126222020270019081) and the National Key R&D Program (grant number 2019YFE0126400).

Data Availability Statement: ESRI_GLC10 comes from the website https://livingatlas.arcgis.com/landcover/ (accessed on 25 November 2023). DW_GLC10 comes from the website https://dynamicworld.app/ (accessed on 25 November 2023). ESA_GLC10 comes from the website https://esa-worldcover.org/en (accessed on 25 November 2023). FROM_GLC10 comes from the website https://data-starcloud.pcl.ac.cn/resource/1 (accessed on 25 November 2023). SinoLC_1 comes from the website https://zenodo.org/records/7711587 (accessed on 25 November 2023). Other data will be made available on request.

Acknowledgments: The authors thank all survey participants and reviewers of the paper. The data used for this paper were provided by ESRI_GLC10 from ESRI, DW_GLC10 from Google, ESA_GLC10 from ESA, FROM_GLC10 from Tsinghua University, and SinoLC_1 from Wuhan University. We greatly appreciate the constructive comments from the anonymous reviewers and the *Remote Sensing* editorial team that helped us improve our paper.

Conflicts of Interest: The authors declare no conflict of interest.

Appendix A

Table A1. The detailed definitions of IPCC land categories.

Land Categories	Detailed Definitions
Forestland	• All land with woody vegetation consistent with thresholds; • Systems with vegetation that are expected to exceed the threshold of the forestland.
Cropland	• Arable and tillage land; • Agro-forestry systems where vegetation falls below the thresholds.
Grassland	• Rangelands and pastureland that is not considered cropland; • Systems with vegetation that fall below the threshold and are not expected to exceed the threshold without human intervention; • All grassland from wildlands to recreational areas, as well as agricultural and silvi-pastural systems.
Wetlands	• Land that is covered by or saturated with water for all or part of the year; • Reservoirs, natural rivers and lakes.
Settlements	• All developed land, including transportation infrastructure and human settlements.
Other land	• Bare soil, rock, ice, and all unmanaged land areas that do not fall into any of the other five categories.

Table A2. Conversion relations between five LC products and the IPCC land cover classification.

IPCC	ESRI_GLC10	DW_GLC10	ESA_GLC10	FROM_GLC10	SinoLC_1
Forestland	Trees	Trees	Tree cover	Forest	Tree cover
	Scrub and shrub		Shrubland	Shrubland	Shrubland
			Mangroves		
Cropland	Crops	Crops	Cropland	Cropland	Cropland
Grassland	Grass	Grass	Grassland	Grassland	Grassland
Wetlands	Water	Water	Permanent water bodies	Waterbodies	Water
	Flooded vegetation	Flooded vegetation	Herbaceous wetland	Wetlands	Wetlands
Settlements	Built area	Built	Built up	Impervious	Roads
					Built up
Other land	Bare ground	Bare	Bare and sparse vegetation	Barren land	Barren and spare vegetation
	Snow and ice	Snow and ice	Snow and ice	Snow and ice	Snow and ice
	Clouds		Moss and Lichen	Tundra	Moss and lichen

Table A3. Stratified sampling calculation parameters.

District	M_l	$\bar{y}_l$	S_l^2	W_l	W_lS_l	m_l
Changping	23,067,539	2.408	1.323	0.082	0.09430	197
Chaoyang	7,925,627	4.401	0.363	0.028	0.01697	35
Daxing	17,461,322	3.221	0.720	0.062	0.05264	110
Dongcheng	713,476	4.879	0.092	0.003	0.00077	2
Fangshan	33,808,138	2.095	1.284	0.120	0.13612	285
Fengtai	5,195,318	4.264	0.478	0.018	0.01276	27
Haidian	7,322,717	3.972	0.679	0.026	0.02145	45
Huairou	36,841,342	1.417	0.896	0.131	0.12396	259
Mentougou	24,734,083	1.269	0.764	0.088	0.07681	161
Miyun	38,531,881	1.742	1.065	0.137	0.14133	295
Pinggu	16,263,416	1.998	1.140	0.058	0.06171	129
Shijingshan	1,436,001	3.957cc	0.763	0.005	0.00446	9
Shunyi	17,292,399	3.288	0.758	0.061	0.05349	112
Tongzhou	15,341,172	3.359	0.695	0.055	0.04544	95
Xicheng	856,827	4.965	0.012	0.003	0.00034	1
Yanqing	34,618,782	1.576	0.861	0.123	0.11415	239

Notes: M_l = the number of pixels in layer l; $\bar{y}_l$ = the sample mean of layer l; S_l^2 = the population variance of layer l; W_l = the weight parameter of layer l; m_l = the number of samples of layer l.

Table A4. Sample numbers for each land cover type in each layer.

District	FL	CL	GL	WL	SL	OL
Changping	125	5	20	3	37	7
Chaoyang	6	1	3	2	22	1
Daxing	38	5	30	0	36	1
Dongcheng	0	0	0	0	2	0
Fangshan	185	4	33	3	54	6
Fengtai	5	2	0	0	17	3
Haidian	15	3	1	1	22	3
Huairou	210	12	15	1	14	7
Mentougou	139	10	4	0	2	6
Miyun	186	17	27	24	20	21

Table A4. *Cont.*

District	FL	CL	GL	WL	SL	OL
Pinggu	73	1	33	9	11	2
Shijingshan	5	0	0	1	3	0
Shunyi	28	6	35	6	32	5
Tongzhou	28	7	24	5	27	4
Xicheng	0	0	0	0	1	0
Yanqing	164	17	27	4	11	16
Total	1207	90	252	59	311	82

Notes: FL = forestland; CL = cropland; GL = grassland; WL = wetlands; SL = settlements; OL = other land.

Table A5. Confusion matrix for ESRI_GLC10 according to the validation dataset.

Classification	FL	CL	GL	WL	SL	OL	Total	P.A. (%)
FL	983	127	0	2	94	1	1207	81.44
CL	51	25	0	1	13	0	90	27.78
GL	25	157	15	0	47	8	252	5.95
WL	7	4	1	42	5	0	59	71.19
SL	5	13	0	0	291	2	311	93.57
OL	35	14	0	4	27	2	82	2.44
Total	1106	340	16	49	477	13	2001	
U.A. (%)	88.88	7.35	93.75	85.71	61.01	15.38		
O.A. (%)				67.87				
Kappa				0.4816				

Note: FL = forestland; CL = cropland; GL = grassland; WL = wetlands; SL = settlements; OL = other land.

Table A6. Confusion matrix for DW_GLC10 according to the validation dataset.

Classification	FL	CL	GL	WL	SL	OL	Total	P.A. (%)
FL	941	131	42	2	89	2	1207	77.96
CL	40	25	5	1	16	3	90	27.78
GL	19	170	1	3	55	4	252	0.40
WL	3	3	0	46	6	1	59	77.97
SL	5	5	0	0	296	5	311	95.18
OL	29	11	0	4	33	5	82	6.10
Total	1037	345	48	56	495	20	2001	
U.A. (%)	90.74	7.25	2.08	82.14	59.80	25.00		
O.A. (%)				65.67				
Kappa				0.4610				

Note: FL = forestland; CL = cropland; GL = grassland; WL = wetlands; SL = settlements; OL = other land.

Table A7. Confusion matrix for ESA_GLC10 according to the validation dataset.

Classification	FL	CL	GL	WL	SL	OL	Total	P.A. (%)
FL	834	112	222	0	18	21	1207	69.10
CL	20	39	20	0	5	6	90	43.33
GL	52	107	67	1	9	16	252	26.59
WL	5	6	3	36	0	9	59	61.02
SL	30	20	1	0	238	22	311	76.53
OL	19	15	13	3	20	12	82	14.63
Total	960	299	326	40	290	86	2001	
U.A. (%)	86.88	13.04	20.55	90.00	82.07	13.95		
O.A. (%)				61.27				
Kappa				0.4118				

Note: FL = forestland; CL = cropland; GL = grassland; WL = wetlands; SL = settlements; OL = other land.

Table A8. Confusion matrix for FROM_GLC10 according to the validation dataset.

Classification	FL	CL	GL	WL	SL	OL	Total	P.A. (%)
FL	922	176	77	1	25	6	1207	76.39
CL	41	31	9	0	9	0	90	34.44
GL	34	182	11	1	24	0	252	4.36
WL	3	15	5	32	4	0	59	54.24
SL	10	55	25	0	217	4	311	69.77
OL	27	25	8	2	19	1	82	1.22
Total	1037	484	135	36	298	11	2001	
U.A. (%)	88.91	6.40	8.15	88.89	72.82	9.09		
O.A. (%)				60.67				
Kappa				0.3894				

Note: FL = forestland; CL = cropland; GL = grassland; WL = wetlands; SL = settlements; OL = other land.

Table A9. Confusion matrix for SinoLC_10 according to the validation dataset.

Classification	FL	CL	GL	WL	SL	OL	Total	P.A. (%)
FL	868	130	123	3	83	0	1207	71.91
CL	47	26	6	0	11	0	90	28.89
GL	45	154	3	1	49	0	252	1.19
WL	30	13	0	9	7	0	59	15.25
SL	16	28	2	0	265	0	311	85.21
OL	32	19	3	0	28	0	82	0
Total	1038	370	137	13	443	0	2001	
U.A. (%)	83.62	7.03	2.19	69.23	59.82	0		
O.A. (%)				58.52				
Kappa				0.3474				

Note: FL = forestland; CL = cropland; GL = grassland; WL = wetlands; SL = settlements; OL = other land.

References

1. Song, X.P.; Hansen, M.C.; Stehman, S.V.; Potapov, P.V.; Tyukavina, A.; Vermote, E.F.; Townshend, J.R. Global land change from 1982 to 2016. *Nature* **2018**, *560*, 639–643. [CrossRef] [PubMed]
2. Stanimirova, R.; Graesser, J.; Olofsson, P.; Friedl, M.A. Widespread changes in 21st century vegetation cover in Argentina, Paraguay, and Uruguay. *Remote Sens. Environ.* **2022**, *282*, 113277. [CrossRef]
3. Li, L.; Zhan, W.; Ju, W.; Peñuelas, J.; Zhu, Z.; Peng, S.; Zhu, X.; Liu, Z.; Zhou, Y.; Li, J.; et al. Competition between biogeochemical drivers and land-cover changes determines urban greening or browning. *Remote Sens. Environ.* **2023**, *287*, 113481. [CrossRef]
4. Wang, J.; Yang, X.; Wang, Z.; Cheng, H.; Kang, J.; Tang, H.; Li, Y.; Bian, Z.; Bai, Z. Consistency Analysis and Accuracy Assessment of Three Global Ten-Meter Land Cover Products in Rocky Desertification Region—A Case Study of Southwest China. *ISPRS Int. J. Geo-Inf.* **2022**, *11*, 202. [CrossRef]
5. Olofsson, P.; Foody, G.M.; Stehman, S.V.; Woodcock, C.E. Making better use of accuracy data in land change studies: Estimating accuracy and area and quantifying uncertainty using stratified estimation. *Remote Sens. Environ.* **2013**, *129*, 122–131. [CrossRef]
6. Zhang, W.; Wang, J.; Lin, H.; Cong, M.; Wan, Y.; Zhang, J. Fusing Multiple Land Cover Products Based on Locally Estimated Map-Reference Cover Type Transition Probabilities. *Remote Sens.* **2023**, *15*, 481. [CrossRef]
7. Comber, A.; Fisher, P.; Brunsdon, C.; Khmag, A. Spatial analysis of remote sensing image classification accuracy. *Remote Sens. Environ.* **2012**, *127*, 237–246. [CrossRef]
8. Chen, J.; Chen, L.; Chen, F.; Ban, Y.; Li, S.; Han, G.; Tong, X.; Liu, C.; Stamenova, V.; Stamenov, S. Collaborative validation of GlobeLand30: Methodology and practices. *Geo-Spat. Inf. Sci.* **2021**, *24*, 134–144. [CrossRef]
9. Wang, Y.; Zhang, J.; Liu, D.; Yang, W.; Zhang, W. Accuracy Assessment of GlobeLand30 2010 Land Cover over China Based on Geographically and Categorically Stratified Validation Sample Data. *Remote Sens.* **2018**, *10*, 1213. [CrossRef]
10. Venter, Z.S.; Barton, D.N.; Chakraborty, T.; Simensen, T.; Singh, G. Global 10 m Land Use Land Cover Datasets: A Comparison of Dynamic World, World Cover and Esri Land Cover. *Remote Sens.* **2022**, *14*, 4101. [CrossRef]
11. Foody, G.M. Explaining the unsuitability of the kappa coefficient in the assessment and comparison of the accuracy of thematic maps obtained by image classification. *Remote Sens. Environ.* **2020**, *239*, 111630. [CrossRef]
12. Bie, Q.; Shi, Y.; Li, X.; Wang, Y. Contrastive Analysis and Accuracy Assessment of Three Global 30 m Land Cover Maps Circa 2020 in Arid Land. *Sustainability* **2022**, *15*, 741. [CrossRef]

13. Pflugmacher, D.; Krankina, O.N.; Cohen, W.B.; Friedl, M.A.; Sulla-Menashe, D.; Kennedy, R.E.; Nelson, P.; Loboda, T.V.; Kuemmerle, T.; Dyukarev, E.; et al. Comparison and assessment of coarse resolution land cover maps for Northern Eurasia. *Remote Sens. Environ.* **2011**, *115*, 3539–3553. [CrossRef]
14. Yang, Y.; Xiao, P.; Feng, X.; Li, H. Accuracy assessment of seven global land cover datasets over China. *ISPRS J. Photogramm. Remote Sens.* **2017**, *125*, 156–173. [CrossRef]
15. Shi, W.; Zhao, X.; Zhao, J.; Zhao, S.; Guo, Y.; Liu, N.; Sun, N.; Du, X.; Sun, M. Reliability and consistency assessment of land cover products at macro and local scales in typical cities. *Int. J. Digit. Earth* **2023**, *16*, 486–508. [CrossRef]
16. Liu, P.; Pei, J.; Guo, H.; Tian, H.; Fang, H.; Wang, L. Evaluating the Accuracy and Spatial Agreement of Five Global Land Cover Datasets in the Ecologically Vulnerable South China Karst. *Remote Sens.* **2022**, *14*, 3090. [CrossRef]
17. Gao, Y.; Liu, L.; Zhang, X.; Chen, X.; Mi, J.; Xie, S. Consistency Analysis and Accuracy Assessment of Three Global 30-m Land-Cover Products over the European Union using the LUCAS Dataset. *Remote Sens.* **2020**, *12*, 3479. [CrossRef]
18. Ye, S.; Pontius, R.G.; Rakshit, R. A review of accuracy assessment for object-based image analysis: From per-pixel to per-polygon approaches. *ISPRS J. Photogramm. Remote Sens.* **2018**, *141*, 137–147. [CrossRef]
19. Kang, J.; Sui, L.; Yang, X.; Wang, Z.; Huang, C.; Wang, J. Spatial Pattern Consistency among Different Remote-Sensing Land Cover Datasets: A Case Study in Northern Laos. *ISPRS Int. J. Geo-Inf.* **2019**, *8*, 201. [CrossRef]
20. Tan, Y.; Shi, Y.; Xu, L.; Zhou, K.; Jing, G.; Wang, X.; Bai, B. An Optimal Transport Based Global Similarity Index for Remote Sensing Products Comparison. *Remote Sens.* **2022**, *14*, 2546. [CrossRef]
21. Brown, C.F.; Brumby, S.P.; Guzder-Williams, B.; Birch, T.; Hyde, S.B.; Mazzariello, J.; Czerwinski, W.; Pasquarella, V.J.; Haertel, R.; Ilyushchenko, S.; et al. Dynamic World, Near real-time global 10 m land use land cover mapping. *Sci. Data* **2022**, *9*, 251. [CrossRef]
22. Zanaga, D.; Van De Kerchove, R.; De Keersmaecker, W.; Souverijns, N.; Brockmann, C.; Quast, R.; Wevers, J.; Grosu, A.; Paccini, A.; Vergnaud, S.; et al. ESA WorldCover 10 m 2020 v100 (Version v100) [dataset]. *Zenodo* **2021**. [CrossRef]
23. Gong, P.; Liu, H.; Zhang, M.; Li, C.; Wang, J.; Huang, H.; Clinton, N.; Ji, L.; Li, W.; Bai, Y.; et al. Stable classification with limited sample: Transferring a 30-m resolution sample set collected in 2015 to mapping 10-m resolution global land cover. *Sci. Bull.* **2017**, *64*, 370–373. [CrossRef] [PubMed]
24. Li, Z.; He, W.; Cheng, M.; Hu, J.; Yang, G.; Zhang, H. SinoLC-1: The first 1-meter resolution national-scale land-cover map of China created with the deep learning framework and open-access data. *Earth Syst. Sci. Data.* **2023**, *15*, 4749–4780. [CrossRef]
25. Karra, K.; Kontgis, C.; Statman-Weil, Z.; Mazzariello, J.C.; Mathis, M.; Brumby, S.P. Global land use/land cover with Sentinel 2 and deep learning. In Proceedings of the 2021 IEEE International Geoscience and Remote Sensing Symposium IGARSS, Brussels, Belgium, 11–16 July 2021; pp. 4704–4707.
26. Friedl, M.A.; Sulla-Menashe, D.; Tan, B.; Schneider, A.; Ramankutty, N.; Sibley, A.; Huang, X. MODIS Collection 5 global land cover: Algorithm refinements and characterization of new datasets. *Remote Sens. Environ.* **2010**, *114*, 168–182. [CrossRef]
27. McCallum, I.; Obersteiner, M.; Nilsson, S.; Shvidenko, A. A spatial comparison of four satellite derived 1km global land cover datasets. *Int. J. Appl. Earth Obs. Geoinf.* **2006**, *8*, 246–255. [CrossRef]
28. Rubner, Y.; Tomasi, C.; Guibas, L.J. The Earth Mover's Distance as a Metric for Image Retrieval. *Int. J. Comput. Vis.* **2000**, *40*, 99–121. [CrossRef]
29. Villani, C. *Optimal Transport: Old and New*; Springer: Berlin/Heidelberg, Germany, 2009.
30. Brenier, Y. Polar factorization and monotone rearrangement of vector-valued functions. *Commun. Pure Appl. Math.* **1991**, *44*, 375–417. [CrossRef]
31. Kolouri, S.; Rohde, G.K.; Hoffmann, H. Sliced Wasserstein Distance for Learning Gaussian Mixture Models. In Proceedings of the 2018 IEEE/CVF Conference on Computer Vision and Pattern Recognition, Salt Lake City, UT, USA, 18–23 June 2018; pp. 3427–3436.
32. Deshpande, I.; Zhang, Z.; Schwing, A. Generative Modeling Using the Sliced Wasserstein Distance. In Proceedings of the 2018 IEEE/CVF Conference on Computer Vision and Pattern Recognition, Salt Lake City, UT, USA, 18–23 June 2018; pp. 3483–3491.
33. Chen, Y.; Li, C.; Lu, Z. Computing Wasserstein-p Distance between Images with Linear Cost. In Proceedings of the 2022 IEEE/CVF Conference on Computer Vision and Pattern Recognition (CVPR), New Orleans, LA, USA, 18–24 June 2022; pp. 509–518.
34. Kolouri, S.; Zou, Y.; Rohde, G.K. Sliced Wasserstein Kernels for Probability Distributions. In Proceedings of the 2016 IEEE Conference on Computer Vision and Pattern Recognition (CVPR), Las Vegas, NV, USA, 27–30 June 2016; pp. 5258–5267.
35. Deshpande, I.; Hu, Y.T.; Sun, R.; Pyrros, A.; Siddiqui, N.; Koyejo, S.; Zhao, Z.; Forsyth, D.; Schwing, A.G. Max-Sliced Wasserstein Distance and Its Use for GANs. In Proceedings of the 2019 IEEE/CVF Conference on Computer Vision and Pattern Recognition (CVPR), Long Beach, CA, USA, 15–20 June 2019; pp. 10640–10648.
36. Wang, H.; Yan, H.; Hu, Y.; Xi, Y.; Yang, Y. Consistency and Accuracy of Four High-Resolution LULC Datasets—Indochina Peninsula Case Study. *Land* **2022**, *11*, 758. [CrossRef]
37. Stehman, S.V. Sampling designs for accuracy assessment of land cover. *Int. J. Remote Sens.* **2009**, *30*, 5243–5272. [CrossRef]
38. Liu, Y.; Shi, W.; Zhang, H.; Zhang, M. A multilevel stratified spatial sampling approach based on terrain knowledge for the quality assessment of OpenStreetMap dataset in Hong Kong. *Trans. GIS* **2023**, *27*, 290–318. [CrossRef]
39. Stehman, S.V.; Foody, G.M. Key issues in rigorous accuracy assessment of land cover products. *Remote Sens. Environ.* **2019**, *231*, 111199. [CrossRef]
40. Tsendbazar, N.E.; de Bruin, S.; Mora, B.; Schouten, L.; Herold, M. Comparative assessment of thematic accuracy of GLC maps for specific applications using existing reference data. *Int. J. Appl. Earth Obs. Geoinf.* **2016**, *44*, 124–135. [CrossRef]

41. Wagner, J.E.; Stehman, S.V. Optimizing sample size allocation to strata for estimating area and map accuracy. *Remote Sens. Environ.* **2015**, *168*, 126–133. [CrossRef]
42. Wickham, J.D.; Stehman, S.V.; Gass, L.; Dewitz, J.; Fry, J.A.; Wade, T.G. Accuracy assessment of NLCD 2006 land cover and impervious surface. *Remote Sens. Environ.* **2013**, *130*, 294–304. [CrossRef]
43. Gong, Y.; Xie, H.; Liao, S.; Lu, Y.; Jin, Y.; Wei, C.; Tong, X. Assessing the Accuracy of Multi-Temporal GlobeLand30 Products in China Using a Spatiotemporal Stratified Sampling Method. *Remote Sens.* **2023**, *15*, 4593. [CrossRef]
44. Li, Z.; Chen, X.; Qi, J.; Xu, C.; An, J.; Chen, J. Accuracy assessment of land cover products in China from 2000 to 2020. *Sci. Rep.* **2023**, *13*, 12936. [CrossRef]
45. Morales-Barquero, L.; Lyons, M.B.; Phinn, S.R.; Roelfsema, C.M. Trends in Remote Sensing Accuracy Assessment Approaches in the Context of Natural Resources. *Remote Sens.* **2019**, *11*, 2305. [CrossRef]
46. Zhao, T.; Zhang, X.; Gao, Y.; Mi, J.; Liu, W.; Wang, J.; Jiang, M.; Liu, L. Assessing the Accuracy and Consistency of Six Fine-Resolution Global Land Cover Products Using a Novel Stratified Random Sampling Validation Dataset. *Remote Sens.* **2023**, *15*, 2285. [CrossRef]
47. Liu, B.; Yang, X.; Wang, Z.; Ding, Y.; Zhang, J.; Meng, D. A Comparison of Six Forest Mapping Products in Southeast Asia, Aided by Field Validation Data. *Remote Sens.* **2023**, *15*, 4584. [CrossRef]
48. Polidori, L.; El Hage, M. Digital Elevation Model Quality Assessment Methods: A Critical Review. *Remote Sens.* **2020**, *12*, 3522. [CrossRef]
49. Liu, Y.; Yu, Y.; Yu, P.; Göttsche, F.M.; Trigo, I.F. Quality Assessment of S-NPP VIIRS Land Surface Temperature Product. *Remote Sens.* **2015**, *7*, 12215–12241. [CrossRef]

remote sensing

MDPI

Article

Finding Misclassified Natura 2000 Habitats by Applying Outlier Detection to Sentinel-1 and Sentinel-2 Data

David Moravec, Vojtěch Barták and Petra Šímová *

Department of Spatial Sciences, Faculty of Environmental Sciences, Czech University of Life Sciences Prague, Kamýcká 129, 165 00 Praha, Czech Republic; dmoravec@fzp.czu.cz (D.M.); bartakv@fzp.czu.cz (V.B.)
* Correspondence: simova@fzp.czu.cz

Abstract: The monitoring of Natura 2000 habitats (Habitat Directive 92/43/EEC) is a key activity ensuring the sufficient protection of European biodiversity. Reporting on the status of Natura 2000 habitats is required every 6 years. Although field mapping is still an indispensable source of data on the status of Natura 2000 habitats, and very good field-based data exist in some countries, keeping the field-based habitat maps up to date can be an issue. Remote sensing techniques represent an excellent alternative. Here, we present a new method for detecting habitats that were likely misclassified during the field mapping or that have changed since then. The method identifies the possible habitat mapping errors as the so-called "attribute outliers", i.e., outlying observations in the feature space of all relevant (spectral and other) characteristics of an individual habitat patch. We used the Czech Natura 2000 Habitat Layer as field-based habitat data. To prepare the feature space of habitat characteristics, we used a fusion of Sentinel-1 and Sentinel-2 satellite data along with a Digital Elevation Model. We compared outlier ratings using the robust Mahalanobis distance and Local Outlier Factor using three different thresholds (Tukey rule, histogram-based Scott's rule, and 95% quantiles in χ^2 distribution). The Mahalanobis distance thresholded by the 95% χ^2 quantile achieved the best results, and, because of its high specificity, appeared as a promising tool for identifying erroneously mapped or changed habitats. The presented method can, therefore, be used as a guide to target field updates of Natura 2000 habitat maps or for other habitat/land cover mapping activities where the detection of misclassifications or changes is needed.

Keywords: Sentinel-1; Sentinel-2; change detection; RADAR; multispectral; nature conservation

1. Introduction

Natura 2000 is a Europe-wide network of protected areas established to protect endangered species and natural and near-natural habitats. The conservation and monitoring of Natura 2000 sites is defined in the European Habitat Directive 92/43/EEC, which states that the conservation status of Natura 2000 sites and habitats must be reported every six years. Because whole-country field investigations are time-consuming and costly [1], satellite imagery has emerged as a promising technology for the monitoring of Natura 2000 sites [2]. A suitable remote sensing method can significantly reduce the costs of Natura 2000 sites monitoring and improve public control over environmental protection [3,4].

So far, many studies have demonstrated the potential of satellite data for general monitoring of Natura 2000 sites or directly for detecting and mapping individual Natura 2000 habitats [5–7]. Studies based on the classification of multispectral (e.g., Rapid Eye, Sentinel-2) data [1,8], hyperspectral data [9–11], or fusion of multispectral and radar data (e.g., fusion of Sentinel-2 and Sentinel-1 [12,13]) achieve promising precision in the identification of individual habitats and encouraging results. However, the studies mostly focus on just a few habitats at the regional level; where a higher number of habitats were classified, the studies were only at a local level. The lack of studies dealing with the classification of a wide range of habitat types within large areas (e. g., whole countries) suggests that the remote-sensing-based classification of all habitats at the national level remains challenging and that field habitat mapping is still valuable for the monitoring of Natura 2000 habitats.

Although field mapping is still an indispensable source of data on the status of habitats and very good field-based data exist in some countries, keeping them up to date without employing remote sensing techniques can be an issue. The Czech Republic is one such example, where detailed (cartographic ratio of 1:10,000) field mapping of all Natura 2000 habitats nationwide was carried out in 2001–2014 [14], based on which a Habitat Layer of the Czech Republic was subsequently created [15,16]. This created a comprehensive data source not only for nature conservation and biodiversity research but also for various sectors of the state administration [14,17,18]. The habitat layer is now being updated, again using field mapping, but ways to simplify and speed up the update using remote sensing are being explored. For example, finding a method capable of identifying habitat segments that were likely misclassified in the original mapping or have changed since the last mapping could make the process more effective, directing field research toward such habitats. Hence, as in some other EU countries, the main challenge remains to keep such field-based data up to date, to clean possible mapping errors (i.e., misidentified habitat classes), and to reveal ongoing gradual or sudden habitat changes as automatically as possible with the least possible expenditure.

Change detection (CD) is a common remote sensing method used to reveal changes in the reflectance characteristics of the mapped surfaces, and many different CD methods have been developed over time (an overview can be found in reference [19]). The CD analysis could be challenging and affected by the detection of unwanted changes such as phenological stages, sudden climatic events (snow, hail, drought, etc.), or differences in observation geometry. It could also be susceptible to errors caused by atmospheric correction or differences in the used sensors [20]. Here, we are introducing a method based on the outlierness of a particular observation in the space of its spectral and other characteristics. The benefit of the proposed method lies in its capability to perform change detection under multi-temporal observation or even using observation from a single time point when comparing the data with existing classification. The robustness of the method consists in using the single-temporal change detection, where effects of climatic events, geometry, atmosphere, or sensors is minimized. The main idea of the method is that changed or misclassified classes should be identifiable as those differing from the correctly classified habitats. Therefore, we identified the possible habitat mapping errors or changes as the so-called "attribute outliers", i.e., outlying observations in the feature space of all relevant (spectral and other) segment characteristics.

Using Sentinel-1 and Sentinel-2 data, this study aimed to develop a repeatable method for the identification of habitats that were potentially misidentified during field mapping or have changed since the mapping. In practical terms, such a method would help to target field efforts for mapping updates to such places.

2. Materials and Methods

2.1. Study Area and Field-Based Habitat Data

We used the Habitat Layer of the Czech Republic (hereinafter Habitat Layer) covering the whole country for our research (see Figure 1). The Habitat Layer was originally created to delineate Special Areas of Conservation (SACs) under the Habitat Directive and its first version was completed in 2005. Since 2006, it has been systematically updated by field mapping led by Nature Conservation Agency, with one-twelfth of the Czech Republic mapped each year. The layer is available under a free open license from the Nature Conservation Agency (https://data.nature.cz, accessed on 5 December 2022).

To find potentially misclassified or substantially changed habitat segments, we used the basic division of the Habitat Catalogue [21], which is compatible with habitat types delimited in Annex I of the Habitats Directive. Hence, the list of Natura 2000 habitats used in our study was as follows: Water (Streams and water bodies); Wetlands (Wetlands and riverine vegetation); Mires (Springs and mires); Screes (Cliffs and boulder screes); Alpines

(Alpine treeless habitats); Grass (Secondary grassland and heathland); Scrubs (Scrubs); and Forests (Woodland). A detailed description of the habitats is available in reference [21].

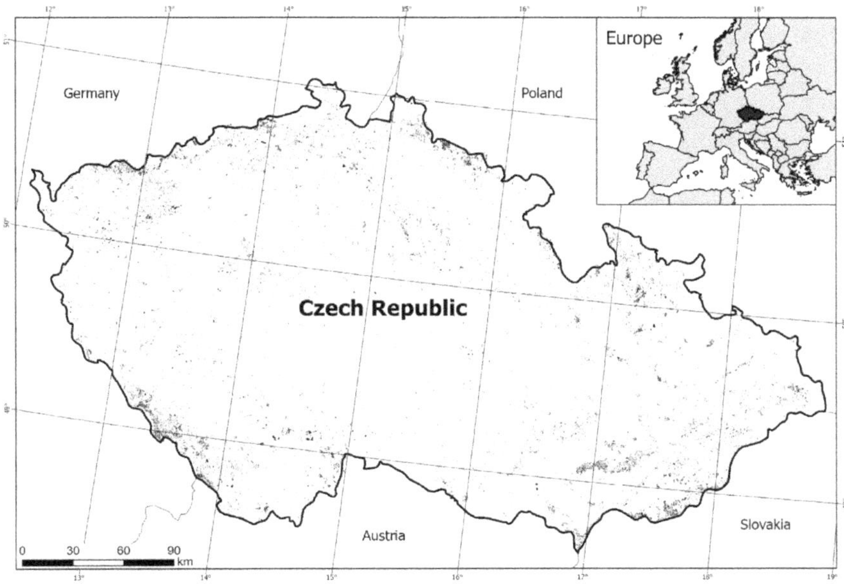

Figure 1. Study area (Czech Republic) with resulting 233,966 habitats used for this study (black color).

2.2. Satellite and DEM Data

We used Sentinel and Digital Elevation Model (DEM) data to find patches (segments) where habitats were potentially misidentified during mapping or have changed since mapping. The Sentinel satellites are operated by the European Commission, which provides the data free of charge. The Sentinel-2 satellite, with its 13 optical bands, is ideal for identifying the species composition of vegetation [22–24]. Despite the many advantages of Sentinel-2 satellites, the applicability in the Czech Republic is limited by cloud cover (Central Bohemia was completely cloud-free only during six satellite flybys in 2018 and only once in 2022). For this reason, we created one representative seamless and cloud-free mosaic of the Czech Republic within the growing season of the years 2018 and 2022 (details in Section 2.3).

To overcome the disadvantages of Sentinel-2 and deliver information about the temporal dynamics of Natura 2000 habitats during the growing season of the year under study, we used Sentinel-1 microwave observation because of its all-weather, day-to-night observation capacity. Since Sentinel-1 has its own illumination source, it allows you to have complete control over the polarization of the transmitted and received signal. Sentinel-1 measures in two polarizations: co-polarization VV (vertical emitting, vertical receiving) and cross-polarization VH (vertical emitting, horizontal receiving). While the source of the VV signal is direct reflections from the earth's surface, or double reflections from mutually perpendicular objects (predominant in cities), the primary source of the VH signal is the so-called volume scattering, in which the signal enters the object and is subsequently reflected many times and interacts with each other. The result is a depolarization of the emitted V-polarized signal and an increase in the received H-polarized signal. The main source of volume scattering (represented by VH polarization) is vegetation. As shown in previous studies, VH polarization could respond well to the vegetation status [25–29] and, therefore, it was used in this study.

The digital elevation model EU DEM version 1.1 (https://land.copernicus.eu/imagery-in-situ/eu-dem/eu-dem-v1.1, accessed on 5 December 2022) was selected to calculate terrain characteristics. The elevation is defined here according to the European Vertical Reference System (EVRS). The spatial resolution is 25 m with an absolute error in elevation of ±7 m RMSE.

2.3. Spatial Data Processing

We used Habitat Layer segments (polygons) that were updated in the field in 2018 and 2022. We chose these two years both to fit the Sentinel-1 and -2 missions and to test the search for outliers on two samples (years) that are as far apart in time as possible. We worked only with natural and near-natural habitats, i.e., we excluded habitats strongly influenced or created by humans (category X according to the Czech Habitat Catalogue). We also excluded so-called mosaics, i.e., polygons consisting of several habitat types. To remove possible edge pixel problems, possible shadowing due to nearby obstacles or possible misregistration of product [30]; we shrunk the remaining polygons by 20 m, which is a spatial resolution in bands essential for vegetation identification of Sentinel-2. The registration of habitat polygons to satellite data was examined visually and no significant shift was observed. Subsequently, all the polygons smaller than the size of one pixel (i.e., 400 m^2) were removed from the analysis. After this, 233,966 polygons remained for the outliers analysis.

We used the Sentinel-2 Global Mosaic service (https://apps.sentinel-hub.com/mosaic-hub/, accessed on 5 December 2022) to create a seamless mosaic over the whole Czech Republic covering the growing season (March-October) in the years 2018 and 2022. The resulting mosaic is formed from medoids of the pixels of each image cleared from clouds, cloud shadows, snow-covered areas, and areas significantly affected by the atmosphere. The medoid is a pixel with the smallest differences from other pixels at the same location but at different times. In this way, the most representative pixel at a given location was always selected [31]. Bands 1, 9, and 10 were omitted because their purpose is atmosphere description and, therefore, they have minimal added value for ground observation; moreover, they have a coarse resolution of 60 m per pixel, which does not correspond to the minimal polygon size. The final band selection is in Table 1.

Table 1. Characteristic of Sentinel-2 Bands used for the final product.

Band Number	2	3	4	5	6	7	8	8a	11	12
Band name	Blue	Green	Red	Red Edge	Red Edge	Red Edge	NIR	Red Edge	SWIR	SWIR
Center (nm)	490	560	665	705	740	783	842	865	1610	2190
Width (nm)	65	35	30	15	15	20	115	20	90	180
Spatial resolution (m)	10	10	10	20	20	20	10	20	20	20

The microwave Sentinel-1 images were processed using the ASF HyP3 tool. The images were radiometrically corrected to the gamma-nought level ($\gamma 0$). The terrain influence was removed using the Copernicus DEM digital terrain model terrain flattening. Additionally, 10 × 2 multi-looking was performed to suppress noise and obtain square pixels. The whole computation was performed using the algorithms of the GAMMA software 5.1.1. More information is available at https://hyp3-docs.asf.alaska.edu/guides/rtc_product_guide/, accessed on 6 December 2022.

The individual images were subsequently used to create nationwide mosaics at a step of one month during the growing season (March–October) for the years 2018 and 2022. We used the Quick Mosaic tool in ENVI 5.6 for the mosaic generation.

We used the EU DEM to account for the effects of elevation, slope, and exposure on cardinal directions of the spectral characteristics of habitats during the growing season. We calculated the average elevation for each polygon of the Natura 2000 habitat, as well as the average slope and the prevailing direction to the cardinal points. This was carried out in ArcGIS Pro 2.7.0 [32], using the Slope and Aspect functions.

The mean, median, and standard deviations were calculated for each mentioned Sentinel-1 and Sentinel-2 band. At the same time, the average slope and elevation were calculated along with the predominant exposure to the cardinal directions and the total polygon size. In total, 55 aggregated values for each of the 233,966 polygons were calculated using QGIS 3.22 [33] multiband zonal statistics. The overview of all the resulting variables is given in Appendix A.

2.4. Spectral Outliers Detection

As the main idea of this study states, the erroneously mapped habitats, or habitats that changed after their mapping, should be identifiable as those that differ in their attribute values from the correctly identified habitats of the same type. Therefore, we identified the possible habitat mapping errors as the so-called "attribute outliers", i.e., outlying observations in the feature space of all relevant (spectral and other) segment characteristics. To characterize individual segments, we used the variables extracted for each polygon as described in the previous parts (see Appendix A). These variables were standardized to unit variance and zero mean. Subsequently, the principal component analysis (PCA) was applied to reduce the dimensionality of the feature space and prevent a redundant use of the same information stored in multiple mutually correlated features. The PCA was applied to each habitat type separately to reflect its specific attribute characteristics. To reduce the dimensionality, we then used the first n principal components that explained together 95% of the variability of the original features. With our data, the mean (±SD) n was 18 (±2.3) (see the biplots of the first two components for each habitat type in Appendix B).

We applied two principally different approaches to identify outliers in the space of the first n principal components. The first approach assumes the compactness of the segment characteristics distribution, i.e., that observations representing correctly mapped habitats (of a given type) lie close together while outliers are observations lying far from the center of this distribution. We used the robust Mahalanobis distance (hereafter referred to as MAH), as implemented in the robust package for R [34]. Mahalanobis distance is the distance of an observation from the center of the distribution relative to the thickness of the covariance ellipsoid in the corresponding direction. The estimation of the covariance ellipsoid itself is sensitive to the presence of outliers, and so we estimated it robustly using the method of Minimum Covariance Determinant [35] implemented in the same package. The result is a robust MAH distance computed for each segment.

The Local Outlier Factor (LOF [36]), implemented in the bigutilsr package for R, was the second approach we applied [37]. It relaxes the compactness assumption and identifies outliers as observations lying in locally "empty" or "sparse" parts of the distribution. To put it simply, each observation is evaluated based on its distance from neighbors relative to the mutual distances among these neighbors. The result is a relative LOF value computed for each observation.

For both approaches, the resulting MAH or LOF value had to be thresholded to decide whether the observation was an outlier or not. Similarly to PCA, this thresholding was performed for each habitat type separately. We applied three different types of thresholds. (1) First, we used the well-known Tukey rule, i.e., an outlier was the observation with the MAH/LOF value higher than the third quartile of all values plus 1.5 times their interquartile range. We used the modified version of this rule implemented in the package bigutilsr, corrected for a possible skewness of the distribution and for multiple comparisons [38]. (2) The second type of threshold we used, implemented in the same package as function hist_out, was based on the identification of a gap in the histogram of MAH/LOF values. The algorithm iteratively searches for an empty class in the histogram that lies above the 80th percentile of the data. If such a gap is found, all observations above this gap are considered outliers and the process is repeated for all the remaining observations. This is iterated until no gap is found. In each iteration, the class breaks are generated using the robust version of Scott's rule [39]. This histogram-based approach has the advantage of being adaptable to the actual distribution of the data. (3) The third type of threshold was applied only to the MAH method and was based on an assumption that the MAH values follow the χ^2 distribution with n degrees of freedom, and identifies outliers as the observations lying above the (1–0.05/N)th quantile, where n is the number of variables, and N is the number of observations. The division by N is the Bonferroni correction for multiple testing. Because LOF values cannot be expected to follow the χ^2 distribution, this approach is not applicable to this method.

The combination of the MAH/LOF methods with the three types of thresholds gave rise to five different methods of identification of outliers, namely MAHTukey, MAHhist, MAHχ^2, LOFTukey, and LOFhist. Apart from these, a combination of all these five methods designated as "Any" was evaluated, in which an observation is classified as an outlier if it was identified as such by any of the remaining five methods.

2.5. Method Validation

We validated our method in two different ways. First, we randomly selected 722 segments and checked their "true" habitat type by visual assessment of the corresponding orthophoto image from Google Earth Pro. Because the assumed prevalence of the truly erroneously mapped segments was very low, we could not apply purely random sampling, as this would probably result in a sample with no true mistakes. Instead, we applied stratified random sampling, trying to equally represent in the validation sample the outliers identified by the individual methods in the individual habitat types, and then added a comparable number of randomly selected inliers (i.e., non-outliers) in each habitat type. Because of the highly unequal prevalence of outliers and habitat types in the dataset, and because the corresponding orthophoto image was not available for some segments and years or the "true" habitat could not be visually identified, the resulting validation sample was not completely balanced (see Table 2). We then compared the truly erroneously mapped segments with the outliers identified using the individual methods using sensitivity, specificity, positive predictive value (PPV), and negative predictive value (NPV), corrected for the partial verification bias (i.e., bias caused by unequal prevalences of true mistakes in the original and validation samples) using the extended Begg and Greenes method [40]. We computed these metrics both for all habitat types together (with applying a correction for their unequal prevalences) and for each habitat type individually. We also computed the bootstrapped 95% confidence intervals for these metrics. We used the PVBcorrect package for R [41] for these computations.

Table 2. Numbers of classification errors and outliers (identified by the "Any" method) in the validation sample for the years 2018 and 2022.

Habitat Type	Mapping Errors 2018		Mapping Errors 2022		Outliers ("Any") 2018		Outliers ("Any") 2022	
	Number	%	Number	%	Number	%	Number	%
Alpines	15	26.79	15	26.79	28	50.00	25	44.64
Scrubs	57	61.29	56	60.22	30	32.26	40	43.01
Forests	41	23.03	70	39.55	78	43.82	104	58.43
Wetlands	67	72.04	69	74.19	47	50.54	54	58.06
Mires	33	68.75	33	68.75	26	54.17	29	60.42
Screes	23	58.97	23	58.97	15	38.46	17	43.59
Grass	32	27.59	41	35.65	68	58.62	65	56.03
Water	14	14.14	23	23.23	70	70.71	72	72.73
Total	282	39.06	15	26.79	331	45.84	406	56.23

Second, we investigated how sensitive the presented methods are to different levels of destruction of Natura 2000 polygons. A typical cause of habitat destruction is the expansion of human activity [42]. For this reason, we identified polygons of natural habitats that are mapped as artificial surfaces in the OpenStreetMap (OSM) database. To achieve this, we extracted polygons of all buildings, quarries, parking, parking multistore, service, bicycle parking, and all roads except bridleways, cycleways, foodways, paths, all grade tracks, and unclassified roads from the OSM database. Since the road network is drawn only as a line, we created the resulting polygons using a 3 m buffer, which is the standard lane size in the country. By subsequently overlaying the polygon layer of natural habitats and the polygon layer of artificial surfaces from OSM, we calculated the percentage destruction for each polygon. For subsequent analysis, we selected only polygons with an overlap greater than 20% and we manually inspected them using orthophoto images to verify if they were really damaged. A total of 59 polygons with verified damage were selected (see Appendix E). We then identified for each polygon whether or not it was marked as an outlier by the "Any" method, which was the method performing best in the previous validation. Finally, we divided the damaged segments into classes based on their percentage destruction and, for each class, we calculated the percentage of damaged segments that were identified as outliers.

3. Results

The results showed that most polygons were probably correctly mapped during the ground survey and all the proposed methods identified only a small fraction (0.06–9.41%) of the input segments as outliers. The total number of identified outliers, however, differed substantially between the methods (see Figure 2). The lowest number of outliers (140 to 290 outliers, i.e., 0.06 to 0.12% of all analyzed segments, respectively) was identified by the LOF methods. The MAHhist and MAHTukey methods identified about 17 times more outliers (from 2377 to 4881, i.e., 1.02 to 2.09%). The largest number of outliers was identified by the MAHχ^2 method, which identified 20,401 (8.72%) and 22,002 (9.40%) outliers in 2018 and 2022, respectively. The results of the "Any" method was, obviously, most influenced by this largest number of outliers identified by the MAHχ^2 method, and, therefore, the resulting number of outliers for ANY was almost identical to MAHχ^2 with an additional 8 and 9 outliers identified by one of the LOF methods in 2018 and 2022, respectively. Compared to the differences between methods, the differences between years were minor (see Figure 2). The numbers of outliers identified in the individual habitat types, however, exhibited much larger interannual variability, and also differed largely between the types, both in totals and percentages (see Appendix C).

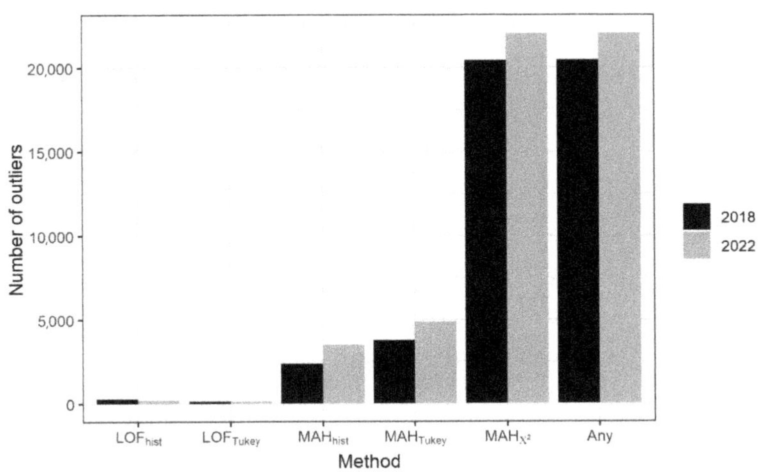

Figure 2. Numbers of outliers identified using different methods (Local Outlier Factor (LOF), Mahalanobis distance (MAH)) in combination with different thresholds (Tukey rule (Tukey), the gap in the histogram (hist), χ^2 distribution (χ^2)) in the years 2018 and 2022.

The structure of the validation sample is described in Table 2. The total number of outliers in particular years was slightly higher than the number of true classification errors. The prevalence of outliers was about 50–55%, while the prevalence of classification errors was about 40–45%. The difference between the number of outliers and classification errors varied a bit more for individual habitat types but, for most of the types, these values were also notably similar (see Table 2).

The sensitivity of individual methods after correcting for the partial verification bias and the unequal prevalence of habitat types (see Figure 3) was very low, especially for the methods using the histogram or Tukey rule thresholds, where it was between 0.1 and 3.9%. The corresponding specificities were above 98%. Applying the χ^2-quantile threshold to the MAH method led to considerably higher sensitivity (17.3 resp. 21.6% in 2018 resp. 2022), at the cost of slightly lower specificity (93.2 resp. 94.3% in 2018 resp. 2022). Unsurprisingly, the "Any" method performed similarly to the MAHχ^2 method. The PPVs were surprisingly similar between the methods (about 40 to 60%), with the values from 2022 being consistently higher than those from 2018. Similarly, the NPV values ranged between 60 to 80% for all methods, with the values from 2022 being lower than those from 2018.

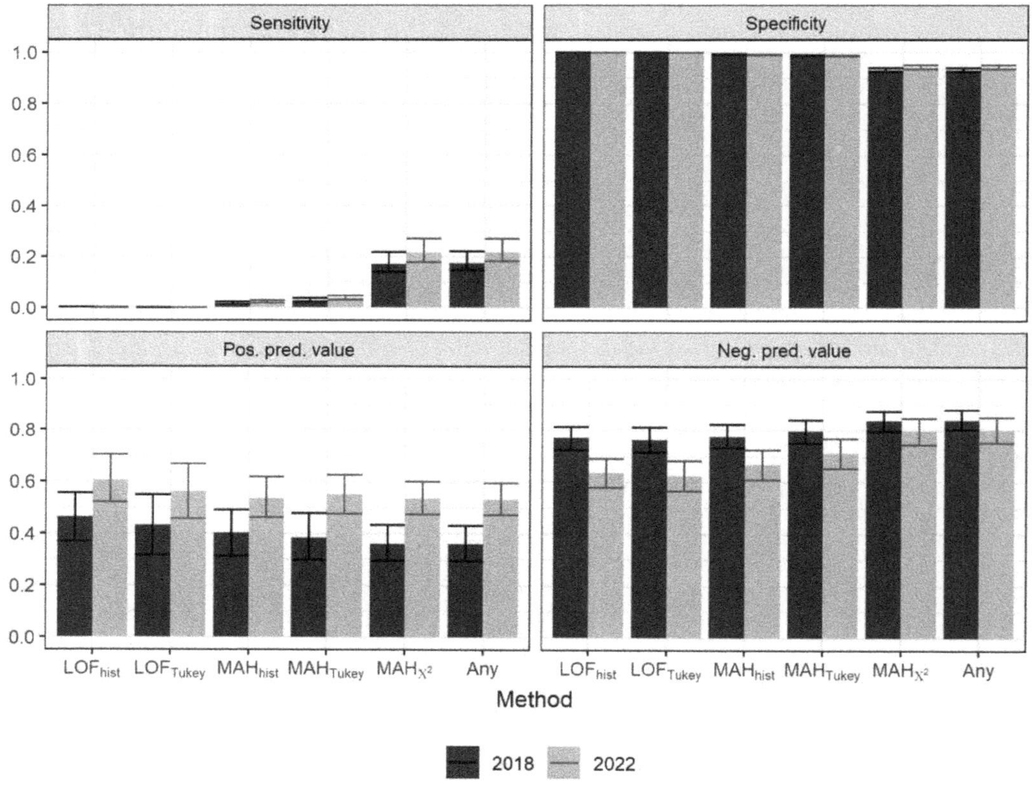

Figure 3. Accuracy metrics of the habitat classification error detection for the individual methods of spectral outlier detection. The method "Any" refers to the case when an outlier is identified if it is marked as an outlier by any of the remaining methods. Values are corrected for verification bias and reflect the unequal characteristics of the individual habitat types. The 95% confidence intervals are based on the percentiles of 999 bootstrap samples.

The variability in accuracy metrics between the individual habitat types (based on the MAHχ^2 method) was considerably higher (Figure 4), with the variability between years remaining low. For water habitats, all 14 and 23 mapping errors in 2018 and 2022, respectively, (see Table 2) were correctly identified as outliers, which led to a sensitivity of 100%. (It was, however, associated with the impossibility of computing confidence intervals due to the perfect separation problem.) The obvious cost of this high sensitivity of water habitats was slightly lower specificity, as the method actually marked 70 and 72 segments in 2018 and 2022, respectively, as outliers (see Table 2). The specificity, however, still remained above 80%. The second-best performance was observed for alpine habitats, with a sensitivity of about 70% and specificity of about 85%, accompanied by a PPV about 50% and NPV about 90%. Sensitivities of about 30–40% were found for forest and grassland habitats, with high specificities (more than 96%), as well as relatively high PPV and NPV. Wetlands showed a sensitivity of about 20%, specificity above 90%, and the highest PPV value (above 80%, although with a cost of NPV being only about 40%). The remaining habitat types (scrubs, mires, and screes) showed very low sensitivity values (about 5–10%).

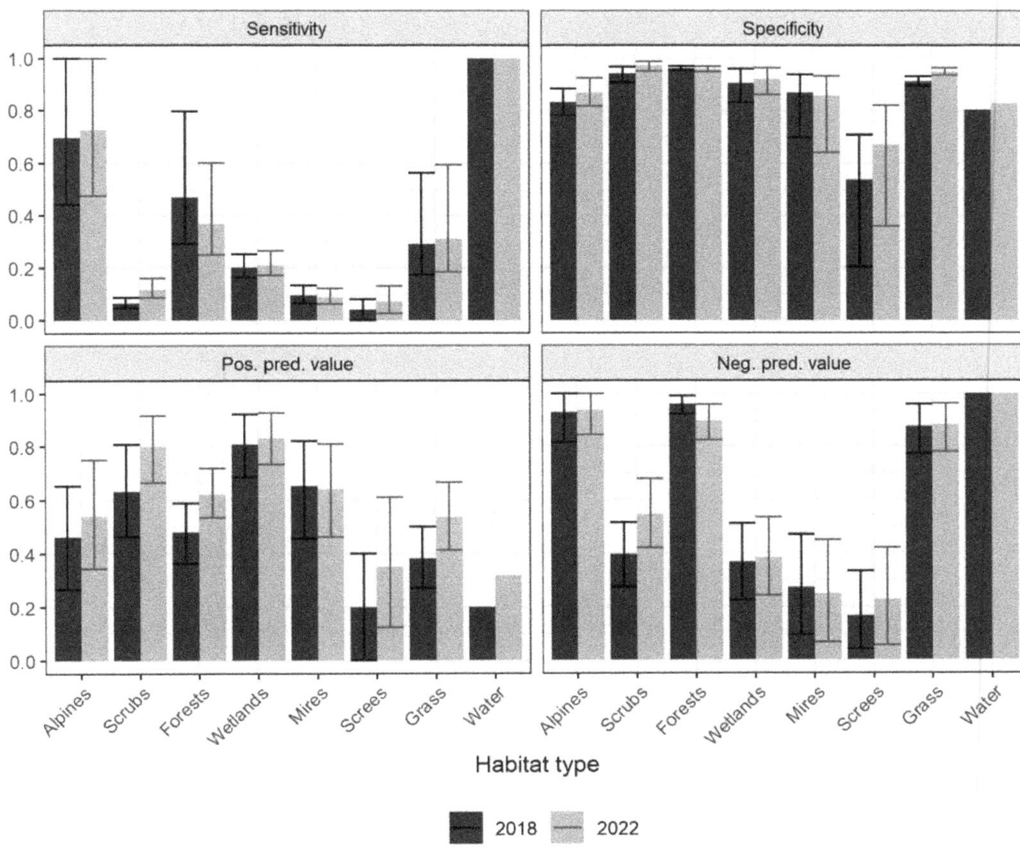

Figure 4. Accuracy metrics of the habitat classification error detection for the individual habitat types, based on the MAHχ^2 method. Values are corrected for verification bias. The 95% confidence intervals are based on the percentiles of 999 bootstrap samples.

Next, we evaluated the performance of the "Any" method with respect to the different levels of habitat destruction (Figure 5). We have found that, with as little as 20% of the area destroyed, the method was able to detect 78% of such degraded polygons. The detection capability increased further with increasing habitat destruction until it stabilized at around 91% for habitats with at least 60% of their area being destroyed.

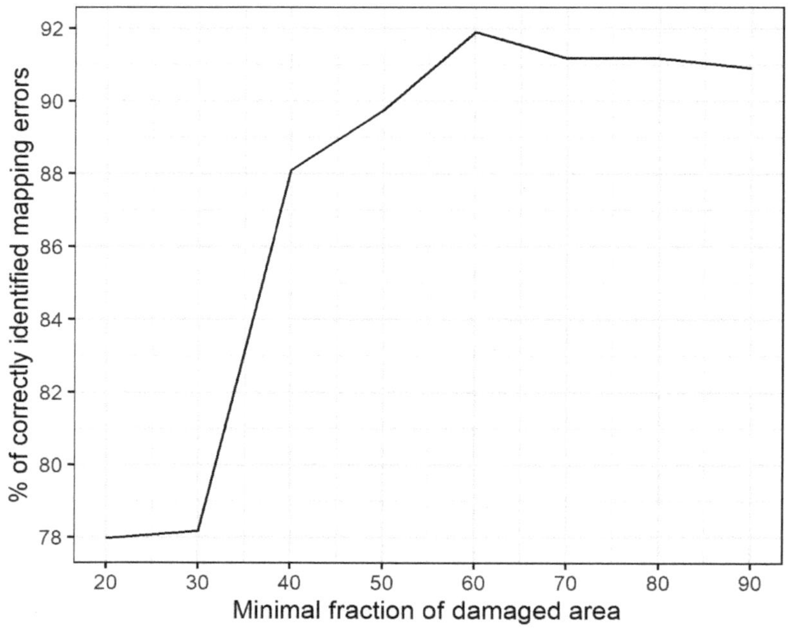

Figure 5. Percentage of correctly identified mapping errors out of 59 segments with a certain fraction of area covered by artificial surface, as a function of this fraction.

4. Discussion

Even though the study area was the whole Czech Republic, with very difficult-to-classify habitats, our results show that the proposed method is partially successful in identifying mapping errors. When all habitat types are analyzed together, the MAHχ^2 method (i.e., the one with the highest sensitivity) would be able to reveal about 20% of real mapping errors, while only about 6% of correctly mapped segments would be misidentified as potential mapping errors. When the method identifies a segment as a possible mapping error, there would be about a 40–50% probability that it is truly a mapping error. The probability that a segment identified by the method as correctly mapped is truly correctly mapped would be about 80%.

Our results also show (see Figure 5) that the proposed method is effective even when only part of the habitat has been altered. For this part of the analysis, we only considered cases where a habitat destruction due to construction occurred. Even in the case of 20% habitat destruction, the proposed method detected a change in 78% of cases. The method's reliability further increased to a maximum of 92% when 60% or more of the habitat was destroyed. This finding has important practical implications, as in most of the observed changes, the habitat is not completely destroyed, but rather gradually or partially transformed.

The proposed method is very flexible and suitable for many different applications. For example, a similar approach [43] was used to solve the one-class classification problem, where the authors also exploited outlier search from the overall hyperspectral data and achieved high classification accuracy that outperformed many other classifications approaches such as one-class random forests or positive and unsupervised learning.

The method showed the best performance for water habitats, where it revealed all mapping errors as outliers. It also performed relatively well for alpines, forests, grasslands, and wetlands, but the performance for other types was rather poor. Except for water, the sensitivities were generally not very high (the maximum was about 70% for alpines). Given the high specificities, however, it does not generate too much of a false signal and so even the relatively small fraction of revealed erroneously mapped habitats might still be valuable. The differences in the method performance between the habitat types can be easily explained by their spectral characteristics [44].

As we can see in Appendix D, some categories had more and some less different spectral characteristics. The water habitats are well separable from other categories (separability was above 1.9 (i.e., good separability) compared to other categories, except for wetlands. This was further reflected in Figure 6, where it can be seen that, in the case of map error occurrences, the habitat will probably appear as a significant attribute outlier. On the other hand, it can be seen from Appendix D that the category scrubs is very poorly separable from the others, except for water and alpines. Therefore, a slight change in the scrubs habitat (e.g., forestation or conversion to grassland) will not be revealed as an attribute outlier and such a change is difficult to detect. Figure 6 also suggests that there is quite clear inter-annual spectral variation within individual Natura 2000 habitats. This could further complicate some of the classical methods for classifying remote sensing data.

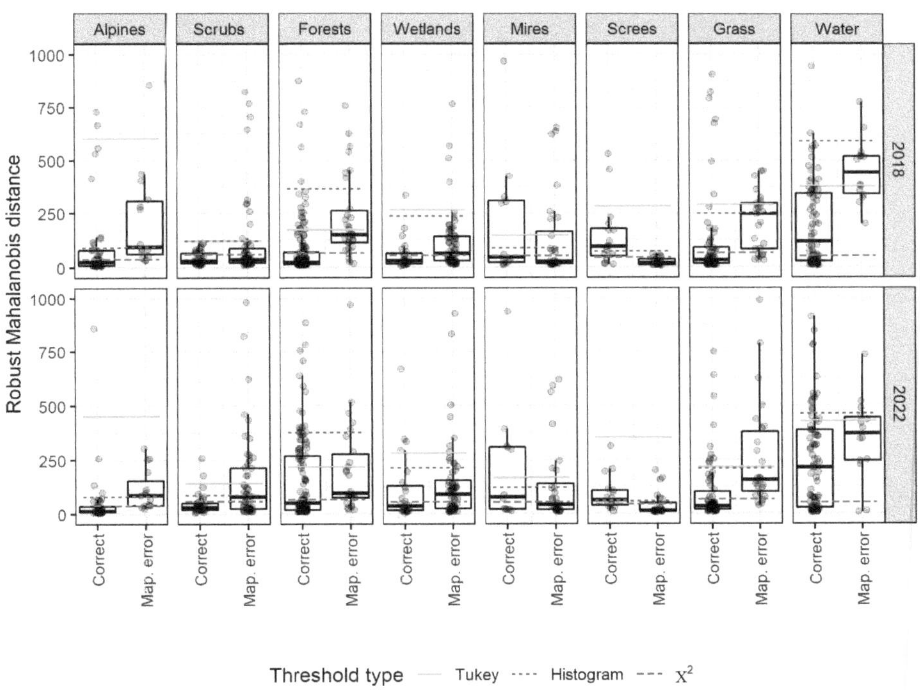

Figure 6. Robust Mahalanobis distance for correctly and erroneously mapped segments based on the validation sample for years 2018 and 2022. The three types of thresholds used to identify outliers are depicted by the horizontal lines. The data points (gray dots) are jittered to avoid overplotting and displayed in light grey under the boxplots.

The low spectral separability of the habitat classes in our study area seems to prevent any reliable classification, which apparently contradicts results of other studies suggesting a high success rate of direct classification of the Natura 2000 habitats from satellite data. Based on Sentinel-2 satellite imagery, Le Dez et al. [5] was able to classify a total of 39 selected habitat types in an alluvial test site in western France covering an area of 260 km^2. They chose the random forest method for classification and the resulting accuracy was up to 98% when using a total of 9 images across the study period. Three selected grassland types were classified in Marcinkowska-Ochtyra [6] with an accuracy of up to 88% on a 25 km^2 test site. A 72% success rate was achieved for a total of 18 different Natura 2000 habitat types in a study by Rapinel [7] on a 20 km^2 test site on the island of Corsica. However, these studies were always carried out on a limited sample of Natura 2000 habitat types and in a limited area.

In contrast to the aforementioned studies, habitat classification could be challenging for remote sensing at a whole country level. One reason is that, over a large area, the same habitats may be in different phenological phases at the same time (i.e., spectrally distinct from each other) and, thus, the classification algorithms may fail to assign them to the same habitat category, even though variables affecting phenology (e.g., elevation and terrain configuration) are also used for the classification. In this study, we suppressed the phenology problem using a medoid mosaic (see Section 2.5). A visual inspection of the results shows that a medoid mosaic effectively suppresses differences in phenology due to altitude and abrupt short-term habitat changes (e.g., water bloom). The second reason is that Natura 2000 habitat categories based on phytocenology are not always following spectral classes detectable by remote sensing. This leads to the definition of classes which do not necessarily correspond well with the remote sensing clusters (such as peatlands overgrown with bushes, dry wetlands, a forest that was in the early stage of development after planting, very sparse grass cover looking like bare soil, or scree completely overgrown with forest, etc.). This could be seen in Appendix D, where almost no or very limited separability of classes occurs over Natura 2000 habitats in the Czech Republic. In such cases, it is obvious that the traditional classification will not be able to classify the given categories correctly at a country level. However, the proposed method proved to be robust enough to be useful in this case as well.

The main limitation of our approach to monitoring Natura 2000 habitat change is that it still relies primarily on the spectral expression of the top layer of land cover. The possible addition of long wavelengths reference [45] or other (in situ sensor measurements) is allowed by the chosen method and could contribute to the resulting accuracy of the classifications. Another limit may occur in areas with very high cloud cover. Here, the medoid mosaic method may not have a sufficient number of cloud-free observations to capture the most representative pixel [46,47].

Despite the above-mentioned limitations, the proposed method proved to be applicable. In the future, it will certainly be interesting to validate this method with a detailed field survey. Unfortunately, it was not possible for the research team to perform a field validation at the level of the whole Czech Republic. However, a validation campaign will take place in the next few years thanks to the Nature Conservation Agency of the Czech Republic. Thanks to these data, we expect the future development and refinement of the method described above.

5. Conclusions

This paper shows that it is possible to use remote sensing, together with the proposed "attribute outliers" method for existing Natura 2000 field-based habitat maps, to detect changes that can then be examined in detail. This method proved to be very reliable in detecting even partial destruction of the habitat. However, it also shows that, in the case of a gradual change between habitat types, many changes remain unidentifiable and hidden in the attribute ambiguity of individual habitats.

The method based on the robust Mahalanobis distance thresholded by the 95% χ^2 quantile proved to be a promising tool for identifying erroneously mapped or changed habitats. Because of its high specificity, it can relatively reliably identify a certain fraction of habitats for which a subsequent terrain revision is likely to confirm a mapping error. In this way, the method can contribute to a significant improvement in the Natura 2000 mapping efforts, although some mapping errors will still remain unrevealed. The method performs best for water habitats, but it is usable for alpines, forests, grasslands, and wetlands as well. Furthermore, the proposed method has broad applicability for detecting misclassifications in remotely sensed data, such as land cover, and its performance is anticipated to improve when the classified classes align with discernible features identifiable from an aerial perspective.

Author Contributions: Conceptualization, D.M., P.Š. and V.B.; methodology, D.M. and V.B.; validation D.M.; formal analysis, D.M. and V.B.; data curation, D.M. and V.B.; writing—original draft preparation, D.M. and V.B.; writing—review and editing, P.Š.; project administration, P.Š.; funding acquisition, P.Š. All authors have read and agreed to the published version of the manuscript.

Funding: This research was funded by a Technology Agency of the Czech Republic (Environment for Life program) project, Possibilities for updating map layers of NATURA 2000 habitats using advanced remote sensing methods, Grant Number: SS01010046, 2021–2023.

Data Availability Statement: The Habitat Layer of the Czech Republic is available from the Czech Nature Conservation Agency (https://data.nature.cz, accessed on 5 December 2022). The digital elevation model EU DEM is available from the Copernicus Programme web page (https://land.copernicus.eu/imagery-in-situ/eu-dem/eu-dem-v1, accessed on 5 December 2022). The Mosaic of Sentinel-2 satellite image can be downloaded from the Sentinel-2 Global Mosaic web page (https://apps.sentinel-hub.com/mosaic-hub/, accessed on 5 December 2022). The processed Sentinel-1 SAR images could be obtained from The Alaska Satellite Facility Vertex (https://search.asf.alaska.edu/, accessed on 5 December 2022).

Acknowledgments: The authors are grateful to the Nature Conservation Agency of the Czech Republic for their consultations during the project. We would also like to thank Jaroslav Janošek for his valuable comments.

Conflicts of Interest: The authors declare no conflict of interest.

Appendix A

Table A1. Characteristics of Sentinel-2 bands used for the final product.

Data Source	Variable	Polygon Aggregation
Sentinel-2	Band n.2—blue	Average, Median, Std. deviation
Sentinel-2	Band n.3—green	Average, Median, Std. deviation
Sentinel-2	Band n.4—red	Average, Median, Std. deviation
Sentinel-2	Band n.5—red edge 1	Average, Median, Std. deviation
Sentinel-2	Band n.6—red edge 2	Average, Median, Std. deviation
Sentinel-2	Band n.7—red edge 3	Average, Median, Std. deviation
Sentinel-2	Band n.8—NIR	Average, Median, Std. deviation
Sentinel-2	Band n.8A—red edge 4	Average, Median, Std. deviation
Sentinel-2	Band n.11—SWIR	Average, Median, Std. deviation
Sentinel-2	Band n.12—SWIR 2	Average, Median, Std. deviation
Sentinel-1	RADAR VH April	Average, Median, Std. deviation
Sentinel-1	RADAR VH May	Average, Median, Std. deviation
Sentinel-1	RADAR VH June	Average, Median, Std. deviation
Sentinel-1	RADAR VH July	Average, Median, Std. deviation
Sentinel-1	RADAR VH August	Average, Median, Std. deviation
Sentinel-1	RADAR VH September	Average, Median, Std. deviation
Sentinel-1	RADAR VH October	Average, Median, Std. deviation
EU DEM	Slope	Average
EU DEM	Altitude	Average
EU DEM	Aspect	Modus
NATURA 2000	Polygon Size	

Appendix B

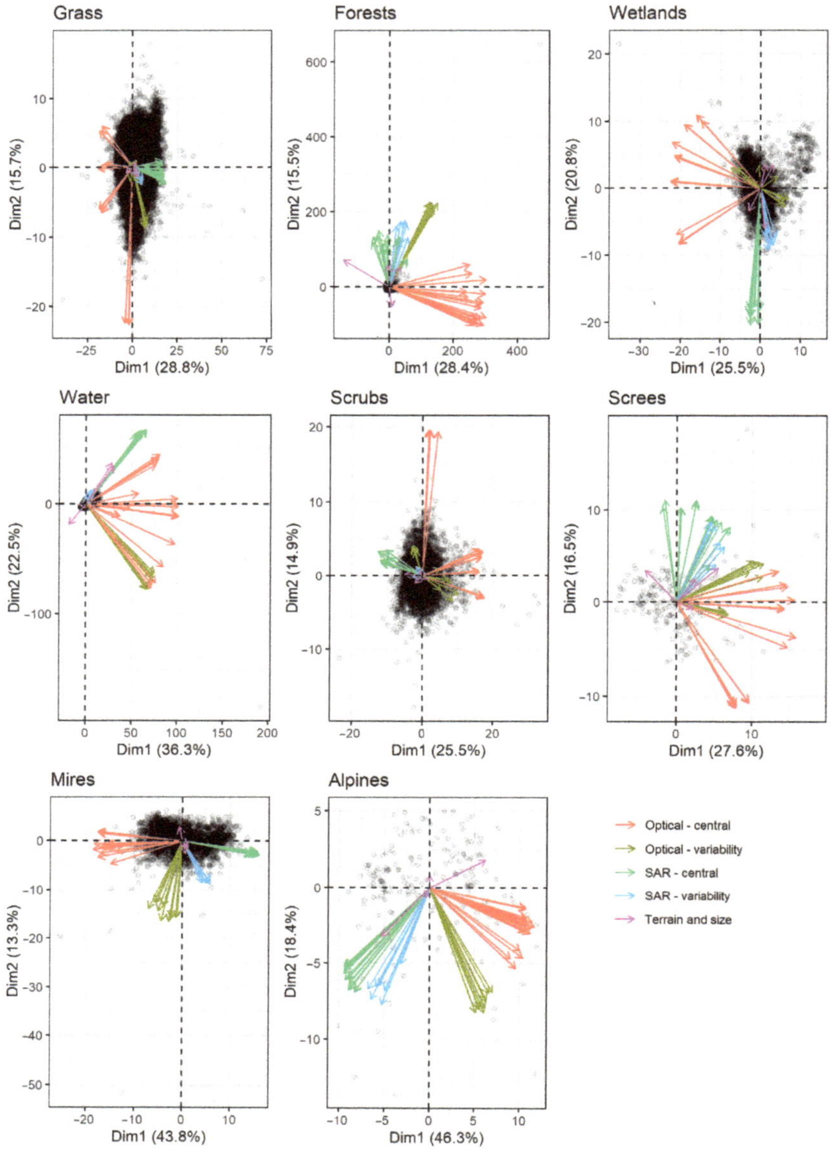

Figure A1. The biplots (gray dots) of the first two principle components for the individual habitat types, based on the features from 2018. The original features are grouped into (1) central measures (means and medians) of the optical bands (red), (2) standard deviations of the optical bands (yellow), (3) central measures (means and medians) of the RADAR bands (green), (4) standard deviations of the RADAR bands (blue), and (5) terrain and geometrical features (violet). The percentage of the data variability explained by each component is shown at the axes labels.

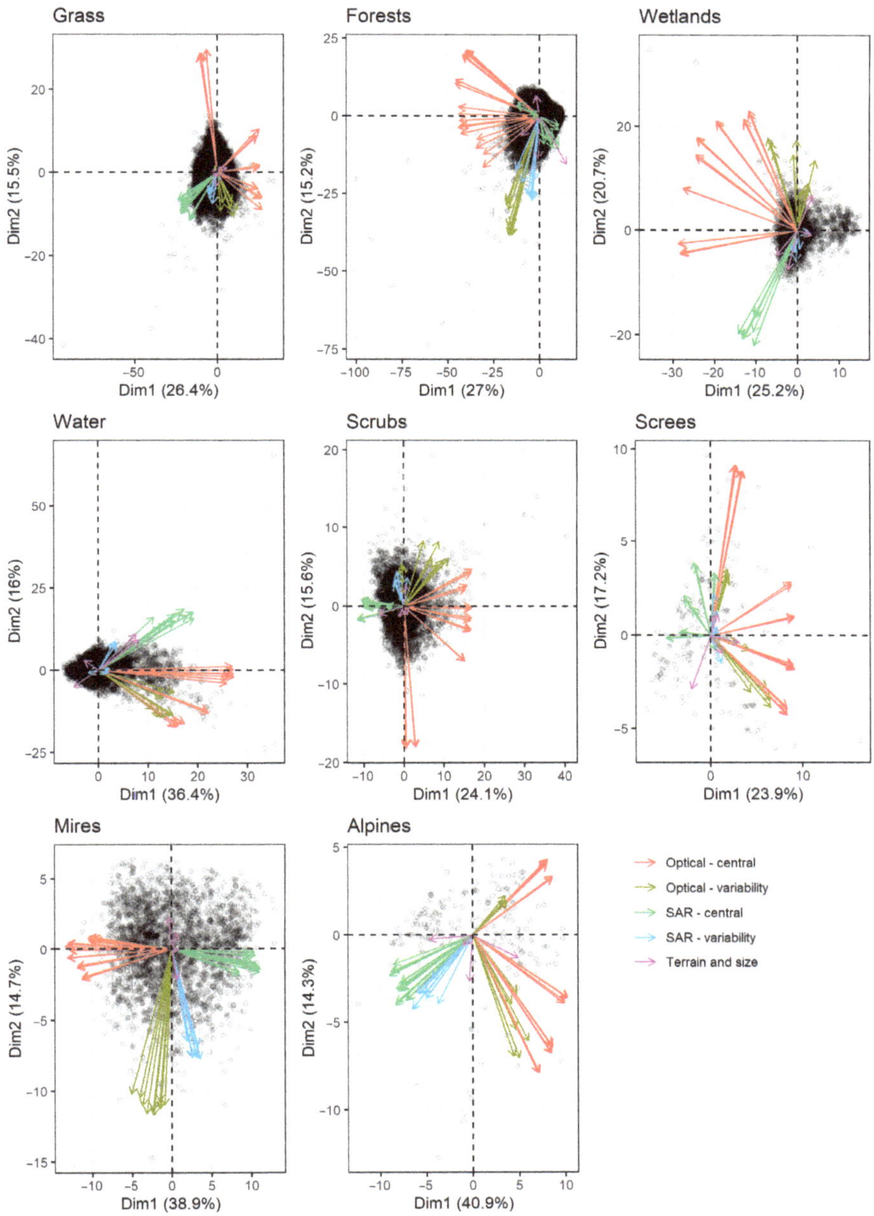

Figure A2. The biplots (gray dots) of the first two principle components for the individual habitat types, based on the features from 2022. The original features are grouped into (1) central measures (means and medians) of the optical bands (red), (2) standard deviations of the optical bands (yellow), (3) central measures (means and medians) of the RADAR bands (green), (4) standard deviations of the RADAR bands (blue), and (5) terrain and geometrical features (violet). The percentage of the data variability explained by each component is shown at the axes labels.

Appendix C

Table A2. Numbers of outliers identified using different methods (Local Outlier Factor (LOF), Mahalanobis distance (MAH)) in combination with different thresholds (Tukey rule (Tukey), the gap in the histogram (hist), X^2 distribution (X^2)) in the year 2018.

Habitat Type	MAH_{hist}		MAH_{Tukey}		MAH_{χ^2}		LOF_{hist}		LOF_{Tukey}		Any	
	#	%	#	%	#	%	#	%	#	%	#	%
Alpines	22	11.5	4	2.1	50	26	0	0	0	0	50	16.2
Scrubs	79	2.01	79	2	246	6.26	16	0.4	8	0.2	246	2.54
Forests	1304	0.84	2749	1.8	10625	6.82	113	0.1	57	0	10630	2.33
Wetlands	61	2.3	57	2.2	439	16.6	16	0.6	6	0.2	439	10.4
Mires	117	5.28	87	3.9	234	10.6	27	1.2	7	0.3	236	7
Screes	15	7.65	2	1	30	15.3	5	2.6	1	0.5	30	9.18
Grass	698	1.1	598	0.9	7537	11.8	96	0.2	46	0.1	7538	6.03
Water	81	1.53	190	3.6	1240	23.5	17	0.3	15	0.3	1240	19

Table A3. Numbers of outliers identified using different methods (Local Outlier Factor (LOF), Mahalanobis distance (MAH)) in combination with different thresholds (Tukey rule (Tukey), the gap in the histogram (hist), X^2 distribution (X^2)) in the year 2022.

Habitat Type	MAH_{hist}		MAH_{Tukey}		MAH_{χ^2}		LOF_{hist}		LOF_{Tukey}		Any	
	#	%	#	%	#	%	#	%	#	%	#	%
Alpines	23	12	2	1	45	23.4	14	7.3	1	0.5	46	24
Scrubs	136	3.46	80	2	276	7.02	23	0.6	12	0.3	276	7.02
Forests	1635	1.05	3166	2	13967	8.96	82	0.1	55	0	13974	8.97
Wetlands	82	3.09	67	2.5	435	16.4	19	0.7	10	0.4	435	16.4
Mires	86	3.88	73	3.3	227	10.3	10	0.5	7	0.3	228	10.3
Screes	21	10.7	0	0	29	14.8	2	1	0	0	29	14.8
Grass	1350	2.12	1324	2.1	5793	9.1	76	0.1	59	0.1	5793	9.1
Water	155	2.93	169	3.2	1230	23.3	8	0.2	8	0.2	1230	23.3

Appendix D

Table A4. Jeffries–Matusita spectral separability of Natura 2000 habitats. Values below 1.9 indicate poor separability.

	Alpines	Scrubs	Forests	Wetlands	Mires	Screes	Grass	Water
Alpines		1.72	1.76	1.65	1.33	1.47	1.68	1.93
Scrubs	1.72		0.93	1.08	1.30	0.99	1.14	1.98
Forests	1.76	0.93		1.45	1.31	1.13	1.77	1.93
Wetlands	1.65	1.08	1.45		1.16	1.32	1.05	1.56
Mires	1.33	1.30	1.31	1.16		1.29	1.16	1.91
Screes	1.47	0.99	1.13	1.32	1.29		1.66	1.89
Grass	1.68	1.14	1.77	1.05	1.16	1.66		2.00
Water	1.93	1.98	1.93	1.56	1.91	1.89	2.00	

Table A5. Transformed divergence spectral separability of Natura 2000 habitats. Values below 1.9 indicate poor separability.

	Alpines	Scrubs	Forests	Wetlands	Mires	Screes	Grass	Water
Alpines		1.90	1.99	1.90	1.60	1.77	1.84	2.00
Scrubs	1.90		1.01	1.39	1.40	1.17	1.21	2.00
Forests	1.99	1.01		1.69	1.58	1.61	1.84	2.00
Wetlands	1.90	1.39	1.69		1.47	1.58	1.58	1.86
Mires	1.60	1.40	1.58	1.47		1.60	1.33	1.99
Screes	1.77	1.17	1.61	1.58	1.60		1.75	1.99
Grass	1.84	1.21	1.84	1.58	1.33	1.75		2.00
Water	2.00	2.00	2.00	1.86	1.99	1.99	2.00	

Appendix E

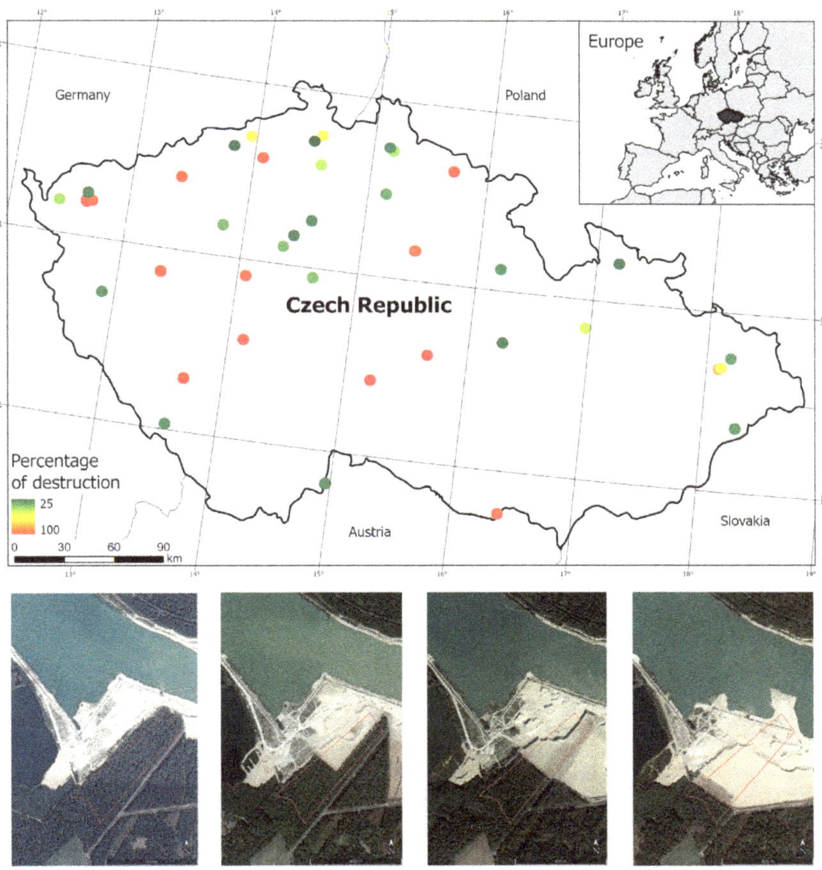

Figure A3. Distribution of verified polygons (red line) and their percentage loss together with an example of the gradual destruction of the forest habitat in 2015, 2017, 2019, and 2022 due to the gradual expansion of the sandpit near Nová Ves nad Lužnicí.

References

1. Feilhauer, H.; Dahlke, C.; Doktor, D.; Lausch, A.; Schmidtlein, S.; Schulz, G.; Stenzel, S. Mapping the local variability of Natura 2000 habitats with remote sensing. *Appl. Veg. Sci.* **2014**, *17*, 765–779. [CrossRef]
2. Lang, S.; Mairota, P.; Pernkopf, L.; Schioppa, E.P. Earth observation for habitat mapping and biodiversity monitoring. *Int. J. Appl. Earth Obs. Geoinf.* **2015**, *37*, 1–160. [CrossRef]
3. Willis, K.S. Remote sensing change detection for ecological monitoring in United States protected areas. *Biol. Conserv.* **2015**, *182*, 233–242. [CrossRef]
4. Vanden Borre, J.; Paelinckx, D.; Mücher, C.A.; Kooistra, L.; Haest, B.; De Blust, G.; Schmidt, A.M. Integrating remote sensing in Natura 2000 habitat monitoring: Prospects on the way forward. *J. Nat. Conserv.* **2011**, *19*, 116–125. [CrossRef]
5. Le Dez, M.; Robin, M.; Launeau, P. Contribution of Sentinel-2 satellite images for habitat mapping of the Natura 2000 site 'Estuaire de la Loire' (France). *Remote Sens. Appl. Soc. Environ.* **2021**, *24*, 100637. [CrossRef]
6. Marcinkowska-Ochtyra, A.; Ochtyra, A.; Raczko, E.; Kopeć, D. Natura 2000 Grassland Habitats Mapping Based on Spectro-Temporal Dimension of Sentinel-2 Images with Machine Learning. *Remote Sens.* **2023**, *15*, 1388. [CrossRef]
7. Rapinel, S.; Rozo, C.; Delbosc, P.; Bioret, F.; Bouzillé, J.B.; Hubert-Moy, L. Contribution of free satellite time-series images to mapping plant communities in the Mediterranean Natura 2000 site: The example of Biguglia Pond in Corse (France). *Mediterr. Bot.* **2020**, *41*, 181–191. [CrossRef]
8. Löfgren, O.; Prentice, H.C.; Moeckel, T.; Schmid, B.C.; Hall, K. Landscape history confounds the ability of the NDVI to detect fine-scale variation in grassland communities. *Methods Ecol. Evol.* **2018**, *9*, 2009–2018. [CrossRef]
9. He, K.S.; Rocchini, D.; Neteler, M.; Nagendra, H. Benefits of hyperspectral remote sensing for tracking plant invasions. *Divers. Distrib.* **2011**, *17*, 381–392. [CrossRef]
10. Middleton, M.; Närhi, P.; Arkimaa, H.; Hyvönen, E.; Kuosmanen, V.; Treitz, P.; Sutinen, R. Ordination and hyperspectral remote sensing approach to classify peatland biotopes along soil moisture and fertility gradients. *Remote Sens. Environ.* **2012**, *124*, 596–609. [CrossRef]
11. Luft, L.; Neumann, C.; Itzerott, S.; Lausch, A.; Doktor, D.; Freude, M.; Blaum, N.; Jeltsch, F. Digital and real-habitat modeling of Hipparchia statilinus based on hyper spectral remote sensing data. *Int. J. Environ. Sci. Technol.* **2016**, *13*, 187–200. [CrossRef]
12. Schmidt, J.; Fassnacht, F.E.; Förster, M.; Schmidtlein, S. Synergetic use of Sentinel-1 and Sentinel-2 for assessments of heathland conservation status. *Remote Sens. Ecol. Conserv.* **2018**, *4*, 225–239. [CrossRef]
13. Erinjery, J.J.; Singh, M.; Kent, R. Mapping and assessment of vegetation types in the tropical rainforests of the Western Ghats using multispectral Sentinel-2 and SAR Sentinel-1 satellite imagery. *Remote Sens. Environ.* **2018**, *216*, 345–354. [CrossRef]
14. Pechanec, V.; Machar, I.; Pohanka, T.; Opršal, Z.; Petrovič, F.; Švajda, J.; Šálek, L.; Chobot, K.; Filippovová, J.; Cudlín, P.; et al. Effectiveness of Natura 2000 system for habitat types protection: A case study from the Czech Republic. *Nat. Conserv.* **2018**, *24*, 21–41. [CrossRef]
15. Härtel, H.; Lončáková, J.; Hošek, M. *Mapování Biotopů v České Republice. Východiska, Výsledky, Perspektivy*; Agentura Ochrany Přírody a Krajiny ČR: Praha, Czech Republic, 2009; ISBN 9788087051368.
16. Divíšek, J.; Chytrý, M.; Grulich, V.; Poláková, L. Landscape classification of the Czech Republic based on the distribution of natural habitats. *Preslia* **2014**, *86*, 209–231.
17. Schneider, J.; Ruda, A.; Kalasová, Ž.; Paletto, A. The forest stakeholders' perception towards the NATURA 2000 network in the Czech Republic. *Forests* **2020**, *11*, 491. [CrossRef]
18. Bastian, O.; Neruda, M.; Filipová, L.; Machová, I.; Leibenath, M. Natura 2000 Sites as an Asset for Rural Development: The German-Czech Ore Mountains Green Network Project. *J. Landsc. Ecol.* **2012**, *3*, 41–58. [CrossRef]
19. Afaq, Y.; Manocha, A. Analysis on change detection techniques for remote sensing applications: A review. *Ecol. Inform.* **2021**, *63*, 101310. [CrossRef]
20. Coops, N.C.; Wulder, M.A.; White, J.C. Identifying and describing forest disturbance and spatial pattern: Data selection issues and methodological implications. In *Understanding Forest Disturbance and Spatial Pattern: Remote Sensing and GIS Approaches*; CRC Press (Taylor and Francis): Boca Raton, FL, USA, 2007.
21. Chytrý, M.; Kučera, T.; Kočí, M. *Katalog Biotopů České Republiky*; AOPK: Praha, Czech Republic, 2001.
22. Forkuor, G.; Dimobe, K.; Serme, I.; Tondoh, J.E. Landsat-8 vs. Sentinel-2: Examining the added value of sentinel-2's red-edge bands to land-use and land-cover mapping in Burkina Faso. *GIScience Remote Sens.* **2018**, *55*, 331–354. [CrossRef]
23. Persson, M.; Lindberg, E.; Reese, H. Tree species classification with multi-temporal Sentinel-2 data. *Remote Sens.* **2018**, *10*, 1794. [CrossRef]
24. Otunga, C.; Odindi, J.; Mutanga, O.; Adjorlolo, C. Evaluating the potential of the red edge channel for C3 (*Festuca* spp.) grass discrimination using Sentinel-2 and Rapid Eye satellite image data. *Geocarto Int.* **2019**, *34*, 1123–1143. [CrossRef]
25. Ferrazzoli, P.; Paloscia, S.; Pampaloni, P.; Schiavon, G.; Solimini, D.; Coppo, P. Sensitivity of Microwave Measurements to Vegetation Biomass and Soil Moisture Content: A Case Study. *IEEE Trans. Geosci. Remote Sens.* **1992**, *30*, 750–756. [CrossRef]
26. Paloscia, S.; Macelloni, G.; Pampaloni, P. The relations between backscattering coefficient and biomass of narrow and wide leaf crops. In Proceedings of the IGARSS '98. Sensing and Managing the Environment. 1998 IEEE International Geoscience and Remote Sensing. Symposium Proceedings. (Cat. No.98CH36174), Seattle, WA, USA, 6–10 July 1998; pp. 100–102.
27. Macelloni, G.; Paloscia, S.; Pampaloni, P.; Marliani, F.; Gai, M. The relationship between the backscattering coefficient and the biomass of narrow and broad leaf crops. *IEEE Trans. Geosci. Remote Sens.* **2001**, *39*, 873–884. [CrossRef]

28. Dobrinić, D.; Gašparović, M.; Medak, D. Sentinel-1 and 2 time-series for vegetation mapping using random forest classification: A case study of northern croatia. *Remote Sens.* **2021**, *13*, 2321. [CrossRef]
29. Kaushik, S.K.; Mishra, V.N.; Punia, M.; Diwate, P.; Sivasankar, T.; Soni, A.K. Crop Health Assessment Using Sentinel-1 SAR Time Series Data in a Part of Central India. *Remote Sens. Earth Syst. Sci.* **2021**, *4*, 217–234. [CrossRef]
30. Cracknell, A.P. Review article Synergy in remote sensing-what's in a pixel? *Int. J. Remote Sens.* **1998**, *19*, 2025–2047. [CrossRef]
31. Kirches, G. Algorithm Theoretical Basis Document Sentinel 2 Global Mosaics Copernicus Sentinel-2 Global Mosaic (S2GM) within the Global Land Component of the Copernicus Land Service. 2018. Available online: https://usermanual.readthedocs.io/en/1.1.2/_downloads/5a2d961d53dea1eb1117ec73e4cbff09/S2GM-SC2-ATBD-BC-v1.3.2.pdf (accessed on 20 December 2022).
32. Esri Inc. ArcGIS Pro 2.7.0. 2020. Available online: https://www.esri.com/ (accessed on 20 December 2022).
33. QGIS 3.22.1. 2021. Available online: https://qgis.org/ (accessed on 20 December 2022).
34. Wang, A.J.; Zamar, R.; Alfiomarazziinsthospvdch, A.M.; Yohai, V.; Salibian-barrera, M.; Maronna, R.; Zivot, E.; Rocke, D.; Martin, D.; Maechler, M. robust: Port of the S+ "Robust Library"; R package version 0.7-1. 2022. Available online: https://cran.r-project.org/package=robust (accessed on 22 January 2023).
35. Rousseeuw, P.J.; Van Driessen, K. A fast algorithm for the minimum covariance determinant estimator. *Technometrics* **1999**, *41*, 212–223. [CrossRef]
36. Breunig, M.M.; Kriegel, H.P.; Ng, R.T.; Sander, J. LOF: Identifying Density-Based Local Outliers. In Proceedings of the SIGMOD '00: Proceedings of the 2000 ACM SIGMOD International Conference on Management of Data, Dallas, TX, USA, 15–18 May 2000; pp. 93–104.
37. Privé, F. Utility Functions for Large-Scale Data; R package version 0.3.4. 2021. Available online: https://cran.r-project.org/package=bigutilsr (accessed on 22 January 2023).
38. Hubert, M.; Vandervieren, E. An adjusted boxplot for skewed distributions. *Comput. Stat. Data Anal.* **2008**, *52*, 5186–5201. [CrossRef]
39. Scott, D.W. On optimal and data-based histograms. *Biometrika* **1979**, *66*, 605–610. [CrossRef]
40. Arifin, W.N.; Yusof, U.K. Correcting for partial verification bias in diagnostic accuracy studies: A tutorial using R. *Stat. Med.* **2022**, *41*, 1709–1727. [CrossRef] [PubMed]
41. Arifin, W.N. PVBcorrect: Partial Verification Bias Correction for Estimates of Accuracy Measures in Diagnostic Accuracy Studies; R package version 0.1.1. 2023. Available online: https://rdrr.io/github/wnarifin/PVBcorrect/man/PVBcorrect.html (accessed on 22 January 2023).
42. Kirschner, V.; Franke, D.; Řezáčová, V.; Peltan, T. Poorer Regions Consume More Undeveloped but Less High-Quality Land Than Wealthier Regions—A Case Study. *Land* **2023**, *12*, 113. [CrossRef]
43. Shi, Z.; Li, P.; Sun, Y. An outlier generation approach for one-class random forests: An example in one-class classification of remote sensing imagery. In Proceedings of the 2016 IEEE International Geoscience and Remote Sensing Symposium (IGARSS), Beijing, China, 10–15 July 2016; pp. 5107–5110.
44. Perrone, M.; Di Febbraro, M.; Conti, L.; Divíšek, J.; Chytrý, M.; Keil, P.; Carranza, M.L.; Rocchini, D.; Torresani, M.; Moudrý, V.; et al. The relationship between spectral and plant diversity: Disentangling the influence of metrics and habitat types at the landscape scale. *Remote Sens. Environ.* **2023**, *293*, 113591. [CrossRef]
45. Zhang, H.; Zhu, J.; Wang, C.; Lin, H.; Long, J.; Zhao, L.; Fu, H.; Liu, Z. Forest growing stock volume estimation in subtropical mountain areas using PALSAR-2 L-Band PolSAR data. *Forests* **2019**, *10*, 276. [CrossRef]
46. Nazarova, T.; Martin, P.; Giuliani, G. Monitoring vegetation change in the presence of high cloud cover with sentinel-2 in a lowland tropical forest region in Brazil. *Remote Sens.* **2020**, *12*, 1829. [CrossRef]
47. Flood, N. Seasonal composite landsat TM/ETM+ Images using the medoid (a multi-dimensional median). *Remote Sens.* **2013**, *5*, 6481–6500. [CrossRef]

remote sensing

MDPI

Article

Multi-Difference Image Fusion Change Detection Using a Visual Attention Model on VHR Satellite Data

Jianhui Luo [1,2], Qiang Chen [1,2,*], Lei Wang [1,2] and Yixiao Huang [1,2]

1 School of Geomatics and Urban Spatial Informatics, Beijing University of Civil Engineering and Architecture, Beijing 102616, China
2 Key Laboratory of Urban Spatial Information, Ministry of Natural Resources, Beijing University of Civil Engineering and Architecture, Beijing 102616, China
* Correspondence: chenqiang@bucea.edu.cn

Abstract: For very-high-resolution (VHR) remote sensing images with complex objects and rich textural information, multi-difference image fusion has been proven as an effective method to improve the performance of change detection. However, errors are superimposed during this process and a single spectral feature cannot fully utilize the correlation between pixels, resulting in low robustness. To overcome these problems and optimize the performance of multi-difference image fusion in change detection, we propose a novel multi-difference image fusion change detection method based on a visual attention model (VA-MDCD). First, we construct difference images using change vector analysis (CVA) and spectral gradient difference (SGD). Second, we use the visual attention model to calculate multiple color, intensity and orientation features of the difference images to obtain the difference saliency images. Third, we use the wavelet transform fusion algorithm to fuse two saliency images. Finally, we execute the OTSU threshold segmentation algorithm (OTSU) to obtain the final change detection map. To validate the effectiveness of VA-MDCD on VHR images, two datasets of Jilin 1 and Beijing 2 are selected for experiments. Compared with classical methods, the proposed method has a better performance with fewer missed alarms (MA) and false alarms (FA), which proves that the method has a strong robustness and generalization ability. The F-measure of the two datasets is 0.6671 and 0.7313, respectively. In addition, the results of ablation experiments confirm that the three feature extraction modules of the model all play a positive role.

Keywords: very high resolution (VHR); change detection; multi-difference image fusion; visual attention model; feature extraction

1. Introduction

In the rapidly changing modern society, phenomena such as soil erosion, geological disasters and deforestation occur from time to time [1]. To build a modern smart city [2,3], it is necessary to monitor information about changes in the land [4,5]. Remote sensing change detection refers to the analysis and identification of changes between remote sensing images using statistics and mathematical models [6]. We can obtain a large amount of ground surface details from very-high-resolution (VHR) remote sensing images, which can be used to compare geographic objects at different stages [7]. Detecting detailed changes in geographic objects has important implications for map updating, urban planning, management and disaster management [8,9].

With the development of satellite technology, the resolution of remote sensing images also increases [10–12]; the improvement in image resolution facilitates the acquisition of detailed information regarding ground objects, including their spatial, contrast and morphological relationships [13–15]. It is important to exploit the correlation between these detailed features. It is easy to describe subject changes in both changing and invariant regions with a single change image, but it is difficult to reveal the details of VHR images. With the development of computer science, computer vision has been applied to many fields of remote sensing [16]. It has become a hot point in remote sensing research of combination VHR image change detection and computer vision [17], which aims to detect changes in multidirectional relationships between images using computer vision and change detection algorithms.

Remote sensing change detection involves primarily two methods: image direct comparison and post-classification comparison [18]. The post-classification comparison method involves the classification of two images of different phases, followed by a comparison of the resulting ground feature types to obtain change detection results. However, this method only considers the current image, disregarding the interconnection between different time phase images, and it is difficult to label samples manually. Consequently, classification errors tend to accumulate and overlap between different objects, leading to inaccuracies during the detection process [19]. In contrast, the direct comparison method is a widely used and straightforward approach for comparing two images [20]. The quality of the direct comparison method relies on obtaining a difference image between images taken at two phases [21,22].

There are a variety of ways to produce difference images, such as the method of taking a logarithmic ratio of remote sensing images [23]. Although the logarithmic method produces relatively good difference images for dual-temporal images, it has been observed that the resulting images lack adequate ground information. Therefore, some researchers have proposed an iterative robust graph-based method for change detection. This method uses the K-nearest neighbor and image cross-mapping techniques to obtain the forward and backward difference images [24]; the Markov model is used to detect changes. Although the robustness is high, the iteration cycles of this method will affect the rate. For classical change detection methods, numerous scholars have conducted relevant research. Some researchers have proposed the use of spectral gradient difference (SGD) to describe the image difference [25]. This method is capable of detecting more prominent feature types. However, it has a disadvantage in that it only considers the spectral characteristics between images and does not take other features into account. Some researchers have proposed an unsupervised change detection method based on the hybrid spectral difference (HSD), which combines the spectral value and spectral shape by fusing the difference images of spectral correlation mapper (SCM) and SGD to describe the change in spectral shape [26]. One advantage of this method is that it can combine more change characteristics of the two images. One disadvantage is that many ground objects cannot be detected using a single pixel. Change vector analysis (CVA) was first used for change detection and applied to forest change [27]. This paper introduces a digital method for change detection using multi-temporal Landsat data. The method calculates the spectral change vectors of two different dates. This method works well in low-resolution images, but its disadvantage is that it only considers a single feature. Since ground objects in VHR images are more complex, applying this method to VHR remote sensing images will result in a lot of noise. To address this limitation, some researchers have proposed a change detection method that combines multiple indexes and uses CVA and SGD to construct a change difference image [28]. However, it has three disadvantages. First, although the two index algorithms are different, they both judge the changes based on image pixels. Second, although multiple index fusion can reduce part of salt and pepper noise, it still cannot use advanced features to express the whole VHR image. Third, multi-index fusion may lead to error superposition of change detection results.

In recent years, with the rapid development of various disciplines, various computer models have made outstanding progress in image processing, for example, machine learning models such as support vector machine (SVM) [29], random forest (RF) [30] and extreme learning machine (ELM) [31]. Visual saliency detection technology has also advanced in leaps and bounds, such as a spectral residual approach (SRA) [32], context-aware saliency detection (CASD) [33] and the richer convolutional features method [34]. Convolutional Neural Networks (CNN) for images have been developed in recent years, including U-Net [35], GoogleNet [36], ResNet [37], etc. These networks show strong effects in various fields of remote sensing. VHR image change detection is roughly divided into three categories. The first category is to extract the image features first and then compare them to get the change detection results. For example, some researchers have used classification CNNs, which are the main method for learning deep features from VHR remote sensing images, to detect building changes from RGB aerial photographs [38]. The second category is to use samples directly to train the change detection neural network model and then output the change image directly. For example, some researchers have used pre-trained image classification CNNs to extract change features [39]. Some researchers have used improved attention mechanisms for training networks to extract change information [40]. The third category uses some visual methods for unsupervised change detection based directly on images. For example, some researchers have proposed a base improved PCA-net model to implement VHR image change detection [41]. It exhibits an excellent performance in the field of VHR change detection. As we all know, most VHR images mainly include four bands. For remote sensing change detection, simply using these visual models to input RGB band information may cause the loss of spectral bands, thus affecting the change detection results. Therefore, it is very important to construct the correlation between remote sensing change detection algorithms and visual models.

From the above analysis, it can be seen that some single feature indexes and their improved versions have different performance improvements in different aspects, but cannot express the global features of the image. The combined multi-index method does not perform well on highly complex VHR images. If a visual model is only used for VHR image change detection, the band information will be lost. Therefore, in order to overcome the problems mentioned above, we combine the visual attention model with difference image fusion and propose a multi-difference image fusion change detection algorithm based on visual attention (VA-MDCD). First, two difference images are calculated using CVA and SGD. Secondly, the two difference images are input into an Ltti visual model to calculate the color, intensity and orientation features of the images [42–44]. These features are combined to produce two saliency feature maps. Thirdly, a fusion result is obtained by performing the wavelet fusion algorithm [45] on two saliency feature images. Finally, the OTSU threshold segmentation algorithm (OTSU) [46] is used to obtain the change detection result. Notably, the Ltti model utilizes visual attention to enhance the efficiency and accuracy of the fusion process, providing a more robust solution for image fusion [42–44]. The VA-MDCD framework is designed to focus visual attention on areas of significant change. VA-MDCD can effectively combine image correlation to extract advanced features of images [47,48]. Before fusing the two indexes, VA-MDCD can significantly reduce the errors of the two indexes and focus on the real changes.

The highlights of this paper as follows: (1) a visual-attention-model-based change detection framework is proposed, which has a higher performance than the traditional multi-difference image fusion change detection method in VHR images with complex features; (2) the framework can detect changes in VHR images without the need for samples and can self-adapt to the extraction of change areas; (3) visual attention is added and a total of 42 change feature maps are computed to accurately capture changes. In total, the model has three change feature extraction modules, which compute 12 color feature maps, 6 intensity feature maps and 24 orientation feature maps (see Section 2.2.2 for specific algorithms).

In the rest of this article, Section 2 describes our proposed VA-MDCD approach and explains how to implement it. Section 3 presents our analysis of the results obtained in the two experiments. Section 4 details the design of ablation experiments to discuss the influence of different model structures on the proposed method, and Section 5 summarizes our conclusions.

2. Methods

The process of the VA-MDCD method proposed in this paper is shown in Figure 1. Firstly, we apply the CVA [27] and SGD [25] change detection algorithms to generate change amplitude images from VHR remote sensing data collected at identical locations during two distinct time periods. Secondly, we utilize a computer vision saliency algorithm to extract the saliency feature of each change amplitude image constructed by CVA and SGD, respectively. We use the Ltti model [42–44] for feature extraction and the Luminance Contrast (LC) algorithm [49,50] as a comparison. Thirdly, we use the wavelet transform [51] to fuse two saliency difference images. Finally, we use the OTSU algorithm to derive the binarized threshold segmentation outcomes and evaluate the change map [46].

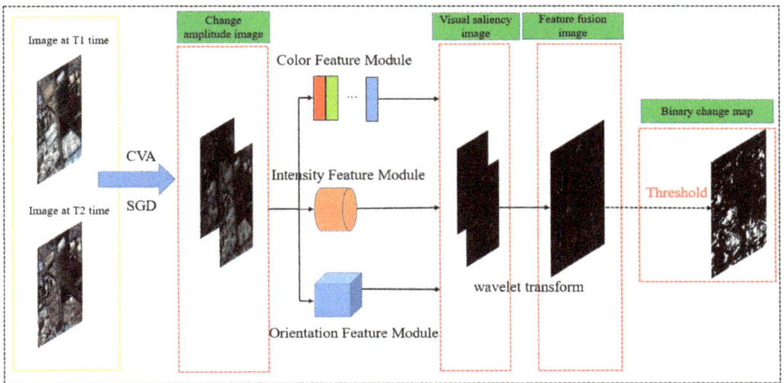

Figure 1. The flow chart of the proposed VA-MDCD.

The proposed VA-MDCD method utilizes the visual attention model to detect changes in images by leveraging color, intensity and orientation features. The fusion results represent the areas where CVA and SGD are salient together. It is also verified that this method has good detection results for complex backgrounds and complex terrain types. The specific algorithms are as follows.

2.1. Construct the Change Intensity Image

2.1.1. Change Vector Analysis

CVA [27,52] can characterize the changes in images regarding both their intensity and direction features. Figure 2 illustrates the direction and magnitude of the change vector, and we have established a separation threshold to delineate the changed and unchanged areas. Different ground objects have varying change angles, so we can classify change types according to these angles.

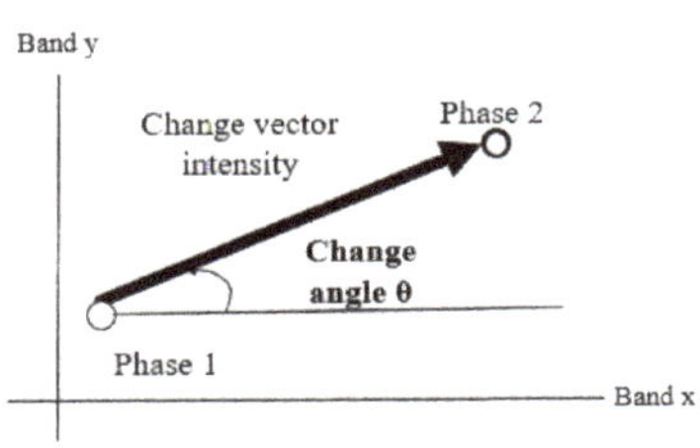

Figure 2. Change vector analysis.

The principle is as follows: Set remote sensing images of phase T_1 and phase T_2 as G_1 and G_2, respectively, and pixel grey values of column j in row i are $G_1 = (X_{ij}^1\ (T_1), X_{ij}^2\ (T_1), \ldots X_{ij}^n(T_1))^T$ and $G_2 = (X_{ij}^1\ (T_2), X_{ij}^2(T_2), \ldots X_{ij}^n(T_2))^T$, respectively. n is the number of selected bands and $X_{ij}^k(T_1)$ is the grey value of the pixel in the T phase of column j in line i of the k band. By calculating the difference between G_1 and G_2, the change vector is obtained as follows:

$$\Delta G = G_1 - G_2 = \begin{bmatrix} X_{ij}{}^1(T_1) - X_{ij}{}^1(T_2) \\ X_{ij}{}^2(T_1) - X_{ij}{}^2(T_2) \\ \cdots\cdots\cdots\cdots\cdots\cdots \\ X_{ij}{}^k(T_1) - X_{ij}{}^k(T_2) \\ \cdots\cdots\cdots\cdots\cdots\cdots \\ X_{ij}{}^n(T_1) - X_{ij}{}^n(T_2) \end{bmatrix} \tag{1}$$

$\mid \Delta G \mid$ is calculated from the following formula which describes information about changes in the entire image:

$$|\Delta G| = \sqrt{\sum_{k=1}^{n} (X_{ij}{}^k(T_1) - X_{ij}{}^k(T_2))^2} \tag{2}$$

2.1.2. Spectral Gradient Difference

SGD [25] is a method that compares the spectral slope of two images captured by satellites in the same area during different time periods. The spectral slope refers to the rate at which the spectral response changes with wavelength [28].

The changing intensity of the spectral slope of the same object at the last two times can be calculated as follows:

$$Dif_K = \sum_{M=1}^{N} \left| \frac{(ref_{i,M+1} - ref_{j,M+1}) - (ref_{i,M} - ref_{j,M})}{\rho'_{M+1} - \rho'_{M}} \right| \tag{3}$$

In this formula, $ref_{i,M+1}$ is the spectral brightness value of the M + 1 band in the T_1 phase, $ref_{i,M}$ is the spectral brightness value of the M band, $ref_{j,M+1}$ is the spectral brightness value of the M + 1 band in the T_2 phase, and $ref_{j,M}$ is the spectral brightness values of the M band. ρ'_{M+1} and ρ'_{M} are the normalized wavelengths of the corresponding bands and Dif_K is the change intensity of the two slope vectors. The larger the value, the more likely the surface cover will change in the area; otherwise, the probability is lower. Finally, it is necessary to set a threshold to determine whether there is a change.

2.2. Visual Attention Model

The Ltti model is a computer vision attention model [42–44]. In this model, three kinds of features are used to represent the image, which are color, intensity and orientation. The model is shown in Figure 3. Firstly, feature pyramids are constructed. Secondly, feature images are calculated. Finally, visual saliency images are calculated.

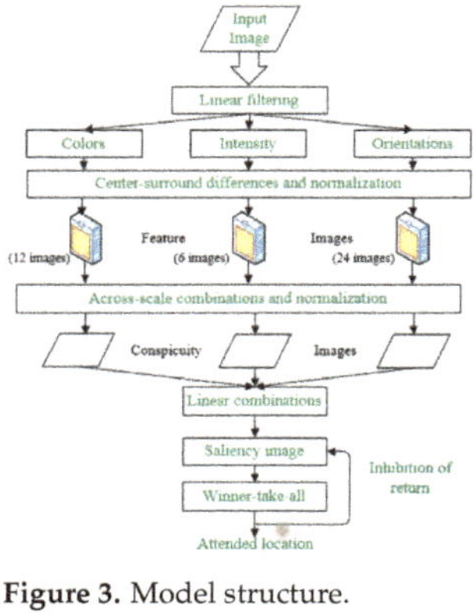

Figure 3. Model structure.

2.2.1. Feature Extraction

An intensity image I is $I = (r + g + b)/3$, where r, g and b are the red, green and blue channels of the input image. I is used to create a Gaussian pyramid $I(\sigma)$, where $\sigma \in [0...8]$ is the scale. The four color channels are $R = r - (g + b)/2$ for red, $G = g - (r + b)/2$ for green, $B = b - (r + g)/2$ for blue and $Y = (r + g)/2 - |r - g|/2 - b$ for yellow. Four Gaussian pyramids, $R(\sigma)$, $G(\sigma)$, $B(\sigma)$ and $Y(\sigma)$, are created from these color channels [53,54].

2.2.2. Feature Image

All features are calculated using a set of methods called center–surround methods (Figure 3). The center is $c \in \{2,3,4\}$ and the surround is $s = c + d$, where $d \in \{3,4\}$. The center and surround values are used to derive the feature image:

$$\begin{cases} I(c,s) = |I(c)\Theta I(s)| \\ RG(c,s) = |(R(c) - G(c))\Theta(G(s) - R(s))| \\ BY(c,s) = |(B(c) - Y(c))\Theta(Y(s) - B(s))| \\ O(c,s,\theta) = |O(c,\theta)\Theta O(s,\theta)| \end{cases} \tag{4}$$

In the above formula, $I\ (c, s)$ is the intensity feature. Above, Θ represents interpolating a coarse scale to a fine scale and subtracting matrix elements. $RG\ (c, s)$ and $BY\ (c, s)$ are used to represent color features. $O\ (c, s, \theta)$ is the orientation feature image, which is generated by comparing the orientation information between the center and surround scales, where the value range of θ is $\theta \in \{0°, 45°, 90°, 135°\}$. A total of 6 intensity features, 12 color features and 24 orientation features were calculated.

2.2.3. Saliency Image

Saliency images are computed on a scale of 4, with $\overline{I}$ representing intensity, $\overline{C}$ representing color and $\overline{O}$ representing orientation. $\oplus$ represents cross-scale addition:

$$\begin{cases} \overline{I} = \bigoplus_{c=2}^{4} \bigoplus_{s=c+3}^{c+4} N(I(c,s)) \\ \overline{C} = \bigoplus_{c=2}^{4} \bigoplus_{s=c+3}^{c+4} [N(RG(c,s)) + N(BY(c,s))] \end{cases} \tag{5}$$

The four orientation values are combined together to obtain the following orientation saliency image:

$$\overline{O} = \sum_{\theta \in \{0°,45°,90°,135°\}} N\left(\bigoplus_{c=2}^{4} \bigoplus_{s=c+3}^{c+4} N(O(c,s,\theta)) \right) \tag{6}$$

The overall saliency image S is:

$$S = \frac{1}{3}(N(\overline{I}) + N(\overline{C}) + N(\overline{O})) \tag{7}$$

2.3. LC Saliency Detection Algorithm

Figure 4 shows the steps of the LC algorithm [49,50].

To calculate the distinct value of a pixel in an image, we compute its global contrast in the entire image. The saliency value of pixels can be obtained using the following formula:

$$SalS(I_k) = \sum_{\forall I_i \in I} |I_k - I_i| \tag{8}$$

In this formula, I_k represents the grey value of pixel k, I_i represents all pixel points, where the value range of i is [1, N], and $SalS(I_k)$ represents the significant value of pixel k.

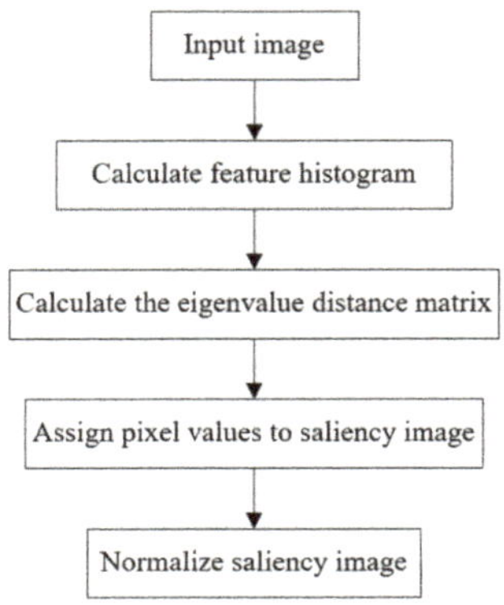

Figure 4. The procedure of LC algorithm.

2.4. Difference Feature Fusion and Segmentation

Figure 5 shows the wavelet fusion method. We establish a wavelet pyramid decomposition of the image by applying the wavelet transform to the saliency image. Then, we fuse the decomposition layers from high to low by using different fusion rules to combine different frequency components in each decomposition layer. Finally, we obtain a fused wavelet pyramid by performing the wavelet inverse transform on the fused pyramid. The resulting image after reconstruction is the fused image [51]. In addition to this, in the process of wavelet decomposition, *L* and *H* stand for low and high frequencies, respectively. The subscripts 1 and 2 represent the first and second level decompositions, and at each decomposition level, the image is decomposed into four frequency bands in vertical and horizontal directions. In addition to this, in the process of wavelet decomposition, *L* and *H* stand for low and high frequencies, respectively. The subscripts 1 and 2 represent the first and second level decompositions, and at each decomposition level, the image is decomposed into four frequency bands in the vertical and horizontal directions. In the wavelet fusion process, low frequencies are directly fused using local variance normalization, while high frequencies are first calculated by the Canny operator to compute the edge information, and then the local variance is computed [55,56].

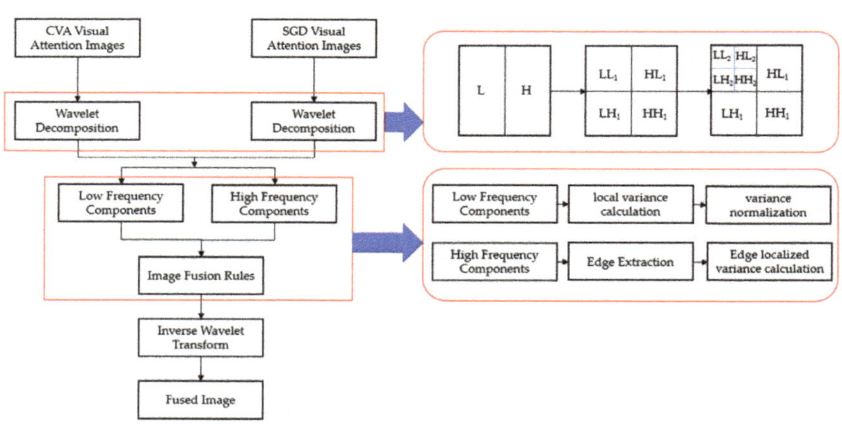

Figure 5. The procedure of wavelet transform fusion.

2.4.1. Wavelet Decomposition

The basic principle of the wavelet transform is that L layer wavelet decomposition is carried out to obtain the $3L + 1$ layer frequency band, including the low frequency baseband C_j and the high frequency sub-band Dh, Dv, Dd of layer $3L$. The original image is represented by $f(x, y)$, denoted by C_0. If the filter coefficient matrix corresponding to the scale coefficient and the wavelet coefficient are H and G, respectively, then the two-dimensional wavelet decomposition algorithm can be described as follows:

$$\begin{cases} C_{j+1} = HC_jH^T \\ D^h_{j+1} = GC_jH^T \\ D^v_{j+1} = HC_jG^T \\ D^d_{j+1} = GC_jG^T \end{cases} \tag{9}$$

where j represents the number of decomposition layers; h, v and d represent horizontal, vertical and diagonal directions, respectively; and H^T and G^T are conjugate transpose matrices of H and G, respectively.

2.4.2. Wavelet Fusion

After wavelet decomposition, the low frequency part of the image is the overview and average characteristics of the image and the high frequency part of the image reflects the details of the image, such as the edge of the image, and regional boundaries. Assuming that the low-frequency components are AA and AB, the local variance $Var(i, j)\ AA$ and $Var(i, j)\ AB$ of all pixels of AA and AB is calculated and normalized. Then, the normalized local variance is used for low-frequency fusion. Assuming that the high-frequency components of images A and B are DLA and DLB, the edge of each high-frequency component is extracted by the Canny operator, then the local variance of each element of the edge image is calculated. The high-frequency coefficient that represents the l*th* layer of the source image processed by the Canny operator is the average value of the l*th* layer of the source image extracted by the Canny operator, which is the local variance obtained by edge extraction of the high-frequency component of the lth layer of the source image. Then, the fused coefficient is reconstructed by a wavelet to obtain the fused image.

The low frequency fusion adopted in this paper adopts the fusion rule of mean square deviation normalization, and the high frequency fusion adopts the Canny operator for edge extraction. The local variance of each pixel is calculated because the high-frequency part is the edge area of the image, so this method is used. The Canny edge detection algorithm is a classic edge detection algorithm used to extract the edge of the image to better highlight high-frequency features.

2.4.3. OTSU Threshold Segmentation Algorithm

After difference image fusion, we use the OTSU [46] segmentation algorithm for image binarization. C_0 and C_1 are two classes whose pixel values have intervals of $[1, \ldots, T]$ and $[T + 1, \ldots, L]$, respectively, with T being the given critical value. ${\sigma_b}^2(T)$ and ${\sigma_w}^2(T)$ represent inter-class variance and intra-class variance, respectively, and the threshold of variance is:

$$T^* = \underset{1 \le T < L}{\arg} \max\{\sigma_b^2(T)\} = \underset{1 \le T < L}{\arg} \min\{\sigma_w^2(T)\} \tag{10}$$

3. Results

To validate the effectiveness of VA-MDCD, we selected two datasets for two experiments. The first dataset was two images taken by the Jilin-1 (JL-1) satellite [57]. The second dataset was two images taken by the Beijing-2 (BJ-2) satellite [58]. Both datasets include multispectral and panchromatic imagery. Before executing our algorithm framework, both datasets underwent precise preprocessing procedures. Firstly, the multispectral imagery underwent radiometric calibration, atmospheric correction and orthorectification. Secondly, the panchromatic imagery was subjected to radiometric calibration and orthorectification. Radiometric calibration was performed to eliminate interference from the sensor itself, the atmosphere, the solar zenith angle and terrain effects. Atmospheric correction was applied to effectively mitigate errors caused by atmospheric scattering, absorption and reflection. Orthorectification was conducted to bring the two temporal images closer to the orthorectified angle. Subsequently, the preprocessed multispectral and panchromatic images were fused using the Gram–Schmidt (G–S) algorithm [59]. Importantly, we finally performed geometric registration on the processed images at different times to ensure the correspondence of pixels between the two images. The land cover types of both images in the JL-1 dataset only included vegetation and impervious surface, while the images in the BJ-2 dataset only included impervious surface.

3.1. Experiment #1

3.1.1. Experimental Data

We have selected JL-1 VHR images as the data of experiment 1. JL-1 satellite remote sensing images have been successfully applied in urban streetlight extraction, surface water resources monitoring, forest ignition recognition, urban real-time traffic monitoring and others. The resolution of panchromatic images taken by the satellite is 0.72 m, and the resolution of multispectral images is 2.88 m. The JL-1 image pair was acquired in October 2018 and November 2019.

3.1.2. Change Detection Results and Statistical Analysis

The images have been cut to 1915 × 2101 pixels in the red, green, blue and near-infrared regions. The cropped images are shown in Figure 6. A reference change image was obtained through detailed field investigations and unmanned aerial vehicle (UAV) measurements, so the visual analysis has some prior knowledge. The black part is the actual change area.

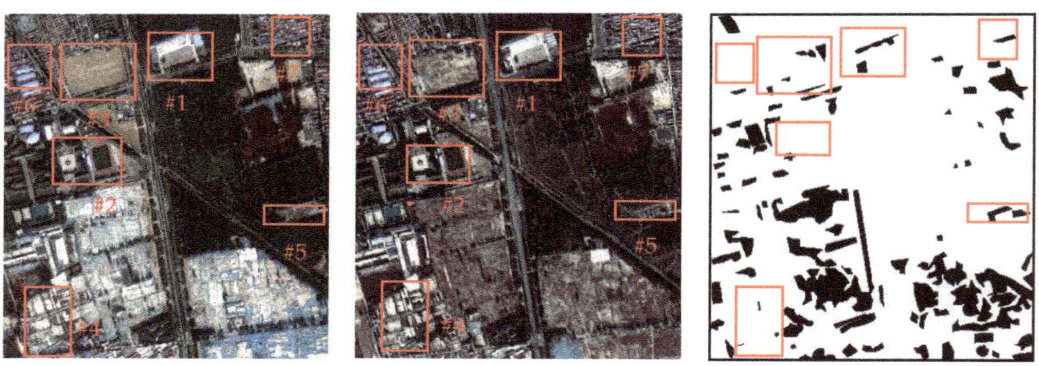

(a) (b) (c)

Change Unchange Sample of selected features

Figure 6. JL-1 remote sensing images (**a**) in 2018; (**b**) in 2019. (**c**) Ground truth image.

In order to evaluate the improvements in the VA-MDCD method on multiple difference image fusion, five pairs of error-prone ground objects marked with red rectangles in the original image were selected as samples for analysis (Figure 7). The spectra of zones #1, 2, 6 and 7 changed in the two images, but they cannot be considered changed. Zones #3, 4 and 5 are vulnerable to rain erosion and cause false detections. They were therefore selected as samples for analysis.

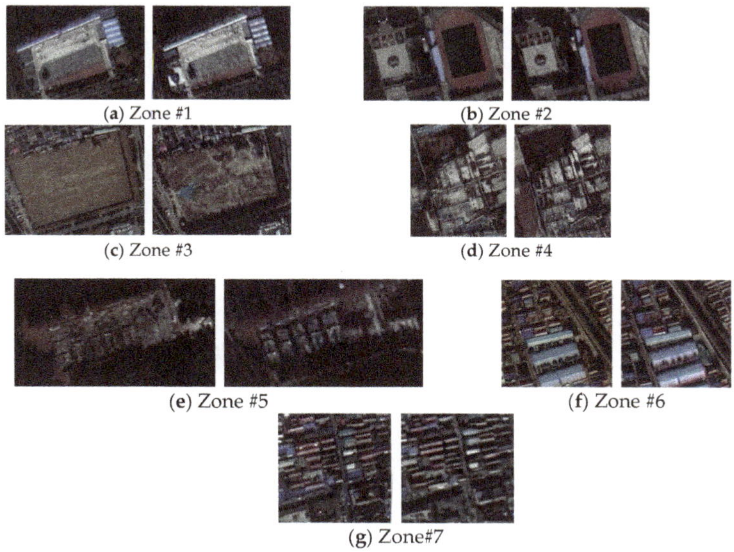

Figure 7. The selected ground object samples.

All experiments were conducted under the same computer environment. To compare with this experiment and verify the reliability of this experiment, in addition to the proposed VA-MDCD method and LC algorithm, another five groups of comparison experiments were designed: the change results of the combination of CVA [27] and OTSU algorithms (CVA-OTSU); the change results of the combination of SGD [25] and OTSU (SGD-OTSU); the change results of CVA-SGD [28] fusion; iterative weighted multivariate [60] change detection (IRMAD-OTSU); and change detection based on PCA kmeans [61] (PCA kmeans). Figure 8a–g shows the change amplitude images of CVA, SGD, CVA-SGD, IRMAD, PCA kmeans, LC algorithm and VA-MDCD, respectively. From VA-MDCD, we can see that our method can well suppress most of the salt-and-pepper noise and spurious changes in CVA and SGD. Influenced by visual attention, our method is closer to human vision, and the magnitude of VA-MDCD changes is closer to the real changes. The change detection results are shown in Figure 9, where the white area is the changed area and the black area is the unchanged area.

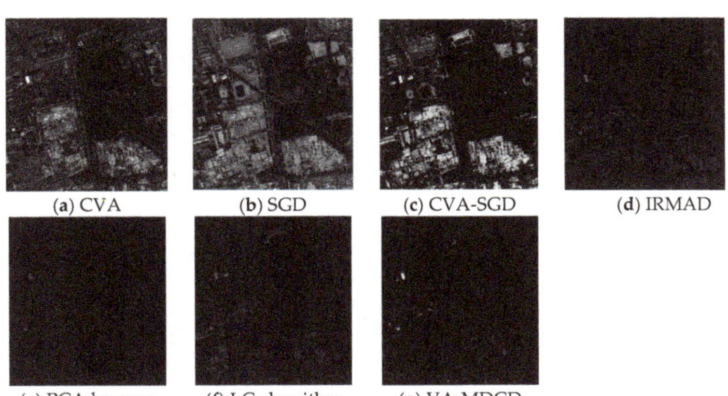

Figure 8. Change magnitude images on the JL-1 dataset.

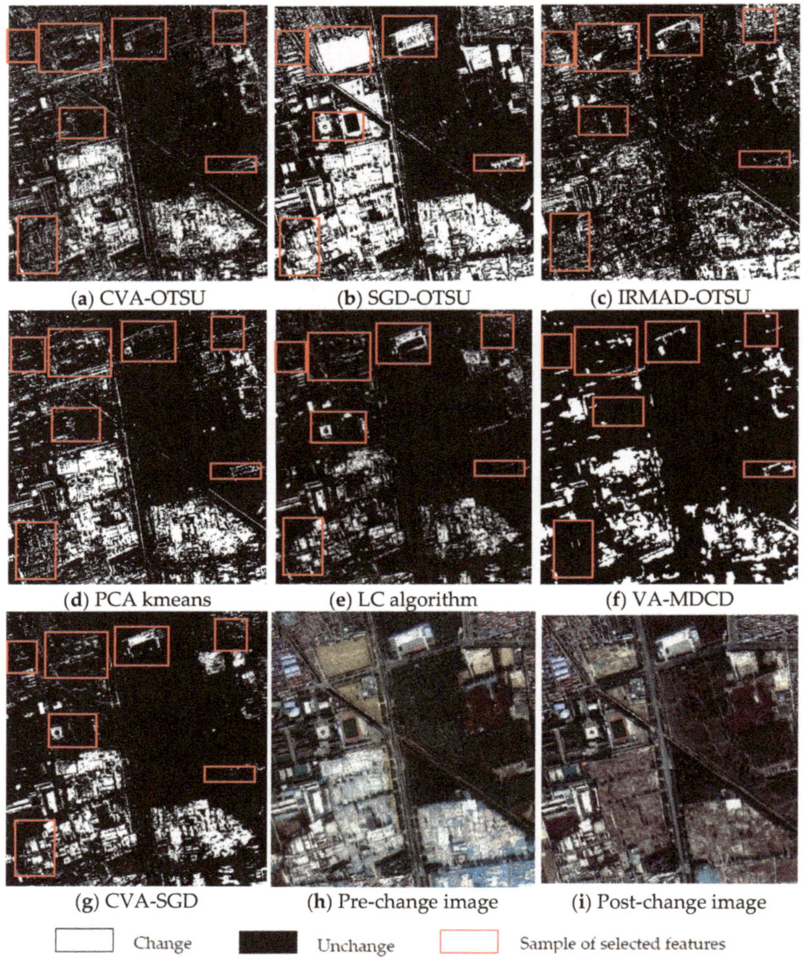

Figure 9. Change detection results. (**a**) CVA-OTSU; (**b**) SGD-OTSU; (**c**) IRMAD-OTSU; (**d**) PCA kmeans; (**e**) LC algorithm; (**f**) VA-MDCD; (**g**) CVA-SGD; (**h**) pre-change image; (**i**) post-change image.

3.1.3. Analysis and Discussion

A confusion matrix was introduced here to evaluate the accuracy. The overall accuracy (*OA*), false alarm rate (*FA*), missed alarm rate (*MA*), *Kappa* and *F-measure* were calculated using the confusion matrix, and the experimental results were quantitatively analyzed and evaluated by these coefficients [62].

The *OA* is the ratio of all correctly classified samples to total samples. The formula is as follows:

$$OA = \frac{TP + TN}{TP + TN + FP + FN} \tag{11}$$

where *TP* and *TN* represent the number of pixels that are actually changed and correctly detected as changed and the number of pixels that are actually unchanged and correctly detected as unchanged, respectively. *FP* and *FN* represent the number of pixels that are actually changed but incorrectly detected as unchanged and the number of pixels that are actually unchanged but incorrectly detected as changed, respectively. *FA* represents the ratio of the unchanged area detected as the changed area to the true unchanged area:

$$FA = \frac{FP}{FP + TN} \tag{12}$$

MA represents the ratio of the undetected area of change to the actual area of change:

$$MA = \frac{FN}{FN + TP} \tag{13}$$

Kappa is used to test the consistency of the results. The higher the *Kappa* value, the more accurate the result. The formula is as follows:

$$Kappa = \frac{(OA - PRE)}{(1 - PRE)} \tag{14}$$

In this formula, *PRE* represents the consistent rate of the theoretical test results. The formula is as follows:

$$PRE = \frac{(TP + FP)(TP + FN) + (TN + FN)(TN + FP)}{(TP + TN + FN + FP)^2} \tag{15}$$

The *F-measure* coefficient is the arithmetic mean divided by the geometric mean; the larger the value, the more accurate the final change detection results. The formula is as follows:

$$F - measure = \frac{TP^2}{TP^2 + FN + FP} \tag{16}$$

According to the accuracy evaluation table (Table 1) and the visualization results in Figure 10, the proposed VA-MDCD had a higher *OA*, *Kappa* and *F-measure* than other comparison methods. For VA-MDCD, they were 94.50%, 72.28% and 66.71%, respectively, and for the LC algorithm, they were 91.54%, 58.74% and 60.18%, respectively. The direct fusion of CVA and SGD could not distinguish spurious changes caused by different sun angles. It should be noted that there were still some noise and spurious variations in the proposed VA-MDCD, which slightly affected the accuracy. In general, based on the analysis of accuracy, we conclude that the VA-MDCD method proposed in this paper is effective and reliable. Furthermore, the *FA* of the proposed VA-MDCD was 8.59%, which was reduced by 8.57% compared to CVA-SGD. Although the proposed method falsely detected some building shadows, it could reduce the effect of CVA and SGD fusion errors.

Table 1. The accuracy table of different change detection methods.

Methods	OA (%)	FA (%)	MA (%)	Kappa	F-Measure
CVA-OTSU	83.47	19.53	23.56	0.1055	0.1922
SGD-OTSU	79.55	26.45	20.22	0.1472	0.1460
CVA-SGD	89.36	17.16	22.47	0.4683	0.3645
IRMAD-OTSU	80.41	30.72	35.34	0.0291	0.0842
PCA kmeans	86.25	13.29	25.92	0.3056	0.2336
LC algorithm	91.54	11.74	27.36	0.5874	0.6018
VA-MDCD	94.50	8.59	23.28	0.7228	0.6671

Table 2 counts the pixel number of the selected five pairs of sample areas, and Table 3 details the false detection pixels of five pairs of samples by CVA, SGD, CVA-SGD and VA-MDCD. In Table 3, the number of false detection pixels of CVA-SGD was higher than that of CVA and SGD, and some were between CVA and SGD, but none of them had a reduced number of false detection pixels. Although the results of the VA-MDCD method also had a lot of pixel misdetection, compared with CVA-SGD, it could effectively improve the fusion effect of CVA and SGD. Although the *FA* of CVA-SGD was lower than that of both CVA and SGD, the number of misdetected pixels in specific areas was very high. The proposed VA-MDCD method reduced the total *FA* and the number of misdetected pixels in specific areas. The spectra of these features in zones #1, #2, #6 and #7 have changed, and we cannot actually consider them as changed zones, so there are only very few erroneous pixels in the VA-MDCD results, which proves that our method is able to remove spurious changes due to spectral effects very well.

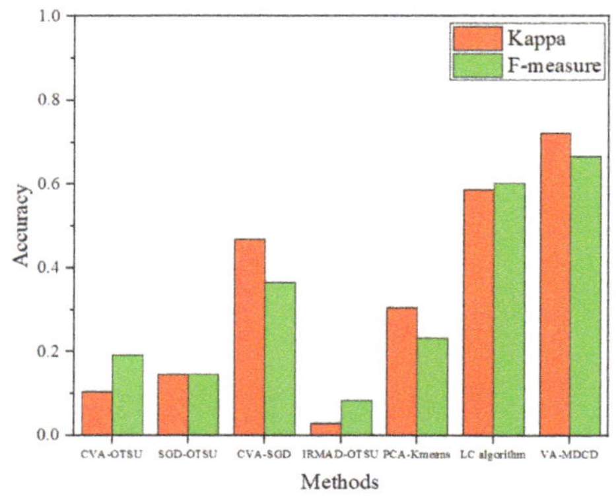

Figure 10. *Kappa* and *F-measure* coefficients of different change detection methods.

Table 2. The number of pixels per sample.

Value	Zone #1	Zone #2	Zone #3	Zone #4	Zone #5	Zone #6	Zone #7
Number of pixels	92,880	61,180	60,532	56,225	6936	40,000	62,500

Table 3. Number of pixels for error detection of each sample by different methods.

	Zone	CVA	SGD	CVA-SGD	VA-MDCD
	Zone #1	541	977	1291	134
	Zone #2	3421	23,823	13,269	0
	Zone #3	25,816	51,185	18,571	1323
Pixel value for error detection	Zone #4	24,594	41,272	29,533	397
	Zone #5	3213	1924	3918	552
	Zone #6	17,327	21,928	3635	0
	Zone #7	19,483	35,167	41,368	1451

3.2. *Experiment #2*

3.2.1. Experimental Data

We have selected BJ-2 VHR images as the data for experiment 2. The satellite consists of three optical remote sensing satellites in the 0.8 m panchromatic band and 3.2 m multi-spectral band, which can provide high-quality remote sensing images. The BJ-2 image pair was acquired in October 2018 and September 2021.

3.2.2. Change Detection Results and Statistical Analysis

The images were cut to 1500 × 1500 pixels in the red, green, blue and near-infrared regions. The cropped image is shown in Figure 11. A reference change detection image was obtained by visual interpretation and geographic analysis, and the black part is the actual changed area.

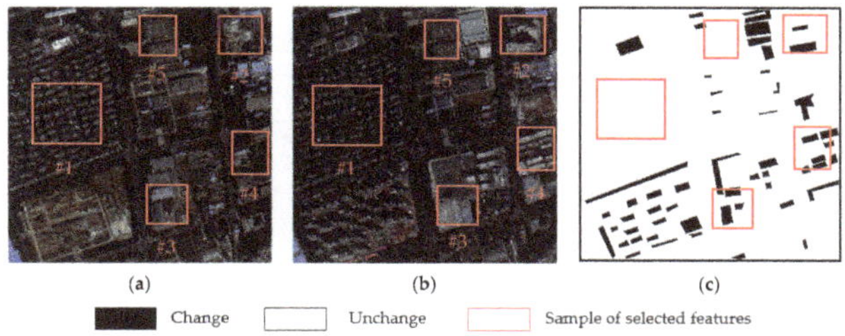

Figure 11. BJ-2 remote sensing images (**a**) in 2018; (**b**) in 2021. (**c**) Ground truth.

In experiment 2, we used red rectangles to select four pairs of ground objects in the original image that were easily affected by the shooting angle of the sensor (Figure 12).

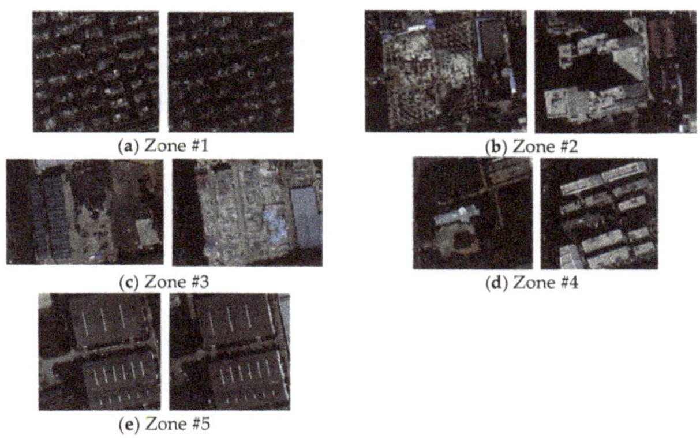

Figure 12. The selected ground object samples.

All experiments were conducted under the same computer environment. To compare with this experiment and verify the reliability of this experiment, in addition to the proposed VA-MDCD method and LC algorithm, another five groups of comparison experiments were designed. The five methods are CVA [27], SGD [25], CVA-SGD [28], IRMAD [60] and PCA kmeans [61]. Figure 13a–g presents the change amplitude images of CVA, SGD, CVA-SGD, IRMAD, PCA kmeans, LC algorithm and VA-MDCD, respectively. From VA-MDCD, we can see that our method effectively suppresses the majority of salt-and-pepper noise and spurious changes, such as urban rooftops. VA-MDCD efficiently focuses attention on areas with salient changes. The change detection results are shown in Figure 14; the white area is the changed area and the black area is the unchanged area.

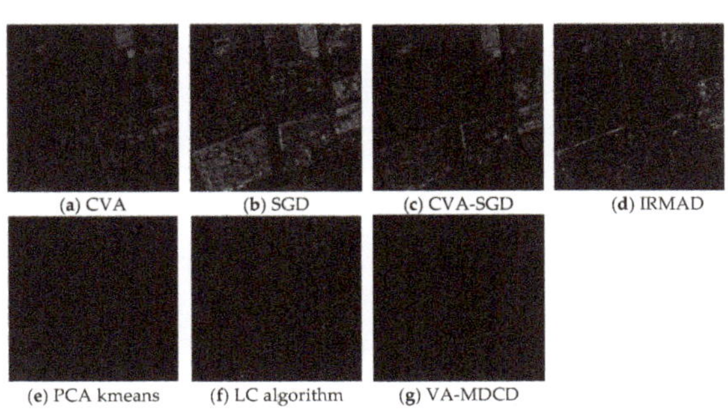

Figure 13. Change magnitude images on BJ-2 dataset.

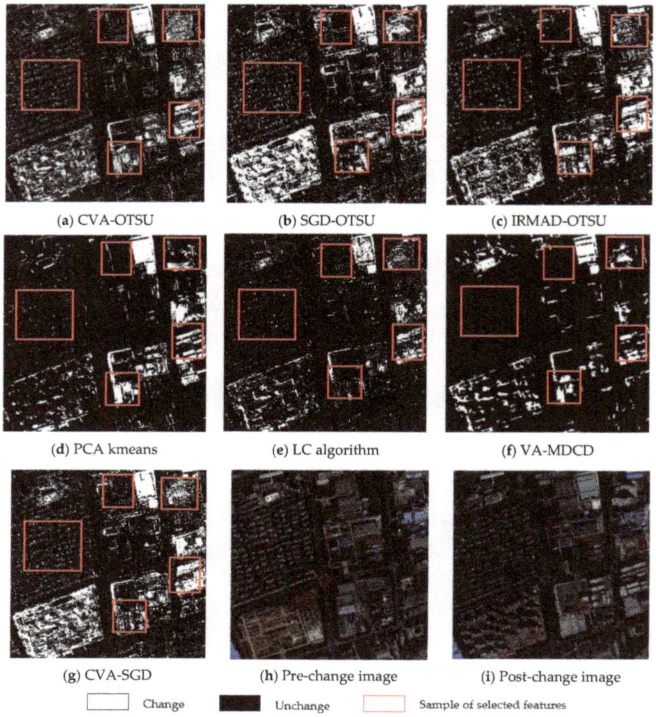

Figure 14. Change detection results. (**a**) CVA-OTSU; (**b**) SGD-OTSU; (**c**) IRMAD-OTSU; (**d**) PCA kmeans; (**e**) LC algorithm; (**f**) VA-MDCD; (**g**) CVA-SGD; (**h**) pre-change image; (**i**) post-change image.

3.2.3. Analysis and Discussion

According to the accuracy evaluation table (Table 4) and the visualization results in Figure 15, the proposed VA-MDCD had a higher *OA*, *Kappa* and *F-measure* than other comparison methods. For VA-MDCD, they were 94.74%, 84.22% and 73.13%, respectively, and for the LC algorithm, they were 91.02%, 59.81% and 54.87%, respectively. In this verification area, the direct fusion method of CVA and SGD could not distinguish the errors caused by different sensor tilt angles. It was worth noting that there was still a small number of false detections in the proposed method, which slightly affected the accuracy of change detection. Overall, based on a quantitative analysis of the results, the *FA* of the proposed method was 6.19%, which was reduced by 9.56% compared to CVA-SGD. We conclude that the proposed VA-MDCD method is also reliable for quasi-urban area detection, and the proposed VA-MDCD can reduce the fusion error of CVA and SGD.

Table 4. The accuracy table of different change detection methods.

Methods	OA (%)	FA (%)	MA (%)	Kappa	F-Measure
CVA-OTSU	85.24	21.36	24.75	0.2258	0.1226
SGD-OTSU	86.58	18.94	17.39	0.1326	0.1865
CVA-SGD	90.21	15.75	22.65	0.5758	0.4903
IRMAD-OTSU	84.92	24.13	21.87	0.3714	0.2642
PCA kmeans	88.76	14.92	26.17	0.3940	0.2036
LC algorithm	91.02	12.76	25.63	0.5981	0.5487
VA-MDCD	94.74	6.19	18.94	0.8422	0.7313

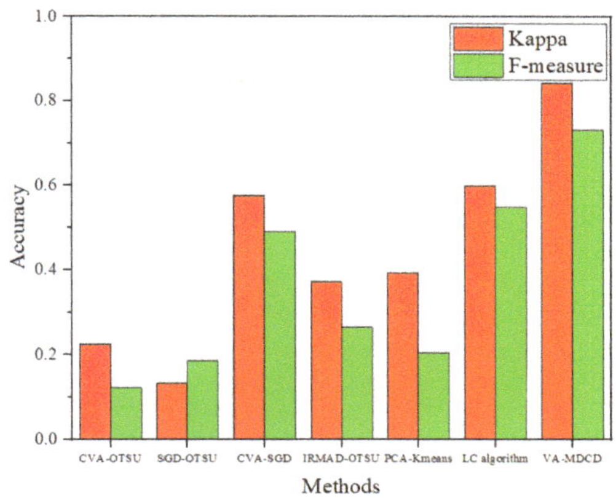

Figure 15. *Kappa* and *F-measure* coefficients of different change detection methods.

Tables 5 and 6, respectively, show the total pixel values of the four pairs of samples and the number of misdetected pixels of CVA, SGD, CVA-SGD and VA-MDCD. Table 6 shows that the proposed VA-MDCD can effectively improve the fusion result of CVA and SGD. Among the four samples, the VA-MDCD method produced a low number of false detection pixels, and one sample even had 0. Although the overall effect CVA-SGD was better than CVA and SGD, it was less effective in some details. Therefore, we conclude that CVA-SGD can remove a lot of noise on the whole. However, because CVA-SGD only calculates spectral features, it could not represent the information of the whole image, so it was not effective to apply it to the VHR image. The proposed VA-MDCD could combine multiple features to effectively improve the performance of CVA-SGD applied to VHR images. It is worth noting that in zones #1 and 5 there is no change in the building morphology. Due to the spectral changes on the roofs, CVA, SGD and CVA-SGD would have detected them as areas of change, which is incorrect. However, VA-MDCD avoids this spurious change by integrating spatial correlation. Therefore, VA-MDCD has a strong performance in solving the problem of spurious changes due to spectral effects.

Table 5. The number of pixels per sample.

Value	Zone #1	Zone #2	Zone #3	Zone #4	Zone #5
Number of pixels	80,089	26,070	21,879	22,194	32,400

Table 6. Number of pixels for error detection of each sample by different methods.

	Zone	CVA	SGD	CVA-SGD	VA-MDCD
Pixel value for error detection	Zone #1	26,323	34,674	27,338	0
	Zone #2	9821	7334	14,935	652
	Zone #3	10,372	16,038	15,973	243
	Zone #4	3679	5614	6385	465
	Zone #5	8154	6012	7305	0

In general, both CVA and SGD could cause false detections on VHR images. CVA-SGD exhibited a slightly better performance compared to CVA and SGD, with a higher *OA*. However, noise remained present, and CVA-SGD did not eliminate the error accumulation of CVA and SGD in certain regions. In the results of IRMAD, both *FA* and *MA* were very high, with numerous errors and significant salt-and-pepper noise. This phenomenon shows that it is not feasible to use spectral features only in VHR images. Due to the consideration of contextual information in the image, PCA kmeans reduced the noise to some extent, but the performance was not efficient. Although the method combined with the LC algorithm eliminated some noise, the overall effect was not good. The reason was that the LC algorithm only calculated the contrast features of the image, which could not fully express the relationship between the pixels of the entire image. VA-MDCD not only had a good detection effect on VHR images, but also eliminated a lot of noise and low building shadows and could reduce the error superposition after the fusion of CVA and SGD, which was especially obvious. It was worth noting that the proposed VA-MDCD did not eliminate the shadows caused by tall buildings. The most important point is that the algorithm framework we designed is unsupervised, which can achieve good change detection results without any samples. Traditional algorithms are mostly unsupervised algorithms. Our algorithm performs better than traditional unsupervised algorithms, and it can be applied to VHR images. Therefore, the novelty of our algorithm is reflected here.

4. Discussion

4.1. *Effect of Different Model Structures*

As we all know, the combination of a computer vision algorithm and a change detection algorithm involves multiple processes, and each stage has uncertainty. We need to design experiments to justify the use of different modules to highlight the benefits of using different modules in the proposed method. We divided the model structure of the proposed method and designed ablation experiments to verify the influence of the color, intensity and orientation modules on the change detection results.

In addition to the proposed methods, the ablation experiments also include methods without color modules, methods without intensity modules and methods without orientation modules. The specific experimental results are summarized in Section 4.2.

4.2. *Ablation Experiments*

The change detection results of the designed experiment on the two datasets are shown in Figure 16. Here, the two datasets are named A and B. As can be seen from the results figure, the results of VA-MDCD show a good detection effect compared with the results with any of the three modules missing, and the textural structure is clear and the fragmented noise processing is better.

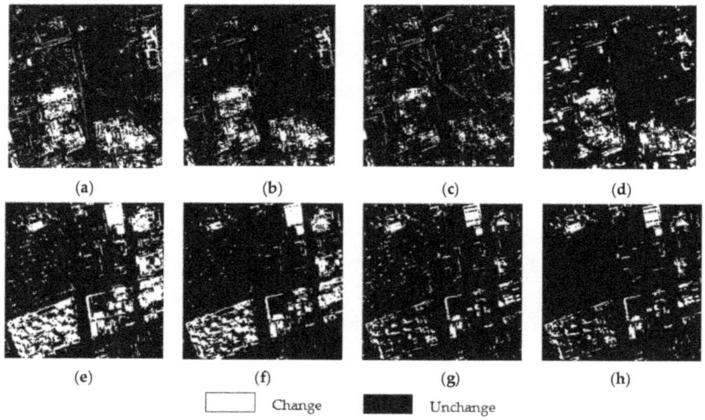

Figure 16. The change detection results of different model structures. (**a–d**) are the four results of dataset A, namely the model without a color module, the model without an intensity module, the model without an orientation module and the proposed VA-MDCD method; (**e–h**) are the four results of dataset B, namely the model without a color module, the model without an intensity module, the model without an orientation module and the proposed VA-MDCD method.

In addition, we also counted the *F-measure* values of the results generated by all model structures. The statistical figure is shown in Figure 17. It can be seen that the *F-measure* value and performance of the proposed VA-MDCD were higher than those produced by the other three model structures regardless of dataset. In dataset A, VA-MDCD improved by 6.33%, 1.18% and 4.41% compared to not using a color module, intensity module or orientation module, respectively. In dataset B, VAMDCD improved by 15.89%, 7.34% and 4.09% compared to not using a color module, intensity module or orientation module, respectively. The color module had the greatest impact on both datasets, since color plays the most important role in the magnitude image. VA-MDCD was affected by the lack of any kind of module, which indicates that the three feature extraction modules introduced can help improve the performance of VHR image change detection.

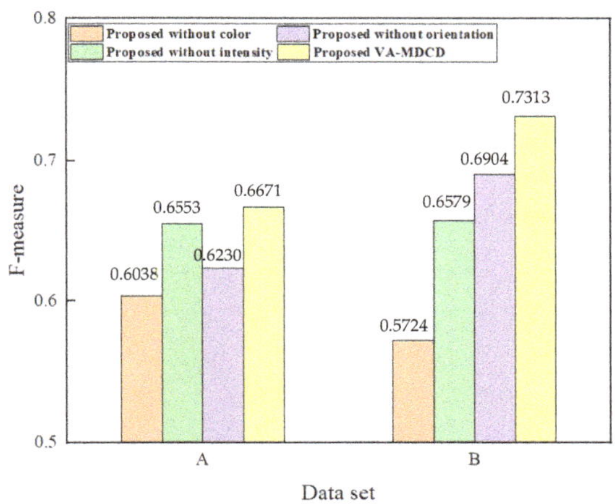

Figure 17. The *F-measure* value of model change detection for different structures used in ablation experiments.

4.3. Comparison with Other Visual Saliency Models

To prove the effectiveness of the proposed method, we compared our method with two other commonly used visual saliency models. They are the Luminance Contrast [63,64] (LC) method and Spectral Residual (SR) method [32,65]. The LC algorithm is an effective method to calculate a spatial saliency map using image color statistics. The algorithm takes into account the computational linear complexity in the number of image pixels. A saliency map of an image is built on the grayscale contrast between image pixels. That is, the sum of the gray value distances between the pixel and all the pixels in the image is taken as the significance value of the pixel. The SR algorithm extracts and emphasizes high-frequency information by transforming the image in the frequency domain and calculating the spectral residue and regards it as salient information.

Figure 18 shows the change detection results of different selected visual saliency models, and Table 7 shows the *F-measure* accuracy and running time of these models. From this, we can see that both LC algorithm and SR algorithm, although their running time is very short, can quickly detect the target area, but they have not eliminated the impact of noise and there are still many false changes. The LC algorithm only includes the relationship between the contrast of each pixel. Although the SR algorithm can be used to quickly calculate the significant region in the image, it may have some limitations in complex scenes, such as an insensitivity to color information and a limited ability to process images with complex backgrounds. Therefore, in relatively complex scenes, our method mimics human vision and combines multi-scale visual features to achieve more effective change detection results.

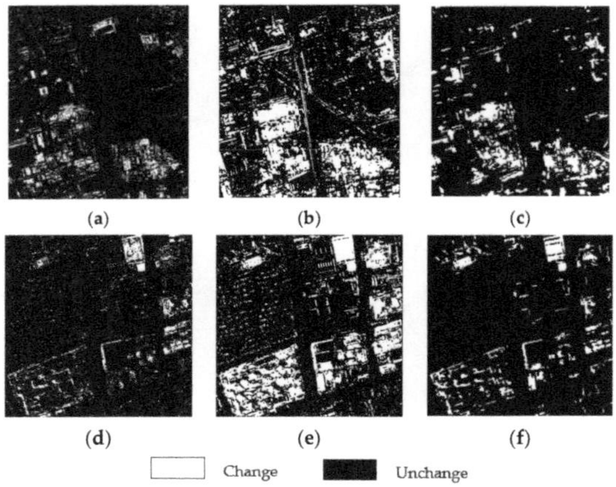

Figure 18. Change detection results based on different visual saliency models. (**a**,**d**) are the change detection results of the LC algorithm for datasets A and B, respectively; (**b**,**e**) are the change detection results of the SR algorithm for datasets A and B, respectively; (**c**,**f**) are the change detection results of the VA-MDCD method for datasets A and B, respectively.

Table 7. *F-measure* values and running times of different visual saliency models.

Dataset	Model	F-Measure	Running Time (s)
A	LC algorithm	0.6018	**73.14**
	SR algorithm	0.6114	83.12
	Proposed VA-MDCD	**0.6671**	135.24
B	LC algorithm	0.5487	**54.28**
	SR algorithm	0.6003	71.06
	Proposed VA-MDCD	**0.7313**	103.48

4.4. Model Complexity

In order to comprehensively evaluate the performance of the proposed framework, we need to compute statistics on the hardware facilities and computing time of the computer. In addition, we also need to evaluate the number of parameters of the whole model. Here, all experiments were conducted on all datasets and all experiments were conducted on an Intel Core i7-11800H CPU @ 2.3 GHz (16 GB RAM) with a NVIDIA GeForce Experience 3060 GPU. The model we developed was mainly implemented in matlab2019b and all the parameters and variables can be debugged on matlab2019b. Table 8 shows the running time and parameter number statistics of the different structural models. In addition, the algorithm running times of the comparison experiment in Section 3 are also listed. It can be seen from the table that the proposed VA-MDCD has more running parameters and takes more time than other models, but this increase can be ignored, which is worthwhile compared with the improvement in the performance of change detection.

Table 8. The number of running parameters of different model structures and the running time of all methods on the two datasets A and B.

Dataset	Model Structure	Number of Running Parameters (k)	Running Time (s)
A	CVA	-	3.21
	SGD	-	2.94
	CVA-SGD	-	6.53
	IRMAD	-	1.79
	LC algorithm	-	73.14
	PCA kmeans	-	49.25
	without color	25.11	126.91

Table 8. *Cont.*

Dataset	Model Structure	Number of Running Parameters (k)	Running Time (s)
	without intensity	23.45	127.25
	without orientation	20.74	113.93
	VA-MDCD	26.32	135.24
B	CVA	-	2.67
	SGD	-	2.14
	CVA-SGD	-	5.33
	IRMAD	-	1.74
	LC algorithm	-	54.28
	PCA kmeans	-	39.58
	without color	25.11	88.72
	without intensity	23.45	97.86
	without orientation	20.74	87.25
	VA-MDCD	26.32	103.48

5. Conclusions

The introduction of this paper summarized the limitations of conventional techniques while suggesting new approaches to overcome these shortcomings. We proposed a novel approach to detect changes in remote sensing images using an attention model in computer vision, which has been called VA-MDCD.

The experimental results obtained from two sets of VHR images validate the efficacy of the proposed VA-MDCD. First, our proposed VA-MDCD method can be used on VHR images and it can effectively identify the changes in VHR images. The experimental results show that this method not only has a higher *F-measure* and *Kappa* compared with other methods, but it can also reduce the *FA* of CVA-SGD. Second, the addition of a visual attention model helps to utilize the overall information of VHR images. Compared with the method using only spectral features, the proposed method can be applied to a wide range of VHR images. It not only contains the spectral information of the image, but also highlights the color, intensity and orientation information of the image. Third, after a statistical analysis of multiple pairs of samples that are easily affected by shooting angles, we conclude that the proposed VA-MDCD can reduce the error caused by the direct fusion of multiple difference images. In addition, we also designed an ablation experiment, and the model included four structures. Four groups of experiments were conducted on two datasets, respectively, and the *F-measure* value was calculated. The experimental results show that the proposed method has a higher *F-measure*, and the change detection effect of VA-MDCD proposed in this paper is better for both dataset compared to not using a color module, intensity module or orientation module.

The method in this paper is based on an improvement of unsupervised algorithms. Due to the lack of training samples, the method does not perform well when applied to extremely complex scenes. Therefore, in future studies, we will consider combining a visual attention model with deep learning and taking the visual attention model as the method of sample generation. This will not only reduce the complexity of manual sample labeling, but also generate reliable training samples.

Author Contributions: Conceptualization, J.L. and Q.C.; methodology, J.L.; software, J.L.; validation, J.L. and Q.C.; formal analysis, Q.C. and L.W.; investigation, J.L.; resources, Q.C.; data curation, J.L.; writing—original draft preparation, J.L.; writing—review and editing, L.W. and Y.H.; visualization, J.L.; supervision, Q.C.; project administration, J.L.; funding acquisition, Q.C. All authors have read and agreed to the published version of the manuscript.

Funding: This research was funded by The Pyramid Talent Training Project of Beijing University of Civil Engineering and Architecture, grant number JDYC20200321; National Natural Science Foundation (NSFC) of China, grant number 42271478; National Natural Science Foundation (NSFC) of China, grant number 41930650.

Data Availability Statement: The data that support the findings of this study are available from the author upon reasonable request.

Acknowledgments: The author expresses gratitude to Changguang Satellite Technology Co., Ltd. for providing Jilin 1 satellite images, and China 21st Century Space Technology Application Co., Ltd. for providing Beijing 2 satellite images.

Conflicts of Interest: The authors declare no conflict of interest.

References

1. Shahat Osman, A.M.; Elragal, A. Smart cities and big data analytics: A data-driven decision-making use case. *Smart Cities* **2021**, *4*, 286–313.
2. Grübel, J.; Thrash, T.; Aguilar, L.; Gath-Morad, M.; Chatain, J.; Sumner, R.W.; Hölscher, C.; Schinazi, V.R. The Hitchhiker's Guide to Fused Twins: A Review of Access to Digital Twins In Situ in Smart Cities. *Remote Sens.* **2022**, *14*, 3095.
3. Zhu, J.; Wu, P. Towards effective BIM/GIS data integration for smart city by integrating computer graphics technique. *Remote Sens.* **2021**, *13*, 1889. [CrossRef]
4. Long, H.; Zhang, Y.; Ma, L.; Tu, S. Land use transitions: Progress, challenges and prospects. *Land* **2021**, *10*, 903.
5. Winkler, K.; Fuchs, R.; Rounsevell, M.; Herold, M. Global land use changes are four times greater than previously estimated. *Nat. Commun.* **2021**, *12*, 2501. [PubMed]
6. Asokan, A.; Anitha, J. Change detection techniques for remote sensing applications: A survey. *Earth Sci. Inform.* **2019**, *12*, 143–160.
7. Chughtai, A.H.; Abbasi, H.; Karas, I.R. A review on change detection method and accuracy assessment for land use land cover. *Remote Sens. Appl. Soc. Environ.* **2021**, *22*, 100482.
8. Du, P.; Liu, S.; Xia, J.; Zhao, Y. Information fusion techniques for change detection from multi-temporal remote sensing images. *Inf. Fusion* **2013**, *14*, 19–27.
9. Zhang, C.; Yue, P.; Tapete, D.; Jiang, L.; Shangguan, B.; Huang, L.; Liu, G. A deeply supervised image fusion network for change detection in high resolution bi-temporal remote sensing images. *ISPRS J. Photogramm. Remote Sens.* **2020**, *166*, 183–200.
10. Liu, Y.; Liu, J.; Ning, X.; Li, J. MS-CNN: Multiscale recognition of building rooftops from high spatial resolution remote sensing imagery. *Int. J. Remote Sens.* **2022**, *43*, 270–298.
11. Xin, J.; Zhang, X.; Zhang, Z.; Fang, W. Road extraction of high-resolution remote sensing images derived from DenseUNet. *Remote Sens.* **2019**, *11*, 2499.
12. Chen, F.; Qin, F.; Peng, G.; Chen, S. Fusion of remote sensing images using improved ICA mergers based on wavelet decomposition. *Procedia Eng.* **2012**, *29*, 2938–2943.
13. Gong, J.; Sui, H.; Sun, K.; Ma, G.; Liu, J. Object-level change detection based on full-scale image segmentation and its application to Wenchuan Earthquake. *Sci. China Ser. E Technol. Sci.* **2008**, *51*, 110–122.
14. Linke, J.; McDermid, G.; Laskin, D.; McLane, A.; Pape, A.; Cranston, J.; Hall-Beyer, M.; Franklin, S. A disturbance-inventory framework for flexible and reliable landscape monitoring. *Photogramm. Eng. Remote Sens.* **2009**, *75*, 981–995. [CrossRef]
15. Blaschke, T. Towards a framework for change detection based on image objects. *Göttinger Geogr. Abh.* **2005**, *113*, 1–9.
16. Ab Wahab, M.N.; Nazir, A.; Ren, A.T.Z.; Noor, M.H.M.; Akbar, M.F.; Mohamed, A.S.A. Efficientnet-lite and hybrid CNN-KNN implementation for facial expression recognition on raspberry pi. *IEEE Access* **2021**, *9*, 134065–134080. [CrossRef]
17. Pires de Lima, R.; Marfurt, K. Convolutional neural network for remote-sensing scene classification: Transfer learning analysis. *Remote Sens.* **2019**, *12*, 86.
18. Khelifi, L.; Mignotte, M. Deep learning for change detection in remote sensing images: Comprehensive review and meta-analysis. *IEEE Access* **2020**, *8*, 126385–126400.
19. Coppin, P.; Jonckheere, I.; Nackaerts, K.; Muys, B.; Lambin, E. Review ArticleDigital change detection methods in ecosystem monitoring: A review. *Int. J. Remote Sens.* **2004**, *25*, 1565–1596. [CrossRef]
20. Celik, T. Change detection in satellite images using a genetic algorithm approach. *IEEE Geosci. Remote Sens. Lett.* **2010**, *7*, 386–390. [CrossRef]
21. Bruzzone, L.; Prieto, D.F. Automatic analysis of the difference image for unsupervised change detection. *IEEE Trans. Geosci. Remote Sens.* **2000**, *38*, 1171–1182. [CrossRef]
22. Lu, D.; Li, G.; Moran, E. Current situation and needs of change detection techniques. *Int. J. Image Data Fusion* **2014**, *5*, 13–38. [CrossRef]
23. Sumaiya, M.; Kumari, R.S.S. Logarithmic mean-based thresholding for SAR image change detection. *IEEE Geosci. Remote Sens. Lett.* **2016**, *13*, 1726–1728. [CrossRef]
24. Sun, Y.; Lei, L.; Guan, D.; Kuang, G. Iterative robust graph for unsupervised change detection of heterogeneous remote sensing images. *IEEE Trans. Image Process.* **2021**, *30*, 6277–6291. [CrossRef] [PubMed]
25. Chen, J.; Lu, M.; Chen, X.; Chen, J.; Chen, L. A spectral gradient difference based approach for land cover change detection. *ISPRS J. Photogramm. Remote Sens.* **2013**, *85*, 1–12. [CrossRef]
26. Yan, Z.; Huazhong, R.; Desheng, C. The research of building earthquake damage object-oriented change detection based on ensemble classifier with remote sensing image. In Proceedings of the IGARSS 2018-2018 IEEE International Geoscience and Remote Sensing Symposium, Valencia, Spain, 22–27 July 2018; pp. 4950–4953.

27. Malila, W.A. Change vector analysis: An approach for detecting forest changes with Landsat. In Proceedings of the LARS Symposia, West Lafayette, IN, USA, 3–6 June 1980; 385p.
28. Xing, H.; Hou, D.; Lu, M.; Chen, J. A Land Cover Change Detection Method Based On Change Difference Map Fusion. *ISPRS Ann. Photogramm. Remote Sens. Spat. Inf. Sci.* **2019**, *IV-2/W5*, 239–243. [CrossRef]
29. Hearst, M.A.; Dumais, S.T.; Osuna, E.; Platt, J.; Scholkopf, B. Support vector machines. *IEEE Intell. Syst. Their Appl.* **1998**, *13*, 18–28. [CrossRef]
30. Breiman, L. Random forests. *Mach. Learn.* **2001**, *45*, 5–32. [CrossRef]
31. Huang, G.-B.; Zhu, Q.-Y.; Siew, C.-K. Extreme learning machine: Theory and applications. *Neurocomputing* **2006**, *70*, 489–501.
32. Hou, X.; Zhang, L. Saliency detection: A spectral residual approach. In Proceedings of the 2007 IEEE Conference on Computer vision and Pattern Recognition, Minneapolis, MN, USA, 17–22 June 2007; pp. 1–8.
33. Goferman, S.; Zelnik-Manor, L.; Tal, A. Context-aware saliency detection. *IEEE Trans. Pattern Anal. Mach. Intell.* **2011**, *34*, 1915–1926. [CrossRef] [PubMed]
34. Liu, Y.; Cheng, M.-M.; Hu, X.; Wang, K.; Bai, X. Richer convolutional features for edge detection. In Proceedings of the IEEE Conference on Computer Vision and Pattern Recognition, Honolulu, HI, USA, 21–26 July 2017; pp. 3000–3009.
35. Ronneberger, O.; Fischer, P.; Brox, T. U-net: Convolutional networks for biomedical image segmentation. In Proceedings of the Medical Image Computing and Computer-Assisted Intervention–MICCAI 2015: 18th International Conference, Munich, Germany, 5–9 October 2015; Proceedings, Part III 18. Springer: Berlin/Heidelberg, Germany, 2015; pp. 234–241.
36. Zhong, Z.; Jin, L.; Xie, Z. High performance offline handwritten chinese character recognition using googlenet and directional feature maps. In Proceedings of the 2015 13th International Conference on Document Analysis and Recognition (ICDAR), Tunis, Tunisia, 23–26 August 2015; pp. 846–850.
37. Szegedy, C.; Ioffe, S.; Vanhoucke, V.; Alemi, A. Inception-v4, inception-resnet and the impact of residual connections on learning. In Proceedings of the AAAI Conference on Artificial Intelligence, San Francisco, CA, USA, 4–9 February 2017.
38. Nemoto, K.; Hamaguchi, R.; Sato, M.; Fujita, A.; Imaizumi, T.; Hikosaka, S. Building change detection via a combination of CNNs using only RGB aerial imageries. In Proceedings of the Remote Sensing Technologies and Applications in Urban Environments II, Warsaw, Poland, 11–12 September 2017; pp. 107–118.
39. Pomente, A.; Picchiani, M.; Del Frate, F. Sentinel-2 change detection based on deep features. In Proceedings of the IGARSS 2018-2018 IEEE International Geoscience and Remote Sensing Symposium, Valencia, Spain, 22–27 July 2018; pp. 6859–6862.
40. Lei, T.; Xue, D.; Ning, H.; Yang, S.; Lv, Z.; Nandi, A.K. Local and global feature learning with kernel scale-adaptive attention network for VHR remote sensing change detection. *IEEE J. Sel. Top. Appl. Earth Obs. Remote Sens.* **2022**, *15*, 7308–7322.
41. Wu, C.; Chen, H.; Du, B.; Zhang, L. Unsupervised change detection in multitemporal VHR images based on deep kernel PCA convolutional mapping network. *IEEE Trans. Cybern.* **2021**, *52*, 12084–12098. [CrossRef]
42. He, D.; Zhang, Y.; Song, H. A novel saliency map extraction method based on improved Itti's model. In Proceedings of the 2010 International Conference on Computer and Communication Technologies in Agriculture Engineering, Chengdu, China, 12–13 June 2010; pp. 323–327.
43. Tang, Z.; Zhang, H.; Pun, C.M.; Yu, M.; Yu, C.; Zhang, X. Robust image hashing with visual attention model and invariant moments. *IET Image Process.* **2020**, *14*, 901–908.
44. Tang, T.; Hu, P.; Wu, G. Influence of promotion mode on purchase decision based on multilevel psychological distance dimension of visual attention model and data mining. *Concurr. Comput. Pr. Exp.* **2021**, *33*, e5587.
45. Li, H.; Manjunath, B.; Mitra, S.K. Multisensor image fusion using the wavelet transform. *Graph. Models Image Process.* **1995**, *57*, 235–245. [CrossRef]
46. Otsu, N. A threshold selection method from gray-level histograms. *IEEE Trans. Syst. Man Cybern.* **1979**, *9*, 62–66. [CrossRef]
47. Jia, Y.; Hao, C.; Wang, K. A new saliency object extraction algorithm based on Itti's model and region growing. In Proceedings of the 2019 IEEE International Conference on Mechatronics and Automation (ICMA), Tianjin, China, 4–7 August 2019; pp. 224–228.
48. Tang, L.; Li, H.; Chen, T. Extract salient objects from natural images. In Proceedings of the 2010 International Symposium on Intelligent Signal Processing and Communication Systems, Chengdu, China, 6–8 December 2010; pp. 1–4.
49. Yan, Q.; Xu, L.; Shi, J.; Jia, J. Hierarchical saliency detection. In Proceedings of the IEEE Conference on Computer Vision and Pattern Recognition, Portland, OR, USA, 23–28 June 2013; pp. 1155–1162.
50. Liu, N.; Cao, Z.; Cui, Z.; Pi, Y.; Dang, S. Multi-layer abstraction saliency for airport detection in SAR images. *IEEE Trans. Geosci. Remote Sens.* **2019**, *57*, 9820–9831.
51. Amolins, K.; Zhang, Y.; Dare, P. Wavelet based image fusion techniques—An introduction, review and comparison. *ISPRS J. Photogramm. Remote Sens.* **2007**, *62*, 249–263. [CrossRef]
52. Nackaerts, K.; Vaesen, K.; Muys, B.; Coppin, P. Comparative performance of a modified change vector analysis in forest change detection. *Int. J. Remote Sens.* **2005**, *26*, 839–852. [CrossRef]
53. Adelson, E.H.; Anderson, C.H.; Bergen, J.R.; Burt, P.J.; Ogden, J.M. Pyramid methods in image processing. *RCA Eng.* **1984**, *29*, 33–41.
54. Lan, Z.; Lin, M.; Li, X.; Hauptmann, A.G.; Raj, B. Beyond gaussian pyramid: Multi-skip feature stacking for action recognition. In Proceedings of the IEEE Conference on Computer Vision and Pattern Recognition, Boston, MA, USA, 7–12 June 2015; pp. 204–212.
55. Ding, L.; Goshtasby, A. On the Canny edge detector. *Pattern Recognit.* **2001**, *34*, 721–725. [CrossRef]

56. Rong, W.; Li, Z.; Zhang, W.; Sun, L. An improved CANNY edge detection algorithm. In Proceedings of the 2014 IEEE International Conference on Mechatronics and Automation, Tianjin, China, 3–6 August 2014; pp. 577–582.
57. Xiao, A.; Wang, Z.; Wang, L.; Ren, Y. Super-resolution for "Jilin-1" satellite video imagery via a convolutional network. *Sensors* **2018**, *18*, 1194. [CrossRef] [PubMed]
58. Fan, Y.; Lou, D.; Zhang, C.; Wei, Y.; Jia, F. Information extraction technologies of iron mine tailings based on object-oriented classification: A case study of Beijing-2 remote sensing images of the Qianxi Area, Hebei Province. *Remote Sens. Nat. Resour.* **2021**, *33*, 153–161.
59. Hoffmann, W. Iterative algorithms for Gram-Schmidt orthogonalization. *Computing* **1989**, *41*, 335–348. [CrossRef]
60. Falco, N.; Marpu, P.R.; Benediktsson, J.A. Comparison of ITPCA and IRMAD for automatic change detection using initial change mask. In Proceedings of the 2012 IEEE International Geoscience and Remote Sensing Symposium, Munich, Germany, 22–27 July 2012; pp. 6769–6772.
61. Li, H.-C.; Longbotham, N.; Emery, W.J. Unsupervised change detection of remote sensing images based on semi-nonnegative matrix factorization. In Proceedings of the 2014 IEEE Geoscience and Remote Sensing Symposium, Quebec City, QC, Canada, 13–18 July 2014; pp. 1289–1292.
62. Stehman, S. Estimating the kappa coefficient and its variance under stratified random sampling. *Photogramm. Eng. Remote Sens.* **1996**, *62*, 401–407.
63. Wang, J.; Liu, J.; Zhang, Y.; Zhu, H.; Han, Y.; Zhang, Y.; Zhou, R.; Hong, Z.; Yang, S. An Image Strategy Based on Saliency Detection Using Luminance Contrast for Artificial Vision with Retinal Prosthesis. In Proceedings of the Sixth International Congress on Information and Communication Technology: ICICT 2021, London, UK, 25–26 February 2021; Springer: Berlin/Heidelberg, Germany, 2022; Volume 3, pp. 273–281.
64. Singh, S.K.; Srivastava, R. A novel probabilistic contrast-based complex salient object detection. *J. Math. Imaging Vis.* **2019**, *61*, 990–1006. [CrossRef]
65. Zhu, M.; Jiao, L.; Liu, F.; Yang, S.; Wang, J. Residual spectral–spatial attention network for hyperspectral image classification. *IEEE Trans. Geosci. Remote Sens.* **2020**, *59*, 449–462.

remote sensing

MDPI

Article

Analysis of Four Decades of Land Use and Land Cover Change in Semiarid Tunisia Using Google Earth Engine

Nesrine Kadri [1,2], Sihem Jebari [1], Xavier Augusseau [3], Naceur Mahdhi [4], Guillaume Lestrelin [3] and Ronny Berndtsson [5,*]

1 National Research Institute for Rural Engineering, Water and Forestry, Tunis 1004, Tunisia; nesriinekadri@outlook.fr (N.K.)
2 National Agronomic Institute of Tunisia, Tunis 1082, Tunisia
3 CIRAD—Centre de Recherche Agronomique pour le Développement, 34090 Montpellier, France
4 Institute of Arid Regions, Medenine 4119, Tunisia
5 Centre for Advanced Middle Eastern Studies & Division of Water Resources Engineering, Lund University, 22100 Lund, Sweden
* Correspondence: ronny.berndtsson@tvrl.lth.se; Tel.: +46-46-2228986 or +46-46-2229609

Abstract: Semiarid Tunisia is characterized by agricultural production that is delimited by water availability and degraded soil. This situation is exacerbated by human pressure and the negative effects of climate change. To improve the knowledge of long-term (1980 to 2020) drivers for Land Use and Land Cover (LULC) changes, we investigated the semiarid Rihana region in central Tunisia. A new approach involving Google Earth Engine (GEE) was used to map LULC using Landsat imagery and vegetative indices (NDVI, MSAVI, and EVI) by applying a Random Forest (RF) classifier. A Rapid Participatory Systemic Diagnosis (RPSD) was used to consider the relation between LULC changes and their key drivers. The methodology relied on interviews with the local population and experts. Focus groups were conducted with practicians of the Regueb Agricultural Extension Services, followed by semi-structured interviews with 52 households. Results showed the following: (1) the RF classifier in Google Earth Engine had strong performance across diverse Landsat image types resulting in overall classification accuracy of $\geq$0.96 and a kappa coefficient $\geq$0.93; (2) rainfed olive land increased four times during the study period while irrigated agriculture increased substantially during the last decade; rangeland and rainfed annual crops decreased by 58 and 88%, respectively, between 1980 and 2021; (3) drivers of LULC changes are predominately local in nature, including topography, local climate, hydrology, strategies of household, effects of the 2010 revolution, associated increasing demand for natural resources, agricultural policy, population growth, high cost of agricultural input, and economic opportunities. To summarize, changes in LULC in Rihana are an adaptive response to these various factors. The findings are important to better understand ways towards sustainable management of natural resources in arid and semiarid regions as well as efficient methods to study these processes.

Keywords: water scarcity; soil erosion; LULC; GEE; random forest; RPSD; semiarid Tunisia

1. Introduction

Land Use and Land Cover (LULC) changes are important for identifying long-term landscape effects caused by natural processes and anthropogenic pressure [1]. These processes have important impacts on the fragile ecology of semiarid regions. Due to increasing human pressure, scarce resources in semiarid Tunisia are frequently overexploited. The pressure has been increasing in recent decades due to a higher occurrence rate of extreme climatic events such as droughts and heavy rainfall, which may force local populations to change crop systems [2]. LULC in semiarid regions has an impact on both local socioeconomic development and the environment, as well as global environmental change [3]. As a result, accurate and long-term information on LULC is a critical tool in natural resource management.

In this regard, Remote-Sensing (RS) imagery is a viable option for detecting changes at various spatial–temporal scales due to detailed resolution and widespread availability [4]. Despite the importance of using RS in long-term change detection analysis, problems arise as data volumes increase and the need to modify and homogenize images from various sensors with varying quality and periods escalates [5,6]. To overcome these problems, Google Earth Engine (GEE) can be used as a cloud-computing platform together with machine-learning algorithms for the analysis of satellite images [7–9]. This tool has recently been recognized as a potential platform for RS analysis [10,11]. Indeed, GEE is a large geospatial, cloud-based platform that provides a powerful tool to process, analyze, and visualize extensive amounts of remote-sensing data [12,13].

As mentioned above, Machine-Learning (ML) techniques can greatly help in the analysis of large data sets, such as in LULC mapping [14]. Common ML techniques include Random Forests (RF), Artificial Neural Networks (ANN), Decision Trees, and Support Vector Machines (SVM) [15,16]. Among these, RF has gained popularity for LULC mapping due to its high accuracy, which has proven to be very efficient in circumstances of large inputs of information, relatively low cost, and the need for a limited number of variables [17,18]. Furthermore, indices such as the Normalized Difference Vegetation Index (NDVI), Normalized Difference Water Index (NDWI), Modified Soil-Adjusted Vegetation Index (MSAVI), Soil-Adjusted Vegetation Index (SAVI), Enhanced Vegetation Index (EVI), and the Normalized Difference Soil Index (NDSI) are effective for mapping of LULC changes [19,20].

Recent studies have demonstrated the effectiveness of using GEE within the RF classifier for automated, long-term, and accurate LULC mapping. For example, [21] utilized Landsat data, an RF classifier in the GEE platform, to map LULC in Côte d'Ivoire for the year 2020, achieving an overall accuracy of over 90%. Another study [22] employed GEE and RF classifiers to monitor annual LULC change in the Tucson Metropolitan Area from 1986 to 2020 using Landsat imagery. Their results consistently demonstrated satisfying classification accuracy, ranging from 90% to 95% for most years. The authors of [23] analyzed spatiotemporal changes in forest cover in the Ashanti region of Ghana for the years 2006, 2011, 2015, and 2020 using GEE and the two classifiers RF and SVM. The RF classifier showed an overall accuracy of 98% and a kappa index of 0.97, while SVM achieved an overall accuracy and a kappa index of 90% and 0.88, respectively. This highlights the superior performance of the RF classifier with low overfitting tendencies. The authors of [24] compared the performance of three machine-learning algorithms, SVM, RF, and CART, for LULC classification for the years 2016, 2018, and 2020. Their findings revealed that RF classifiers in the GEE platform consistently outperformed SVM and CART classifiers in terms of accuracy for both Landsat-8 and Sentinel-2 imagery.

Scholars using satellite imagery data have, to some extent, studied LULC changes in Tunisia [25]. For instance, [26] used Landsat imagery and maximum likelihood classification to examine the changes in LULC between 2007 and 2017. In another study, [27] employed the Normalized Difference Vegetation Index (NDVI) derived from the Very High-Resolution Radiometer (AVHRR) to assess the spatial and temporal patterns of land use change in North and West Africa spanning the period from 1985 to 2015. The authors of [28] focused on mapping irrigated and rainfed areas in semi-arid regions of Tunisia between 2015 and 2017 using Sentinel 1 and Sentinel 2 data. The authors of [29] analyzed vegetation dynamics over the central region of Tunisia during drought events by utilizing a multi-temporal series of NDVI derived from SPOT-VEGETATION and TERRA-MODIS satellite data. However, there has been a lack of studies conducted in Tunisia that specifically employ the GEE platform to accurately map and analyze LULC changes. Also, there is a lack of research that integrates local knowledge to understand the factors that influence farmers' land use decision making at a local scale during a long period corresponding to decades.

In view of the above, we investigated the long-term changes in LULC for a representative region in central semiarid Tunisia to improve the knowledge of possible natural and socioeconomic drivers. Hence, we assessed the decade-wise changes in LULC between 1980 and 2021 using Landsat imagery and RF classifier on the GEE platform. The NDVI, EVI, and SAVI indices were selected to characterize LULC, see, e.g., [30]. Rapid Participatory Systemic Diagnosis (RPSD) was used to clarify the relationships between LULC changes and their drivers. The objectives were the following: Firstly, to investigate whether the GEE and RF methods can provide automated, long-term, and improved accuracy of LULC mapping. Secondly, to examine the spatiotemporal patterns of LULC change in semi-arid central Tunisia. Thirdly, to identify the major natural socioeconomic drivers that influence LULC changes.

2. Study Area

The focus of this study was the Rihana region located in central Tunisia (Regueb delegation, Sidi Bouzid governorate) and one of the involved sites of the PACTE program «Climate change adaptation program for vulnerable rural territories of Tunisia» (PACTE: https://pacte.tn/) (accessed on 15 January 2019)). The Rihana region covers about 10,000 ha (Figure 1) with an annual average temperature of about 20 °C and a summer temperature that can reach 47 °C. The annual rainfall is irregular, with a mean of 225 ± 129 mm (Figure 2). Rihana agriculture is characterized by small farms, often less than 10 ha, with low economic productivity. The landscape is dominated by rainfed olive production in the up- and middle stream parts and irrigated crops in the downstream plain. Rihana is among Tunisia's environmentally and socioeconomically most vulnerable territories [31]. The vulnerability is essentially characterized by degrading natural resources, such as the loss of a large part of its production potential and the overexploitation of hydro-agricultural resources and rangelands. The vulnerability is worsened by large temporal and spatial variability of rainfall, which results in alternating periods of drought, flooding, and erosion. The risk of flooding is high, and heavy rains lead to significant surface runoff, resulting in floods in almost all wadis in the region [32]. Figure 2a,b display the annual rainfall depth as well as the annual average together with mean, maximum, and minimum temperature. The data were obtained from the General Directorate of Water Resources [33] covering the period between 1980 and 2021. Historical rainfall data since 1980 show several annual extremes, e.g., the wet year 2013/2014 (395.5 mm). About ten deficit years, with a minimum of 50 mm for the year 2000/2001, occurred in the period (Figure 2).

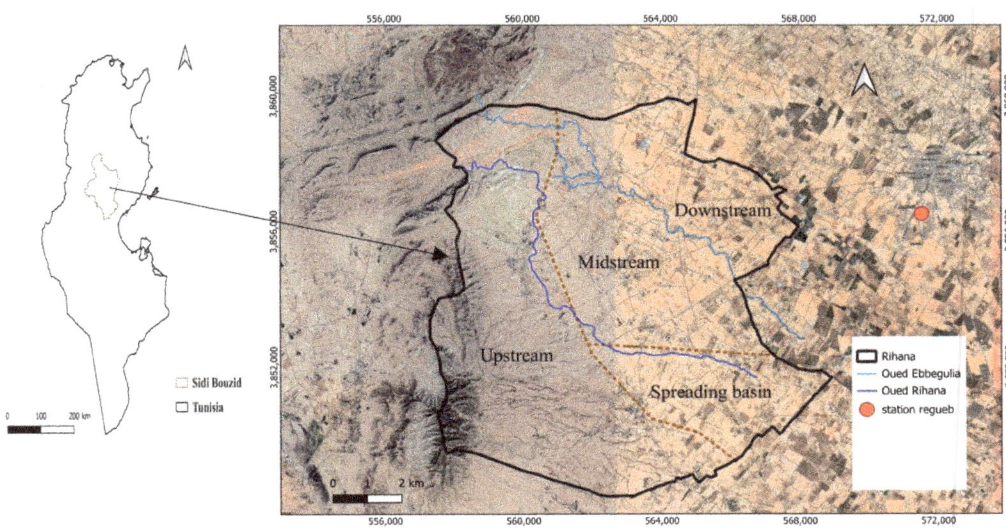

Figure 1. Geographic location of the Rihana study area in Tunisia.

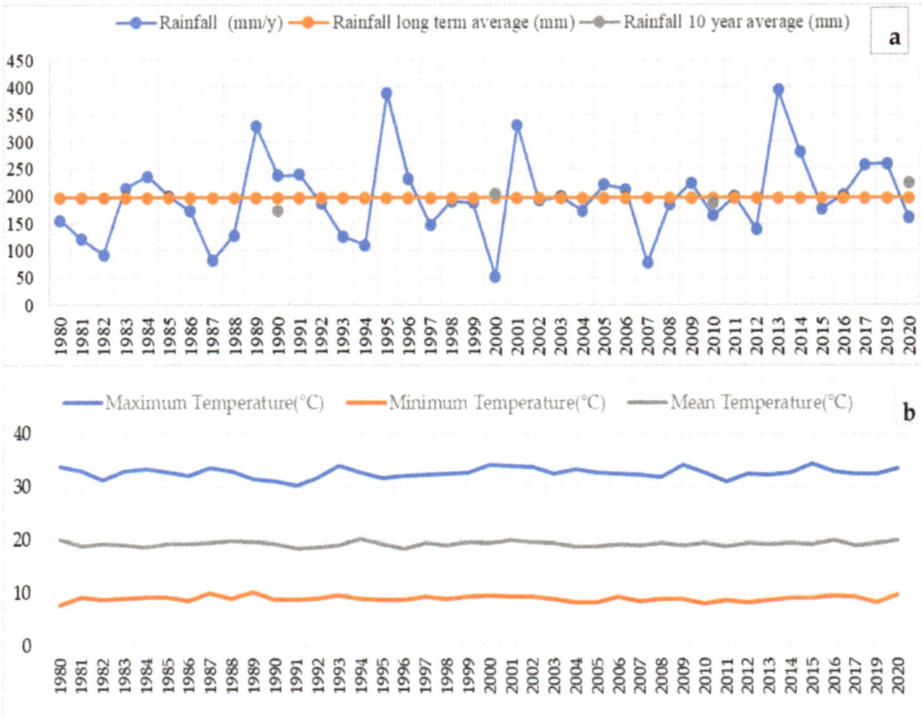

Figure 2. (**a**) Average annual rainfall; (**b**) temperature (mean, maximum, and minimum) observed at the Regueb climate station.

3. Data and Methods

3.1. Datasets

3.1.1. Satellite Imagery

Landsat 3/5/7/8 data archived by the United States Geological Survey (USGS) (http://glovis.usgs.gov/ (accessed on 15 October 2022)) for mapping LULC changes were used in this study. Table 1 presents all details of the selected images. Surface Reflectance data were utilized from Landsat-5 TM, Landsat-7 ETM+, and Landsat 8 OLI between 1 September and 31 August (hydrological year) corresponding to the years 1990, 2000, 2010, and 2020. All datasets were used and directly integrated into the GEE platform. For Landsat 3 MSS, images were corrected before being used because available corrected images for the year 1980 did not cover our area of interest.

Table 1. Used satellite imagery description.

Data Sets	Acquisition Date	Resolution (m)
USGS Landsat 3 MSS	01 September 1980/31 August 1981	60
USGS Landsat 5 Tier 1 Surface Reflectance	01 September 1990/31 August 1991	30
USGS Landsat 7 Tier 1 Surface Reflectance	01 September 2000/31 August 2001 01 September 2010/31 August 2011	30
USGS Landsat 8 Tier 1 Surface Reflectance	01 September 2020/31 August2021	30

3.1.2. Training and Validation

Training and validation are critical steps in LULC mapping. Thus, data were collected via field observations, farmer surveys, and use of Google Earth Pro. Field observations were performed in the study area during 2020 (September, October, and December) and 2021 (January, March, June, and July), where 320 polygons were selected. For 1980, 1990, 2000, and 2010, samples of polygons were collected via farmer surveys. Together with stakeholders, we identified polygons that cover the period for each LULC class. For example, field visits were performed to areas that were covered by olive trees in 1980, 1990, 2000, and 2010, and similar field visits were conducted for other classes. In general, 240 ground-truth polygons were collected. For 2010, we also used Google Earth Pro. For reference data collection, true and false color composites of Landsat imagery were used. The generated polygons were evenly distributed over the Rihana area and had a homogeneous spatial distribution. It is worth noting that former studies [34], as well as the local population, helped to define the four classes, as shown in Table 2, representing the overall land cover of the study area except for the most upstream part. The latter, which represents a nature reserve and scrubland, is considered a stable class that has not changed over the past 40 years. Table 3 displays the number of polygons for each year and LULC class, along with the average surface area.

Table 2. Description of Land Use and Land Cover classes.

LULC Category	Description
Rangeland	Grazed land characterized by natural vegetation.
Rainfed olive	Olives cultivated in solely rainfed areas.
Rainfed annual crops	Annual crops that rely on rainfall for water, specifically cereals and forage crops
Irrigated crops	Includes areas with at least one crop cycle using irrigation as a water source

Table 3. Training and validation data.

		Rangeland	Rainfed Olive	Rainfed Annual Crops	Irrigated Crops
1980 (Landsat 3)	Polygons	16	10	18	-
	Surface (pixels)	75	42	92	-
1990 (Landsat 5)	Polygons	15	7	10	-
	Surface (pixels)	233	67	89	-
2000 (Landsat 7)	Polygons	21	23	16	-
	Surface (pixels)	444	689	144	-
2010 (Landsat 7)	Polygons	23	67	8	6
	Surface (pixels)	756	2644	311	111
2020 (Landsat 8)	Polygons	92	76	6	146
	Surface (pixels)	1156	2433	233	633

3.2. Methodology

We utilized the GEE platform to classify and detect changes in LULC. The adopted approach is shown in Figure 3. Further explanation follows below.

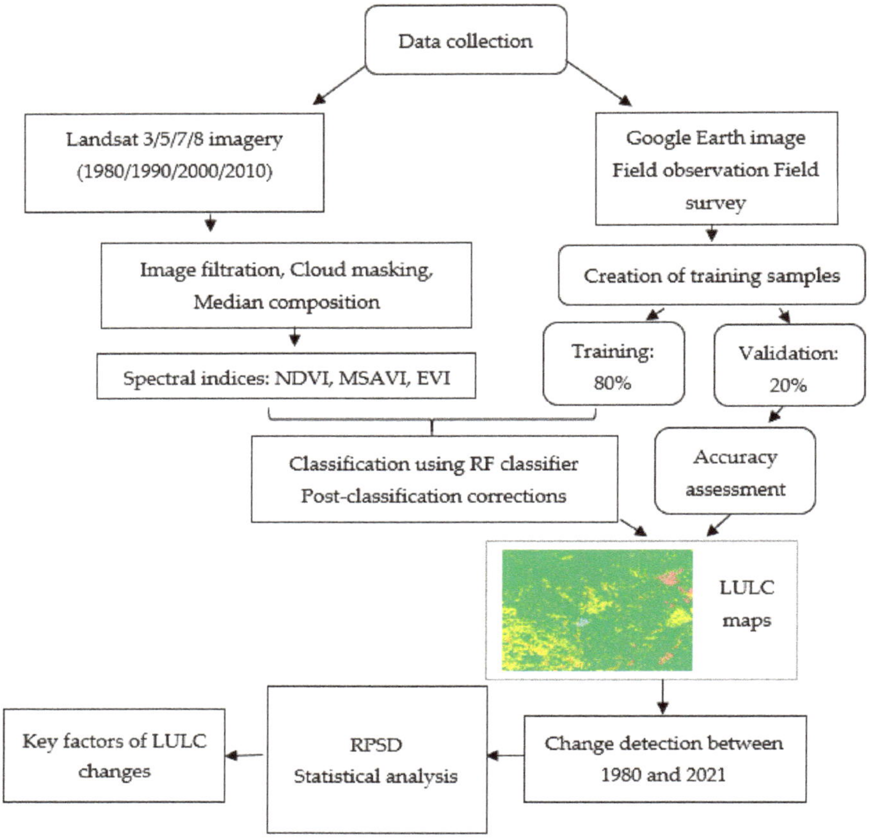

Figure 3. Description of the methodology used in this present study.

3.2.1. Classification and Post-Processing

The input data (Landsat images) were processed using the GEE platform, as shown in Figure 3. It is preferable if the imagery used for LULC change detection contains as few clouds as possible. As a result, images were filtered with 10% cloudiness for the selected years 1990, 2000, 2010, and 2020. For 1980, images were filtered with 25%, which represents the low-cloudiness percentage available for the study area. Then, we combined each year's dataset (Landsat imagery) based on the median reducer function to produce one single image from the image collection. For the entire stack of images, a median value was used for each pixel.

To improve the classification accuracy, NDVI, MSAVI, and EVI indices were derived and calculated from all data used. The definitions for these are as follows [35]:

$$\text{NDVI} = (\text{nir} - \text{red})/(\text{nir} + \text{red}) \quad (1)$$

$$\text{MSAVI} = (2 \times \text{nir} + 1 - \text{sqrt}((2 \times \text{nir} + 1)2 - 8 \times (\text{nir} - \text{red})))/2 \quad (2)$$

$$\text{EVI} = 2.5 \times (\text{nir} - \text{red})/(\text{nir} + 6.0 \times \text{red} - 7.5 \times \text{blue} + 1) \quad (3)$$

where nir, red, and blue represent spectral reflectance in the near-infrared, red, and blue bands, respectively.

After that, and as mentioned before, we used RF machine-learning algorithm to train the classification for Landsat images. Once the classification was completed, an evaluation was carried out to determine the accuracy level of each classified image using a confusion matrix. This algorithm built on GEE depicts the relationship between ground truth and classification results [36]. We trained and validated the algorithm with 80% and 20%, respectively. To validate classification, both kappa coefficient and overall accuracy were adopted [37]. The final product was chosen based on the RF with the highest overall accuracy and kappa value. We used post-classification corrections based on the neighborhood algorithm in GEE to replace isolated pixels with surrounding values to guarantee smoother appearance of the classified results.

3.2.2. Drivers of LULC Change Based on the RPSD Methodology

To comprehend the main LULC transitions that occurred in Rihana over the past four decades and to investigate the key drivers behind these changes, a Rapid Participatory Systemic Diagnosis (RPSD) was conducted with local farmers and experts. The RPSD is a participatory and multi-actor method. It is based on semi-open interviews with the local population, either individually or collectively, complemented by interviews with local experts. Together, it allows for the co-construction of a common understanding of the functioning and development issues of territory with local actors [34,38]. Initially, focus groups were initiated with the help of experts from the Regueb Agricultural extension for different specialties: crop production, water resources, soil resources, and water and soil conservation. During these discussions, local experts provided valuable insights and analysis of the local context of Rihana dynamics and the changing availability and use of water and soil resources. Furthermore, these groups reviewed the main changes in LULC over time. We then had detailed discussions with farmers to better understand these changes and the underlying factors that drove them in both current and historical situations. The discussions were conducted in two focus groups and via field surveys with 52 farmers. Participants were selected from various locations in Rihana, including the upstream, midstream, and downstream basin parts (Figure 4). The participating farmers had an average age of 59 years, with a majority between the ages of 41 to 64 years. Only 11% of them were below the age of 40. They owned an average farmland area of 18 ha, which is usually composed of 3 parcels. The surveyed farms were asked to list the LULC changes they had observed. In addition, participants were also asked to identify the drivers behind these changes and rank them from most to least significant. This approach allowed for a deeper understanding of the LULC dynamics and the factors contributing to them, as explained by local population and experts. During the fieldwork, a variety of tools was utilized; among them were satellite images of the region and a tablet equipped with GPS. Coordination with the local authorities and involvement of researchers, associations, and practitioners were ensured.

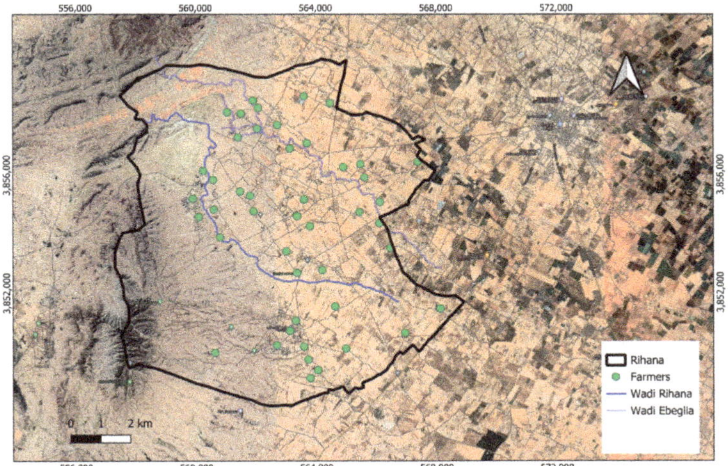

Figure 4. Geographic distribution of surveyed farms.

4. Results and Discussion

4.1. Classification Performance

An accuracy assessment was considered for the five classified maps to verify that mapped content corresponds to ground truth in the study area. The overall accuracy (OA), kappa index (K), and user and producer's accuracy were performed to identify class performance and misclassifications. Results show that the use of the RF classifier in GEE provided an OA $\geq$ 0.96 and a K $\geq$ 0.93. For Landsat 3 (1980), Landsat 5 (1990), and Landsat 7 (2000), the mean OA and K were 0.99 and 0.98, respectively. The OA for Landsat 7 (2010) and Landsat 8 (2020) was 0.96 and 0.97, respectively, with individual K equal to 0.93 and 0.95, respectively. Thus, results were generally of high quality and adequate for the subsequent analysis and change detection [39].

The classification accuracy results obtained for the different classes for every date are shown in Table 4. UA and PA did not show a big difference between rangeland and rainfed olive. They were classified with higher accuracies for all years, with producer accuracy (PA) and user accuracy (UA) $\geq$ 95%. For rainfed annual crops, they also showed a good performance for the first three years, unlike for 2010 (PA = 96%) and 2020 (PA = 94%). This can be explained by the confusion between rainfed annual crops and irrigated crops, which could be explained as a seasonal effect. For irrigated crops, the performance was better for the year 2020 than for 2010, with PA = 97% and 71%, respectively. The relatively lower classification shows that irrigated areas are more easily detected with Landsat 8 than with Landsat 7. Refs. [40,41] note that the low accuracy for some classes is due to the reduced samples of training that significantly affects the classification results, which is true, especially for the irrigated crops class for the year 2010.

Table 4. Accuracy assessment of the different LULC classes.

		Rangeland	Rainfed Olive	Rainfed Annual Crops	Irrigated Crops
1980 (Landsat 3)	PA	0.98	0.96	0.98	–
	UA	0.97	0.97	0.99	–
	OA	0.98			
	K	0.97			
1990 (Landsat 5)	PA	0.99	0.96	0.99	–
	UA	0.98	0.98	0.99	–
	OA	0.99			
	K	0.98			
2000 (Landsat 7)	PA	0.98	0.99	0.97	–
	UA	0.99	0.98	0.98	–
	OA	0.99			
	K	0.98			
2010 (Landsat 7)	PA	0.97	0.99	0.96	0.71
	UA	0.95	0.96	0.98	0.99
	OA	0.96			
	K	0.93			
2020 (Landsat 8)	PA	0.97	0.97	0.94	0.97
	UA	0.95	0.97	0.97	0.99
	OA	0.97			
	K	0.95			

PA: Producer's Accuracy, UA: User's Accuracy, OA: Overall Accuracy, K: Kappa index.

4.2. LULC Classification Analysis

Multi-temporal LULC for 1980, 1990, 2000, 2010, and 2020 are shown in Figure 5. Figure 6 shows the areal statistical distribution of LULC and their ratios for the respective period. Results indicate that in 1980, most of the study area was rangeland, and rainfall annual/cereal crops covered about 36.2% and 28.1%, respectively. Rainfed olive covered an area of 1507.7 ha (15.4%). Rangeland is distributed across the territory, with significant presence in the upstream and piedmont parts. As for rainfed annual crops, they have mainly been concentrated in the riverbed of wadi Rihana and in the spreading basin. On the other hand, in 1990, the most dominant land classes were rainfed olive and rangeland, which covered 38.8% and 30.1%, respectively. Rainfed olive cultivation was widespread across the entire study area, while rangeland was primarily concentrated in the upstream and midstream regions. The least aerial coverage was rainfed annual crops, which accounted for only 10.8%, mainly located in the spreading basin. The analysis of LULC for the year 2000 revealed that rainfed olive cultivation occupied approximately 47% of the total area, with a notable increase observed in the midstream and downstream regions. Rangeland, on the other hand, experienced a decrease and accounted for 25.8% of the area. Rainfed annual crops also decreased in various parts of the study area, except for the spreading basin, where they remained dominant, covering 6.3% of the area. In the year 2010, notable changes occurred in the LULC of Rihana. The introduction of irrigation was observed in the downstream plain, covering approximately 13.2 hectares (0.13%) of the total area. Rainfed olive cultivation continued to expand and emerged as the dominant land cover, encompassing more than 53% of the study area. Rangeland experienced a further decline, representing 22.7% of the total area, primarily in the midstream region. Rainfed annual crops witnessed a significant decrease, particularly in the spreading basin, accounting for only 3.2% of the total area. Looking at the year 2020, the most remarkable change was observed in the expansion of irrigated crop areas, which covered approximately 6.4% of the total area. Rainfed olive cultivation continued to constitute the largest proportion, although with a slight increase compared to previous years, accounting for 55%. Conversely, rangeland experienced a further decline, representing 15% of the total area. However, there was an increase in rainfed annual crops, which now accounted for 3.4% of the area.

4.3. Drivers of LULC Change

The participating local farmers provided insights into the various factors driving the changes in LULC. They reported that topography, soil type, accessibility, hydrology, local climate, population growth, individual and collective farmer strategies, socio-economic conditions of farmers, new policies, high cost of agricultural inputs, economic opportunities/financial incentives, the Jasmin revolution, and cultural factors all play significant roles in LULC dynamics in the study area. They indicated that each decade of LULC change had its unique drivers (Figures 6 and 7). Table 5 shows the main drivers of LULC changes, and their ranking as considered by the local population.

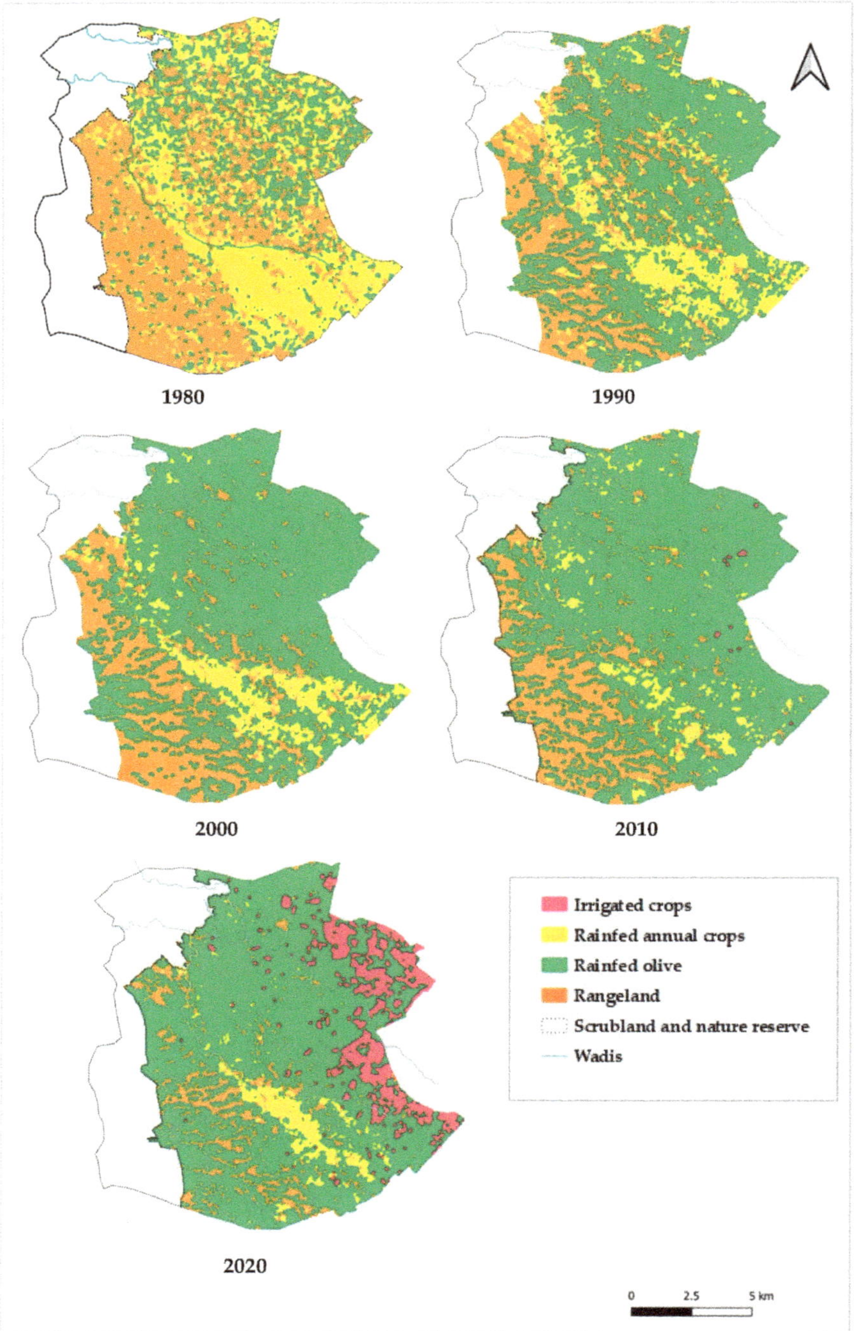

Figure 5. LULC change in Rihana for the years 1980, 1990, 2000, 2010, and 2020.

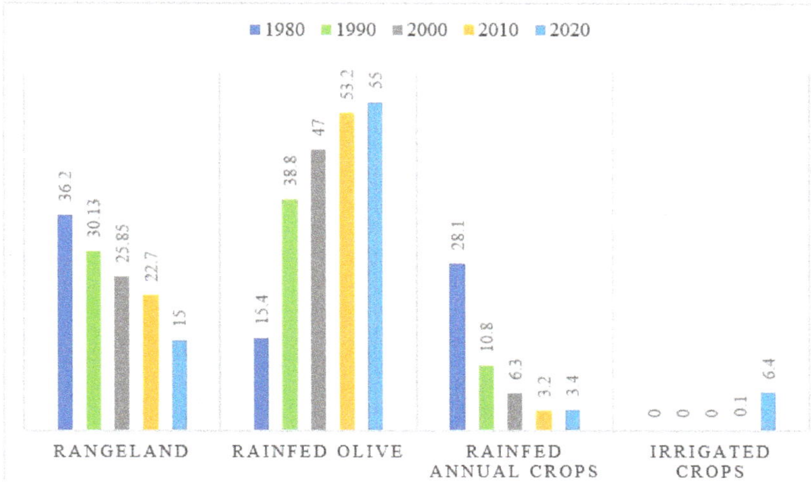

Figure 6. Successive LULC changes in Rihana from 1980 to 2020 (%).

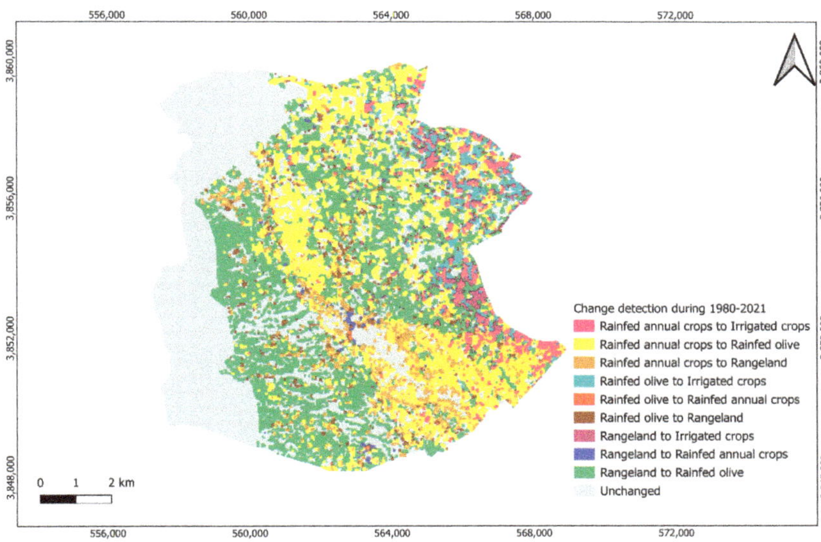

Figure 7. LULC change detection during 1980–2021.

From 1980 to 1990, major changes were observed for the rainfed annual crops and rainfed olive classes. Indeed, the annual rainfed crops experienced a massive decline in the area corresponding to 61%. This decline was a result of expanding the rainfed olive area, which increased by 152% throughout the study area. Moreover, the rangeland area, particularly in the upstream part, declined by about 17%. There were no differences among the interviewees in their responses regarding the LULC change that occurred in this period. They all indicated that there was a significant decline in rainfed annual croplands and a corresponding significant increase in rainfed olive lands. According to the local population of Rihana, the consequences of the rapid increase in the population since independence in 1956 had a significant impact on the change of LULC. This has resulted in an increase in the number of douars (small nomadic villages), with populations varying greatly from one locality to another. The main activity of this population was based on grazing sheep resulting in the large use of rangelands and cereal crops. Rainfed agriculture usually gives low yields [42]. A total of 89% of the participants indicated that the sedentarization policy led to the growth of rainfed olive crop areas between 1980 and 1990. This policy consisted of the promotion of olive and almond plantations and the creation of the first Soil and Water Conservation (SWC) techniques [43]. Consequently, the settling of nomadic populations led to a major change in their relationship with the land. The prospects of privatization made them attribute a value to the land for its own sake, whereas previously, it had only been used as a support for livestock feeding. Additionally, 60% of the interviewed local population mentioned that seasonal migration plays a significant role in the changes that occurred. These additional incomes allowed them to continue to invest in rainfed orchards and livestock. Furthermore, all farmers stressed the considerable influence of the 1987 and 1988 droughts (Figure 2). They caused the destruction of many almond trees, which were later replaced by more drought-resistant olive trees. Farmers have long developed their own SWC techniques like the mgoud and tabias for floodwater harvesting and distribution [44]. Investment in rainfed olives was favored by the operation of land registration organized by the state between 1980 and 1990. Finally, the evolution of the production system in this period is the result of the local climate, the local population strategies, and the agricultural policies [45], which were confirmed by local experts.

Table 5. Drivers of LULC change and their relative ranking according to the local population.

Period	Drivers	Rank
1980–1990	Sedentarization policy	1
	Decreased sheep grazing	2
	Severe droughts	3
	Increased implementation of SWC techniques by the local population	4
	Seasonal migration	5
	Land registration operation	6
1990–2000	the implementation of the first national strategy of SWC	1
2000–2010	Severe droughts	1
	The widespread use of irrigation in Regueb	2
	Topography and soil fertility	3
	Availability of larger farms	4
2010–2020	The revolution resulting in the weak administrative control	1
	Topography and soil fertility	2
	Occurrence of larger farms	3
	Extended drought periods	4
	High cost of forage grass	5
	Increased cost of fuel and electricity	6
	Tradition of cultivating rainfed olive	7

For the period from 1990 to 2000, results show that the rainfed olive area continued to increase by about 21%. On the other hand, rangeland and rainfed annual crops continued to decrease by 14% and 42%, respectively. The decrease in rangeland was specifically observed in the midstream part, while the decrease in the rainfed annual crops was particularly obvious in the upper areas of the downstream part. However, these crops are still cultivated in the spreading basin. All interviewees pointed out that the main driving force behind these changes is closely linked to state policies and strategies. These changes are likely to be associated with the implementation of the first national strategy of SWC in Tunisia

(1990–2000). The strategy relied on different techniques of SWC designed to limit erosion and gullying. In 1990, different techniques were implemented in Rihana, such as tabias, flood spreading, and recharging wells in the wadi's upstream part. Under this strategy, the Ministry of Agriculture continued to provide Rihana farmers, landowners with which these techniques were implemented, with olive plants between 1990 and 1995. This encouraged them to plant rainfed olives that are better adapted to the soil and water availability.

During the period 2000–2010, a continuously increasing rainfed olive crop area by about 13% is observed. Rainfed annual crops experienced a massive decline of about 49%. They covered 625.2 ha of the study area in 2000 and decreased to 319 ha in 2010. The decline was primarily observed in the spreading area. Additionally, the rangelands witnessed a decrease of 12% throughout the study area. In 2010, we noticed the introduction of irrigation covering about 13.2 ha (0.13%) in the plain area. A total of 98% of the interviewees stated that these LULC changes may be primarily due to extreme climatic events observed in this specific period. Most of the territory falls under high exposure to drought, especially from 2000 to 2007 (Figure 2). The local population responded to droughts by cultivating rainfed olives. It was already mentioned [46] that drought poses the most significant threat to the socioeconomic stability of small farmers. In addition, all local experts indicated that the irrigation from drilled wells which became widespread in Regueb, motivated the local population to start the irrigation of crops. The local population situated in the plain opts for irrigation agriculture due to favorable topography, convenient access to water through wells, and the occurrence of larger farms (5 to 50 ha) compared to the upstream part. The topography of these areas is characterized by steep slopes and rocky, shallow soils and low fertility with farms that are smaller than 5 ha. Additionally, farmers consider irrigation a viable adaptation strategy to cope with the impact of climate change [47].

Considering the LULC changes between 2010 and 2020, we note the extension of the irrigated areas in the plain. They increased from 13.2 ha in 2010 to 634 ha in 2020. In addition, rainfed annual crops increased from 319 ha in 2010 to 335 ha in 2020; this trend is especially evident in the spreading basin area. During the same period, rainfed olives showed a small increase of 3.38%, particularly in the midstream part. In contrast, rangeland showed a reverse trend, decreasing by 40%. It is worth noting that since the outbreak of the Tunisian revolution in 2011, the region, as well as the country, experienced a period of transformation that induced significant environmental changes [48]. These changes are marked by the weakening of administrative control that has led to an increase in the number of illicit drillings, as reported by the experts. The local population has reported an increase in the number of illicit well drilling in Rihana, which has risen from 35 in 2019 to 70 in 2020. We also notice that the use of solar panels has recently soared in Rihana and in the semiarid region of Tunisia in general due to the increasing costs of fuel and electricity [49]. About 25% of the farmers who practice irrigation use solar panels. Additionally, another important factor reported by 90% of the interviewees is the extended drought period followed by wet years that lead to a decrease in the rangeland area (Figure 2). Moreover, 83% of the respondents mentioned that with the high cost of forage grass, they returned to cultivating rainfed annual crops for the feeding of their livestock. Furthermore, 60% of the farmers indicated that the expansion of rainfed olive lands is due to the long-standing tradition of cultivating rainfed olives, which has become an integral part of their culture and livelihoods.

Finally, it is worth noting that Sidi Bouzid, and particularly Rihana, have experienced a significant transformation regarding its agrarian system. Over the 40 past years, rainfed olive land and irrigated land increased in area by 4 times between 1980 and 2021 and 60 times between 2010 and 2021 (Figure 7). The rangeland and rainfed annual crop area decreased by 58% and 88%, respectively. It is mainly the transition from a system dominated by pastoralism based on breeding, transhumance, and rangeland to an agricultural system dominated based on irrigated and rainfed olive crops in dry conditions associated with episodic cereal cultivation and decreasing rangeland area [50]. These changes are influenced by a variety of factors, such as economic opportunities and policies, etc., but they are also highly dependent on the local context and directly linked to the livelihoods of the local population. The changes in LULC can be explained as an adaptive response to these various factors. As a result, even in small areas, there is a great diversity of agricultural situations.

5. Conclusions

Accurate and long-term LULC data are crucial for the natural resources' sustainable management in the semiarid regions of Tunisia. In this study, we started with the analysis of the LULC changes in Rihana from 1980 to 2020 using the Random Forest algorithm (RF) in Google Earth Engine (GEE) platform, along with Landsat 3, 5, 7, and 8 imageries. We characterized the dynamic LULC change using RPSD. We used time series of satellite images with an RF classifier in the GEE platform for the semiarid region. Change detection analysis enabled us to compare the spatial-temporal patterns of LULC changes. The convenience of the used approach can be considered in the future as a reference for conducting similar studies in other regions. We found that the Rihana study area has undergone major LULC changes during the past 40 years. The observed trend is an increase in rainfed olives, especially in the upstream part, and a corresponding fluctuating trend in irrigated crops in the downstream part. We noted a decrease in rangeland and annual rainfed crops over the same period. In general, the observed landscape transformations are mainly linked to local factors followed by economic opportunities and political factors. These changes increased the pressure on groundwater resources. Hence, there is a risk of a decline in rangeland and rainfed land area soon. Finally, the findings of this research have shown the necessity for raising awareness toward implementing a comprehensive assessment of human activities along with LULC management practices within the Rihana study area. Therefore, it is crucial to embrace the management and sustainability of these LULC practices, water and soil conservation measures, participatory approaches, and offering alternative livelihoods to reverse the non-desired consequences related to LULC changes in semiarid regions of Tunisia.

Author Contributions: Conceptualization, N.K., S.J., X.A., N.M. and G.L.; formal analysis, N.K.; funding acquisition, G.L. and R.B.; methodology, N.K., S.J., X.A., N.M. and G.L.; writing original draft, N.K.; writing review and editing, S.J. and R.B. All authors have read and agreed to the published version of the manuscript.

Funding: This research was supported by the climate change adaptation program for vulnerable rural territories of Tunisia (PACTE), a program implemented by the Tunisian Ministry of Agriculture with funding from the Agence Française de Développement (AFD) and the Fonds Français pour l'Environnement Mondial (FFEM).

Data Availability Statement: Data are available on request from the authors.

Acknowledgments: R.B. & S.J. acknowledge the support of the Strategic Research Area: The Middle East in the Contemporary World (MECW) at the Centre for Advanced Middle Eastern Studies, Lund University, Sweden. as well as the European Union Horizon 2020 program FASTER project, grant agreement No. [810812]. All the authors would also like to thank the national experts, the staff of the Regional commission for agricultural development at Sidi Bouzid (CRDA) including the territorial extension unit of Regueb (CTV), as well as the local population of Rihana.

Conflicts of Interest: The authors declare no conflict of interest.

References

1. Quintero-Gallego, M.E.; Quintero-Angel, M.; Vila-Ortega, J.J. Exploring land use/land cover change and drivers in Andean mountains in Colombia: A case in rural Quindio. *Sci. Total Environ.* **2018**, *634*, 1288–1299. [CrossRef] [PubMed]
2. Mariem, J.; Burte, J.; Biard, Y.; Benaissa, N.; Amara, H.; Sinfort, C. A framework for coupling a participatory approach and life cycle assessment for public decision-making in rural territory management. *Sci. Total Environ.* **2019**, *655*, 1017–1027. [CrossRef]
3. Yan, X.Y.; Zhang, Q.; Yan, X.M.; Wang, S.; Ren, X.Y.; Zhao, F.N. An overview of distribution characteristics and formation mechanisms in global arid areas. *Adv. Earth Sci.* **2019**, *34*, 826–841. [CrossRef]
4. Zhang, F.; Kung, H.T.; Johnson, V.C. Assessment of Land-Cover/Land-Use Change and Landscape Patterns in the Two National Nature Reserves of Ebinur Lake Watershed, Xinjiang, China. *Sustainability* **2017**, *9*, 724. [CrossRef]

5. Lasaponara, R.; Abate, N.; Fattore, C.; Aromando, A.; Cardettini, G.; Di Fonzo, M. On the Use of Sentinel-2 NDVI Time Series and Google Earth Engine to Detect Land-Use/Land-Cover Changes inFire-Affected Areas. *Remote Sens.* **2022**, *14*, 4723. [CrossRef]
6. Phan, T.N.; Kuch, V.; Lehnert, L.W. Land Cover Classification Using Google Earth Engine and Random Forest Classifier—The Role of Image Composition. *Remote. Sens.* **2020**, *12*, 2411. [CrossRef]
7. Sazib, N.; Mladenova, I.; Bolten, J. Leveraging the Google Earth Engine for drought assessment using global soil moisture data. *Remote. Sens.* **2018**, *10*, 1265. [CrossRef]
8. Felegari, S.; Sharifi, A.; Moravej, K.; Amin, M.; Golchin, A.; Muzirafuti, A.; Tariq, A.; Zhao, N. Integration of Sentinel 1 and Sentinel 2 Satellite Images for Crop Mapping. *Appl. Sci.* **2021**, *11*, 10104. [CrossRef]
9. Ngongo, Y.; deRosari, B.; Basuki, T.; Njurumana, G.N.; Nugraha, Y.; Harianja, A.H.; Ardha, M.; Kustiyo, K.; Shofiyati, R.; Heryanto, R.B.; et al. Land Cover Change and Food Security in Central Sumba: Challenges and Opportunities in the Decentralization Era in Indonesia. *Land* **2023**, *12*, 1043. [CrossRef]
10. Amani, M.; Ghorbanian, A.; Ahmadi, S.A.; Kakooei, M.; Moghimi, A.; Mirmazloumi, S.M.; Moghaddam, S.H.A.; Mahdavi, S.; Ghahremanloo, M.; Parsian, S.; et al. Google Earth Engine Cloud Computing Platform for Remote Sensing Big Data Applications: A Comprehensive Review. *IEEE J. Sel. Top. Appl. Earth Obs. Remote Sens.* **2020**, *13*, 5326–5350. [CrossRef]
11. Kruasilp, J.; Pattanakiat, S.; Phutthai, T.; Vardhanabindu, P.; Nakmuenwai, P. Evaluation of Land Use Land Cover Changes in Nan Province, Thailand, Using Multi-Sensor Satellite Data and Google Earth Engine. *Environ. Nat. Resour. J.* **2023**, *21*, 186–197. [CrossRef]
12. Kolli, M.K.; Opp, C.; Karthe, D.; Groll, M. Mapping of major land-use changes in the Kolleru Lake freshwater ecosystem by using Landsat satellite images in Google Earth Engine. *Water* **2020**, *12*, 2493. [CrossRef]
13. Basheer, S.; Wang, X.; Farooque, A.A.; Nawaz, R.A.; Liu, K.; Adekanmbi, T.; Liu, S. Comparison of Land Use Land Cover Classifiers Using Different Satellite Imagery and Machine Learning Techniques. *Remote Sens.* **2022**, *14*, 4978. [CrossRef]
14. Xie, S.; Liu, L.; Zhang, X.; Yang, J.; Chen, X.; Gao, Y. Automatic land-cover mapping using Landsat time-series data based on google earth engine. *Remote Sens.* **2019**, *11*, 3023. [CrossRef]
15. Jamali, A. Land use land cover modeling using optimized machine-learning classifiers: A case study of Shiraz, Iran. *Model. Earth Syst. Environ.* **2020**, *7*, 1539–1550. [CrossRef]
16. Abdi, A.M. Land cover and land use classification performance of machine learning algorithms in a boreal landscape using Sentinel-2 data. *GISci. Remote Sens.* **2020**, *57*, 1–20. [CrossRef]
17. Thanh Noi, P.; Kappas, M. Comparison of Random Forest, k-Nearest Neighbor, and Support Vector Machine Classifiers for Land Cover Classification Using Sentinel-2 Imagery. *Sensors* **2018**, *18*, 18. [CrossRef]
18. Naushad, R.; Kaur, T.; Ghaderpour, E. Deep Transfer Learning for Land Use and Land Cover Classification: A Comparative Study. *Sensors* **2021**, *21*, 8083. [CrossRef] [PubMed]
19. Pham, T.D.; Yokoya, N.; Bui, D.T.; Yoshino, K.; Friess, D.A. Remote Sensing Approaches for Monitoring Mangrove Species, Structure, and Biomass: Opportunities and Challenges. *Remote Sens.* **2019**, *11*, 230. [CrossRef]
20. Maurya, K.; Mahajan, S.; Chaube, N. Remote Sensing Techniques: Mapping and Monitoring of Mangrove Ecosystem—A Review. *Complex. Intell. Syst.* **2021**, *7*, 2797–2818. [CrossRef]
21. Kouassi, C.J.A.; Qian, C.; Afridi, D.K.; Suika, A.L.; Kebin, Z.; Omifolaji, K.; Yang, X. Goofle Earth Engine for Landsat Image Processing and Assessing LULC Classification in Southwestern CÔTE D'IVOIRE. *Geod. Cartogr.* **2023**, *49*, 37–50. [CrossRef]
22. Dubertret, F.; Le Tourneau, F.M.; Villarreal, M.L.; Norman, L.M. Monitoring Annual Land Use/Land Cover Change in the Tucson Metropolitan Area with Google Earth Engine (1986–2020). *Remote Sens.* **2022**, *14*, 2127. [CrossRef]
23. Agyekum Codjoe, K.; Afrifa Acheampong, A. Spatio Temporal Analysis in Forest Cover Using Google Earth Engine in Ashanti Region, Ghana. *Ajgis* **2022**, *11*, 41–50. [CrossRef]
24. Loukika, K.N.; Keesara, V.R.; Sridhar, V. Analysis of Land Use and Land Cover Using Machine Learning Algorithms on Google Earth Engine for Munneru River Basin, India. *Sustainability* **2021**, *13*, 13758. [CrossRef]
25. Negm, A.M.; Khebour Allouche, F. Introduction to "Environmental Remote Sensing and GIS in Tunisia". In *Environmental Remote Sensing and GIS in Tunisia*; Springer Nature: Berlin/Heidelberg, Germany, 2021; pp. 3–14. Available online: https://link.springer.com/chapter/10.1007/978-3-030-63668-5_1 (accessed on 8 June 2023).
26. Boussema, S.; Allouche, F.; Ajmi, R.; Chaabane, B.; Gad, A. Assessing and monitoring the effects of land cover changes in biodiversity. Case study: Mediterranean coastal region, Sousse, Tunisia. *Egypt. J. Remote Sens. Space Sci.* **2023**, *26*, 185–196. [CrossRef]
27. Henchiri, M.; Liu, Q.; Essifi, B.; Javed, T.; Zhang, S.; Bai, Y.; Zhang, J. Spatio-Temporal Patterns of Drought and Impact on Vegetation in North and West Africa Based on Multi-Satellite Data. *Remote Sens.* **2020**, *12*, 3869. [CrossRef]
28. Bousbih, S.; Zribi, M.; El Hajj, M.; Baghdadi, N.; Lili-Chabaane, Z.; Gao, Q.; Fanise, P. Soil Moisture and Irrigation Mapping in A Semi-Arid Region, Based on the Synergetic Use of Sentinel-1 and Sentinel-2 Data. *Remote Sens.* **2018**, *10*, 1953. [CrossRef]
29. Zribi, M.; Dridi, G.; Amri, R.; Lili-Chabaane, Z. Analysis of the Effects of Drought on Vegetation Cover in a Mediterranean Region through the use of SPOT-VGT and TERRA-MODIS Long Time Series. *Remote Sens.* **2016**, *8*, 992. [CrossRef]
30. Da Silva, V.S.; Salami, G.; Da Silva, O.; Silva, E.A.; Junior, J.J.M.; Alba, E. Methodological evaluation of vegetation indexes in land use and land cover (LULC) classification. *Geol. Ecol. Landsc.* **2020**, *4*, 159–169. [CrossRef]
31. DGACTA. *Rapport de Programme d'Adaptation au Changement Climatique des Territoires Tunisiens: Cadre de Gestion Environnemental et Social*; DGACTA, Ed.; DGACTA: Tunis, Tunisie, 2018; p. 95.

32. Hajjem, A. Etude sur l'hydrogéologie de la plaine de Bled Regueb. In *Rapport de la Direction Générale des Ressources en eau (DGRE)*; DGRE: Tunis, Tunisie, 1999; p. 50.
33. DGRE. *Rapport Annuaire Hydrologique*; Internal Report; Regional Directorate of Water Resources of Tunis, Ed.; DGRE: Tunis, Tunisie, 2022; p. 35.
34. Ghanmi, K.; Kadri, S.; Morardet, S.; Younsi, S.; Nsiri, N.; Saïdi, A.; Omrani, K.; Kadri, N. *Programme d'Adaptation au changement Climatique des Territoires de Tunisie. Diagnostic Territorial, Participatif et Systémique: Zone d'intervention de Sidi Bouzid Rihana*; PACTE program, Ed.; PACTE program: Tunis, Tunisie, 2020; p. 64.
35. Muchsin, F.; Dirghayu, D.; Prasasti, I.; Rahayu, M.I.; Fibriawati, L.; Pradono, K.A.; Mahatmanto, B. Comparison of atmospheric correction models: FLAASH and 6S code and their impact on vegetation indices (case study: Paddy field in Subang District, West Java). *IOP Conf. Ser. Earth Environ. Sci.* **2019**, *280*, 012034. [CrossRef]
36. Belay, T.; Mengistu, D.A. Land use and land cover dynamics and drivers in the Muga watershed, Upper Blue Nile Basin, Ethiopia. *Remote Sens. Appl. Soc. Environ.* **2019**, *15*, 100249. [CrossRef]
37. Liu, C.; Frazier, P.; Kumar, L. Comparative assessment of the measures of thematic classification accuracy. *Remote Sens. Environ.* **2007**, *107*, 606–616. [CrossRef]
38. Burte, J. *Diagnostic Rapide Participatif Systémique: Guide Pratique*; CIRAD-ES-UMR GEAU: Tunis, Tunisia, 2016; p. 19.
39. Lee, E. Analysis of MODIS 250 m NDVI Using Different Time-Series Data for Crop Type Separability. Ph.D. Thesis, University of Kansas, Lawrence, KS, USA, 2014; pp. 1–62.
40. Htitiou, A.; Boudhar, A.; Lebrini, Y.; Hadria, R.; Lionboui, H.; Elmansouri, L.; Tychon, B.; Benabdoulwaheb, T. The performance of random forest classification based on phenological metrics derived from Sentinel-2 and Landsat 8 to map crop cover in an irrigated semi-arid region. *Remote Sens. Earth Syst. Sci.* **2019**, *2*, 208–224. [CrossRef]
41. Huang, Y.; Li, J.; Li, X. Comparison of the Accuracy of Three Commonly Used Landsat Imagery Classification Methods. *Remote Sens.* **2018**, *10*, 542. [CrossRef]
42. Abdallah, H.; Gammar, A.M. Steppisation et exploitation du bois dans le secteur de Soughas (versant sud de la Dorsale tunisienne). In *Climat, Société et dynamique des Paysages Ruraux en Tunisie*; Faculté des Sciences Humaines et Sociales de Tunis: Tunis, Tunisie, 2010; pp. 142–164.
43. Attia, H. Les Hautes Steppes Tunisiennes. De la Société Pastorale à la Société Paysanne. Ph.D.Thesis, l'Université Paris VII, Paris, France, 1977; p. 78.
44. El Amami, S. *Traditional Water Management in Tunisia*; Centre de Recherche de Génie Rural: Tunis, Tunisie, 1984; p. 194.
45. Abaab, A. La Modernisation Agricole et ses Effets sur les Systèmes de Production Agricole. Cas de la région de Sidi Bouzid en Tunisie Centrale. Ph.D. Thesis, Université de Ghent, Ghent, Belgique, 1999.
46. Jaramillo, S.; Graterol, E.; Pulver, E. Sustainable Transformation of Rainfed to Irrigated Agriculture Through Water Harvesting and Smart Crop Management Practices. *Front. Sustain. Food Syst.* **2020**, *4*, 437086. [CrossRef]
47. Li, X.; Troy, T.J. Changes in rainfed and irrigated crop yield response to climate in the western US. *Environ. Res. Lett.* **2018**, *13*, 064031. [CrossRef]
48. Achour, H.; Toujani, A.; Rzigui, T.; Faiz, S. Forest cover in Tunisia before and after the 2011 Tunisian revolution: A spatial analysis approach. *J. Geovis. Spat. Anal.* **2018**, *2*, 10. [CrossRef]
49. CRDA Sidi Bouzid. *CRDA Sidi Bouzid. Gestion des ressources naturelle à Sidi Bouzid*; Internal report; Regional Agricultural Development Comisssion, Ed.; CRDA: Tunis, Tunisie, 2021.
50. Elloumi, M. L'agriculture tunisienne dans le contexte de la libéralisation. *Libéralisation Agricole Pays Développement (Revue Rég. Développement)* **2006**, *23*, 129–159. Available online: https://regionetdeveloppement.univ-tln.fr/wp-content/uploads/R23_Elloumi.pdf (accessed on 5 November 2022).

remote sensing MDPI

Article

Advancing High-Resolution Land Cover Mapping in Colombia: The Importance of a Locally Appropriate Legend

J. Camilo Fagua [1,2,*], Susana Rodríguez-Buriticá [3] and Patrick Jantz [4]

1 Grupo de Biodiversidad, Biotecnología y Conservación de Ecosistemas, Departamento de Biología, Facultad de Ciencias, Universidad Nacional de Colombia—Sede Bogotá, Bogota D.C. 111321, Colombia
2 Centro de Excelencia en Computación Científica, Universidad Nacional de Colombia—Sede Bogotá, Bogotá D.C. 111321, Colombia
3 Spatial Ecology Group, Biodiversity Sciences Program, Alexander von Humboldt Institute for Research on Biological Resources, Bogota D.C. 110311, Colombia
4 Global Earth Observation & Dynamics of Ecosystems Lab (GEODE), School of Informatics, Computing, and Cyber Systems (SICCS), Northern Arizona University, Flagstaff, AZ 85123, USA
* Correspondence: jcfaguag@unal.edu.co; Tel.: +57-318-550-2326

Abstract: Improving the remote sensing frameworks related to land cover mapping is necessary to make informed policy, development, planning, and natural resource management decisions. These efforts are especially important in tropical countries where technical capacity is limited. Land cover legend specification is a critical first step when mapping land cover, with consequences for its subsequent use and interpretation of results. We integrated the temporal metrics of SAR (Synthetic Aperture Radar) and multispectral data (Sentinel-1 and Sentienel-2) with visual pixel classifications and field surveys using five machine learning algorithms that apply different statistical methods to assess the prediction and mapping of two different land cover legends at a high spatial resolution (10 m) in a tropical region with seasonal flooding. The evaluated legends were CORINE (Coordination of Information on the Environment) and ECOSO, a legend that we defined based on the ecological and socio-economic conditions of the study area. Compared with previous studies, we obtained high accuracies for land cover modeling (kappa = 0.82) and land cover mapping (kappa = 0.76) when using ECOSO. We also found that the CORINE legend generated lower accuracies than the ECOSO legend (kappa = 0.79 for land cover modeling and kappa = 0.61 for the land cover mapping). Although CORINE was developed for European environments, it is the official land cover legend of Colombia, a South American country with tropical ecosystems not found in Europe. Therefore, some of the CORINE classes have ambiguous definitions for the study area, explaining the lower accuracy of its modeling and mapping. We used free and open-access data and software in this research; thus, our methods can be applied in other tropical regions.

Keywords: land cover; spatial modeling; Random Forest; wetlands; CORINE; Sentinel-1; Sentienel-2

1. Introduction

Land cover changes continuously transform the Earth from local to global scales [1–3]. Land cover changes also are the cause and consequence of climatic change, biodiversity loss, hydrologic alteration, soil degradation, and loss of ecosystem services [4–6]. Thus, developing accurate methods for land cover mapping is crucial to generate detailed information to monitor and mitigate the current environmental crisis as well as to implement international agreements addressing sustainable development goals [7,8]. Earth observations from satellite-based sensors provide accurate and consistent data for mapping and monitoring land cover in large areas [9–11]. Consequently, land cover mapping via satellite has been a central topic in remote sensing for decades [12,13].

The reliability of representing biophysical conditions using a thematic land-cover legend affects the accuracy of land-cover mapping [8,14,15]. Thematic legends are developed by a variety of organizations and for specific objectives, sometimes generating discrepancies among them [8,16]. Some assessments have shown differences in accuracies among land cover maps with different legends when common areas are evaluated, even after legend harmonization [15,17,18]. Additional issues can emerge when thematic legends are developed for specific environments and then transferred to new environments [16], such as the case of the CORINE (Coordination of Information on the Environment) land cover legend [19,20]. CORINE was developed for environmental purposes in Europe based on their climatic, geological, and socioeconomic conditions [21]. Some scholars have criticized the Mediterranean bias of the CORINE nomenclature [22,23] and its errors at local scales, and have identified challenges for its use in detailed landscape analysis [23,24]. Despite these issues in European environments, CORINE has been transferred to Colombia, a South American country with tropical conditions, and has been used as the official land cover legend of this country since 2010 [25].

Mapping areas with mixed land cover is a well-studied problem in land cover and land use change research [26]. In tropical environments, areas where wetlands, forests, and other land cover classes converge are especially difficult to map due to seasonal water-level changes in wetlands and because of the lack of standardized criteria by which wetlands should be identified [27–29]. Forests and wetlands are under increasing threat due to land-use changes in many tropical countries in South America [30–32]. In particular, deforestation and the draining of wetlands have increased in Colombia following the peace agreement in 2016 between the FARC, the largest guerrilla group in the country, and the national government [33,34]. The loss of these ecosystems has been alarmingly high in recent years, even within national parks [35,36]. Thus, land cover legends that accurately describe such ecosystems are a necessary first step in efforts to map their extent and change over time.

Some cloud platforms now offer open access to massive and systematic satellite remote sensing data [37], potentially allowing the development of more accurate and advanced methodologies for land cover detection and mapping. The estimation of multi-temporal metrics is one of these methodologies [38–40] where time series data of multispectral imagery (e.g., MODIS, Landsat, Sentinel-2) have helped to overcome the inter-annual or seasonal inconsistencies produced by atmospheric contamination (e.g., clouds, shadows, and water vapor) [7,39,41]. Once high-quality stacks of imagery are obtained, temporal metrics [32,38,42–44] or time series statistics (change detection algorithms such as CCDC—Continuous Change Detection and Classification [45], LandTrendr [46], VCT—Vegetation Change Tracker [47], DTW—Dynamic Time Warping [48], and BFAST—Breaks For Additive Seasonal and Trend [49]) of individual bands and spectral indices can be used as predictors to capture the phenological characteristics that increase the accuracy of the land cover mapping [7,30,40,50] or the spatial modeling of land cover attributes [51–53].

The integration of multispectral and SAR (Synthetic Aperture Radar) imagery is another of the methodologies facilitated by cloud platforms to improve land cover mapping [54–56]. Multispectral and SAR metrics together detect more regions of the electromagnetic spectrum, offering a larger set of predictors related to phenological and structural components which can improve the accuracy of land cover maps [7,52,57]. For instance, the integration of Sentinel-2 (multispectral) and Sentinel-1 (SAR) has allowed important progress in land cover detection due to the higher spatial, spectral, radiometric, and temporal resolutions of these Sentinel data compared with previously launched multispectral and SAR instruments [7,54–56,58].

Calibration and validation techniques also are essential components of the methodological framework for mapping land cover. Sample data (i.e., calibration data or training data) are required to apply machine learning algorithms, which are among the best-performing methods for developing maps, while external validation data are used to assess the final accuracy of the land cover maps [59–61]. Methodological independence between the sample and validation data can mitigate inflated map accuracy statistics [62,63]. Sample data based on visual classifications of high-resolution optical imagery (e.g., WorldView, Ikonos, QuickBird, and GeoEye) can be obtained from open cloud platforms, allowing researchers to acquire the abundant sample data required for the calibrations of learning algorithms [40,60,64,65]. Imagery spatial resolution is another key component of the methodological framework for land cover mapping because it influences the detail level of land-cover classes and thus the accuracy of the resulting maps [66]. Coarser spatial resolutions tend to mix different land cover classes in individual pixels (mixed pixels), reducing the accuracy of the classifications [67]. Detailed spatial resolutions conversely reduce the prevalence of mixed pixels by decreasing the inclusion of fractional areas of land cover classes within pixels in the landscape [68].

Here, our main objective is to compare land cover prediction and mapping from a regional legend developed for temperate and Mediterranean environments to a legend developed specifically for the study area, a tropical environment. Additional objectives are to integrate high spatial resolution (10 m), multitemporal, multispectral, and SAR data to improve the prediction and mapping of land cover in a seasonally flooded tropical region where wetlands, forests, and other land cover converge. We developed land cover maps using a set of sample data agreeing with two legends: (1) the CORINE legend adapted to Colombia [25] and (2) a second legend, ECOSO, that we defined using the ECOlogical and SOcioeconomic conditions of the study area. We hypothesized that these legends should show different accuracies for resulting land cover maps due to their differences in representing the biophysical conditions of the study area. We also hypothesized that the temporal metrics (seasonal and annual) of multispectral and SAR data together should increase the sensitivity of machine learning algorithms to discriminate land covers and thus the accuracy of the resulting land cover maps. We used free and open-access data and software; therefore, our methods can be readily adopted in other tropical regions.

2. Materials and Methods

2.1. Study Area

Our analysis was focused on a tropical seasonally flooded area that is part of the Magdalena-Urabá Moist forests [69]. The study area is also included in the Caribbean region, one of the five primary natural regions of the country identified by the Colombian environmental authorities [70] (Figure 1). Annual rainfall ranges from 2095 mm to 3119 mm throughout the study area. The rainy season is bimodal, with maximum rainfall from April to May and September to October; January and February are the driest months of the year [71]. The Magdalena River, the most important river in Colombia in terms of transportation, crosses the study area from south to north, forming seasonal wetlands where the water level changes through rainy and dry months. Land cover changes generated by human influence, such as deforestation and the draining of wetlands, have generated mosaics of native forests, wetlands, and land cover of anthropic origin in the study area.

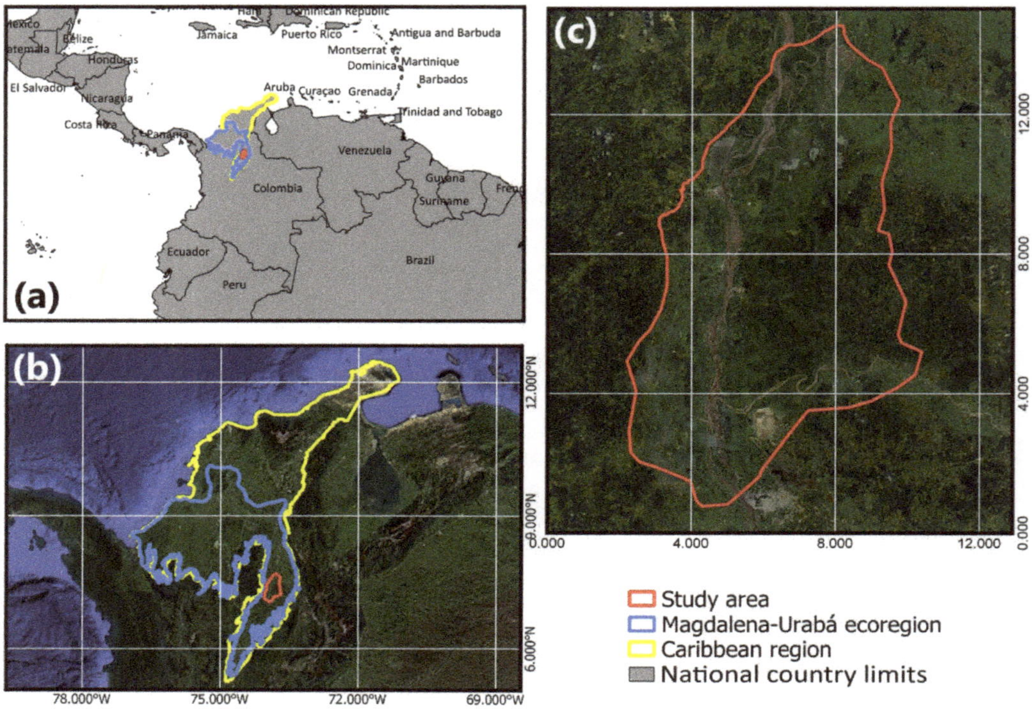

Figure 1. Study area. (**a**) Location of the study area in South America. (**b**) Location of the study area in the Magdalena-Urabá ecoregion of the global map of terrestrial ecoregions [69] and in the Caribbean region of Colombia [70]; the grid is shown in decimal geographic degrees. (**c**) A detailed scale of the study area. Magdalena River is observed crossing the study area from south to north. The grid is shown in meters.

2.2. Response Variables and Sample Data

We used two sets of land cover legends as categorical response variables. (1) The first land cover legend was the CORINE second level, which is formed by ten classes (Table 1). In Colombia, CORINE is structured in five hierarchical levels where the first two levels are the most general and present the same land cover classes of Europe [25]. After assessing field observations and high spatial resolution imagery, we found that the land cover classes of the CORINE second level adjust better to represent the biophysical conditions of the study area than the other CORINE levels. (2) We also developed a second legend, termed ECOSO, tailored to the ECOlogical and SOcioeconomic conditions of the study area. The ECOSO legend consists of eight classes representing the three main natural covers of the study area, moist forest, wetlands, and areas dominated by natural herbs and shrubs. We represented the primary agricultural activities in the study area with two classes: palm plantations and grassland, the latter of which is used for cattle grazing. We divided palm plantations into young and mature classes because the field surveys of some animal taxa (insects, birds, and mammals) have shown differences in diversity and composition when these two palm plantation ages are sampled. The definitions of each land cover of ECOSO are presented in Table 1.

Table 1. Summary of the thematic land cover legends modeled in this study. (1) CORINE second level and (2) ECOSO legend. Land cover codes are in parentheses after the land cover names. Each ECOSO class is matched to the CORINE class with which it has the greatest correspondence.

CORINE LEGEND	ECOSO LEGEND
Forest (C1): Areas occupied by forests and woodlands with a vegetation pattern composed of native or exotic coniferous and/or broad-leaved trees.	**Tropical moist Forest (E1):** Evergreen vegetation dominated by trees that reach over 30 m in height. These forests are the primary natural non-flooding vegetation cover type of the study area [72–75].
Continental humid areas (C2): Areas flooded or liable to flooding during a great part of the year by fresh, brackish, or standing water with specific vegetation coverage made of low shrub, semi-ligneous, or herbaceous species.	**Wetlands (E2):** Swamps and shallow lakes where water saturates the soil to generate a particular type of evergreen vegetation that varies from shrubs to trees. These areas are located along rivers, and their water levels vary during the year according to rainfall [74,75].
Areas with herbaceous and/or shrub vegetation (C3): Areas covered by natural grassland, moors, sclerophyllous vegetation, and transitional woodland/shrub.	**Herbaceous and Shrubland (E3):** Areas dominated by natural herbaceous vegetation and shrubs. Introduced forbs can be present [75].
Grassland (C4): Dense grass cover dominated by graminaceae not under a rotation system. Mainly for grazing, but the folder may be harvested mechanically. Includes areas with hedges.	**Grassland (E4):** Introduced grass species that are used primarily for cattle grazing [75].
Urban areas (C5): Any urban surface.	**Infrastructure (E5):** Human infrastructures, such as urban areas, buildings, concrete, and others.
Permanent crops (C6): All surfaces occupied by permanent crops, not under a rotation system.	**Mature palm plantations (E6):** plantations of African palm (Elaeis Guineensis Jacq). These cultivated areas are relatively stable vegetation because palm requires three years to mature and produce oil and its useful life is about 25 years, at which point individuals are replanted with younger palms [76].
	Young palm plantations (E7): First three years of a new plantation of African palm [76].
Water (C7): Bodies of continental water.	**Water (E8):** Bodies of continental water.
Temporary crops (C8): All surfaces occupied by crops in rotation systems.	
Heterogeneous agricultural areas (C9): Areas covered by annual crops associated with permanent crops and complex cultivation. Land principally occupied by agriculture, with significant areas of natural vegetation or Agro-forestry areas.	
Open Areas with little or no vegetation (C10): Areas covered by beaches, bare rock, sparsely vegetated areas, burnt areas, glaciers, and perpetual snow.	

To obtain the sample data, the study area was divided into square sample areas of 10 m × 10 m that match the Sentinel imagery pixel size. We then visually identified and selected the square sample areas with 100% of any of the land cover classes of the two legends for the year 2020, using the Google Satellite Plugin of QGIS 3.4 Madeira. As with other GIS applications (e.g., Google Earth and the Basemap of ArcGIS), the Madeira Plugin grants the visualization of high spatial resolution imagery (e.g., WorldView, Ikonos, QuickBird, and GeoEye) to experienced analysts who later select ideal areas of each land cover [40,64,65]. We randomly selected 49,500 of these square sample areas as sample data, using a spatial filter of 20 m^2 to reduce spatial autocorrelation effects. We used a sample size of 49,500 because it was the sample size where classification algorithms started to achieve confident landcover classification (see land cover modeling section).

2.3. Predictor Variables

We built mosaics of temporal mean metrics using the pixel values of all imagery of Sentinel-1 (SAR data) and Sentinel-2 (multispectral data) available between 1 January 2020 and 31 December 2020 in the Google Earth Engine [37]. These mosaics were used as predictor variables and were built for the backscatter coefficients, bands, and indices of Table 2. These mosaics also were built for three different periods: (1) The annual mean using all imagery of 2020, (2) the dry-season mean using the imagery of the two driest months of 2020, and (3) the rainy-season mean using the imagery of the five rainiest months of the year.

Table 2. Summary of the multispectral and SAR data used to build the temporal mean (mean-X) mosaics. The first letter and second letter in the SAR data (H or V) refer to the transmit and return signals, where H stands for horizontal and V for vertical polarization.

Satellite (Data Type)	Band, Index Name, or Backscatter Coefficient	Wavelength or Definition
Sentinel-1 (SAR)	VV of C band	5.6 cm (5.405 GHz)
	VH of C band	5.6 cm (5.405 GHz)
	VH/VV of C band	5.6 cm (5.405 GHz)
	VV-VH of C band	5.6 cm (5.405 GHz)
Sentinel-2 (Multispectral)	Blue	492.1–496.6 nm
	Green	559–560 nm
	Red	664.5–665 nm
	Red edge 1	703.8–703.9 nm
	Red edge 2	739.1–740.2 nm
	Red edge 3	779.7–782.5 nm
	Near Infrared—NIR	835.1–833 nm
	Red edge 4	864–864.8 nm
	Short wave infrared 1—SWIR1	−1610.4–1613.7 nm
	Short wave infrared 2—SWIR2	2185.7–2202.4 nm
	NDVI—Normalized Difference Vegetation Index	(NIR − Red)/(NIR + Red)
	EVI—Enhanced Vegetation Index	G * (NIR − Red)/(NIR + C1 * RED − C2 * Blue + L)
	SAVI—Soil Adjusted Vegetation Index	(1 + L) * (NIR − Red)/NIR + Red + L)
	RNDVI—Red edge Vegetation Index	(NIR − Red Edge 2)/(NIR + Red Edge 2)

To create the Sentinel-1 mosaics, we used the product Sentinel-1 SAR GRD (C-band Synthetic Aperture Radar Ground Range Detected) of Google Earth Engine-GEE [77]. This product provides calibrated and ortho-corrected Sentinel-1 data. We also applied an angular-based radiometric slope correction to this product [78], using the correction model used for vegetation covers and the ALOS Global Digital Surface Model (AW3D30) to estimate surface slope. To create the Sentinel-2 mosaics, we used the products Sentinel-2 MSI (Multispectral Instrument-Level-2A) [79] and Sentinel-2—Cloud Probability of GEE [80]. Sentinel-2 MSI provides the corrected BOA (Bottom Of Atmosphere) reflectance of the Sentinel-2 images. Sentinel-2—Cloud Probability provides information to mask pixels with a high probability of cloud using the LightGBM library. We combined these two Sentinel-2 products following Braaten (2022) for masking clouds and cloud shadows [81]. Clouds were identified from the Sentinel-2—Cloud Probability dataset and shadows are defined by a cloud projection intersection with low-reflectance near-infrared (NIR) pixels.

We also used two geomorphological auxiliary predictors generated from the ALOS Global Digital Surface Model (AW3D30), topographic slope and topographic position index (TPI) [82]. This type of geomorphological data is useful in assisting the mapping of wetlands because the landforms constrain the wetland distribution [27]. The topographic slope was defined as the degree of inclination between the surface normal and a horizontal plane. TPI is an index of terrain classification where the altitude of each pixel is evaluated against its neighborhood pixels. If a pixel is higher than its surroundings, the TPI will be positive, for instance on ridges and hilltops; TPI will be negative for low-lying pixels, such as those corresponding to valleys where wetlands are more common [27,83]. After evaluating the effects of the predictors on the land cover mapping, we excluded the TPI of the final modeling because the TPI did not affect the prediction of the land cover (see the section on land cover mapping).

2.4. Land Cover Discrimination by Temporal Mean Metrics

We selected 400 sample data per land cover of the CORINE and the ECOSO legend to perform one-way ANOVAS for evaluating the land cover classifications generated by the three temporal mean metrics (annual, dry-season, and rainy-season periods) of each SAR and multispectral datum. Significant differences would show that the backscatter coefficients, bands, or spectral indices behave differently depending on the period, providing support that metrics could be used as predictor variables to help to discriminate between land covers.

2.5. Land Cover Modeling

We modeled the land covers identified in the sample data as response variables and used their corresponding values of the three temporal mean metrics of Sentinel (Sentinel-1 and Sentinel-2) and the auxiliary data as predictor variables. This procedure was performed for both legends. We assessed five learning algorithms that apply to different statistical methods to predict the response variable: (1) Bootstrap aggregating trees or Bagging (BAG), (2) Random Forest (RF), (3) Linear support vector machine (SVM_L), (4) Radial support vector machine (SVM_R), and (5) Multivariate Adaptive Regression Splines (MARS). We tuned the parameters of the learning algorithms to achieve the most accurate predicted models following Boehmke and Greenwell (2019) [84]. We performed the learning algorithms in different sample sizes from 1500 to 49,500 (1500, 3000, 4500, . . . , and 49,500 data samples) using the R Package 'caret' [85]. We used the Overall Accuracy (OA) and Cohen's kappa coefficient (kappa) estimated in five-cross validations (CV) to measure the accuracy of the models.

To compare the effect of the three types of temporal mean metrics (annual, dry-season, and rainy-season metrics) and the three types of remote sensing data (Sentinel-1, Sentinel-1, and geomorphological auxiliary) on the accuracy of the land cover classifications, we calculated the predictors' importance for the learning algorithms at a sample size of 49,500 sample data using the R package caret [85]. This package applies the methods of each learning algorithm to estimate predictor importance and scales the maximum importance to the value of 100, allowing the comparison of the importance among different algorithms. For BAG and RF classifications, the prediction accuracy of the out-of-bag portion of the data is recorded for each tree. Then, the same is repeated after permuting each predictor variable. The difference between the two accuracies is then averaged over all trees and normalized by the standard error [85,86]. For MARS, a backward elimination feature selection routine that looks at reductions in the generalized cross-validation estimate of error is performed. The function tracks the changes in model statistics for each predictor and accumulates the reduction in the statistic when each predictor's feature is added to the model. This total reduction is used as the variable importance measure [87]. SVM_L and SVM_R lack a reliable methodology to estimate the importance of the predictors.

2.6. Land Cover Mapping

We generated land cover maps for the CORINE and ECOSO legends using the learning algorithm that produced the highest accuracy measures using the previous modeling of the sample data. Before the creation of these maps, we performed a Boruta analysis [88] to identify the importance of the predictors on the land cover modeling and eliminate possible predictors with no importance. The Boruta algorithm compares predictor importance with shadow predictors (predictor copies generated by random permutations of their own values) in numerous classifications, 100 in our case. Predictors with a significantly larger or smaller importance than shadow predictors are declared as important or unimportant for the modeling. The result of the Boruta analyses showed that all the predictors, except

TPI, presented significant effects on the modeling of the land cover for the ECOSO and CORINE legends (Figures A1 and A2). Therefore, the final modeling excluded TPI.

We then measured map accuracies by estimations of OA and kappa, using the predicted land covers of the maps against the validation data obtained from 2131 field surveys. We also estimated sensitivity (the proportion of testing data of a land cover correctly classified or true positive rate), specificity (the proportion of testing data of a land cover incorrectly classified as another land cover or true negative rate), and F1 score (the harmonic-mean of precision and recall for the minority positive class) to evaluate the accuracy of the classifications per land cover. These accuracy metrics were estimated for the CORINE and ECOSO legends by partitioning the sample data into training (70%) and testing (30%) sets.

3. Results

Five-cross validations showed that the learning algorithms used in the land cover modeling produced more accurate classifications when the land covers of the ECOSO legend were used as response variables compared with the CORINE legend for the different sample sizes evaluated (Paired T-Tests values: For OA; $T > 10.02$ and p-value < 0.001. For kappa; $T > 7.1$ and p-value < 0.001) (Figures 2 and 3).

Most temporal means of the multispectral and SAR data showed significant differences in the same land cover when annual, dry-season, and rainy-season periods were compared for the ECOSO legend and CORINE legend (Tables A1 and A2). The backscatter coefficients VV, VH, and VV—VH of Sentinel-1 and most bands and indices of Sentinel-2 were different within each land cover (p-value < 0.04; $F = 3.06$), excepting land cover corresponding to infrastructure and water bodies of the ECOSO legend. This also occurred within land cover corresponding to open areas with little or no vegetation, urban zones, and water bodies of the CORINE legend. The VV/VH backscatter ratio and the blue and green bands tended to present the lowest variations within each land cover.

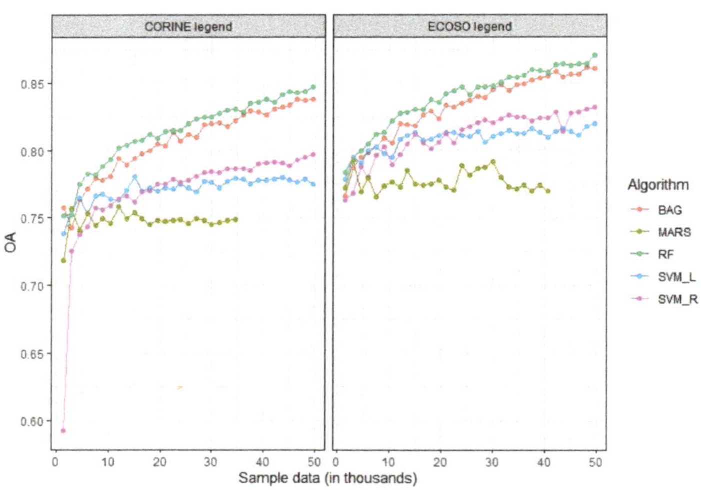

Figure 2. The Overall Accuracy (OA) of CORINE legend and ECOSO legend land cover modeling using different sample data sizes and five learning algorithms: Bootstrap aggregating trees or Bagging (BAG), Random Forest (RF), Linear support vector machine (SVM_L), Radial support vector machine (SVM_R), and Multivariate Adaptive Regression Splines (MARS).

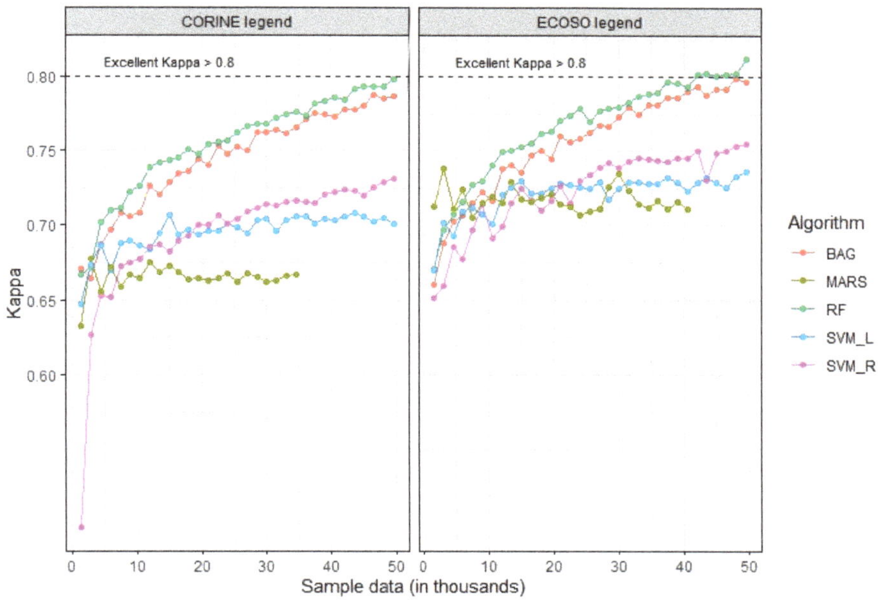

Figure 3. Cohen's kappa coefficient (kappa) generated by CORINE legend and ESOCO legend land cover modeling using different sample data sizes and five learning algorithms: Bootstrap aggregating trees or Bagging (BAG), Random Forest (RF), Linear support vector machine (SVM_L), Radial support vector machine (SVM_R), and Multivariate Adaptive Regression Splines (MARS).

The RF algorithm generated the most accurate models across the different sample sizes for both land cover legends, the ECOSO (paired T-test values: For OA; T > 11.82 and p-value < 0.001. For kappa; T > 6.22 and p-value < 0.001) and CORINE (paired T-test values: For OA; T > 10.72 and p-value < 0.001. For kappa; T > 10.62 and p-value < 0.001). We also found that only the RF algorithm generated excellent classifications (kappa > 0.8) for the land cover of the ECOSO legend when the sample size was over 42,000. No algorithms generated classifications with kappa > 0.8 for the CORINE legend (Figures 2 and 3).

The predictor importance generated by the three types of temporal metrics varied for the ECOSO (X^2 = 10.38, df = 3, p-value = 0.01) and the CORINE (X^2 = 9.97, df = 3, p-value = 0.01) legends in the classifications generated by the BAG, RF, and MARS algorithms. Dry-season metrics presented the highest importance (~65%) compared with the annual (~37%) and rainy-season (~7.6%) metrics for both legends (ECOSO: Z > 2.03, p-value < 0.03; CORINE: Z > 2.15, p-value < 0.03) (Figure 4). In addition, the three types of remote sensing data showed different predictor importance for the ECOSO (X^2 = 6.48, df = 2, p-value = 0.03) and CORINE (X^2 = 7.2, df = 2, p-value = 0.02) legends in the three learning algorithms evaluated. Metrics generated by Sentinel-2 presented the highest importance (~75.1%) compared with the metrics generated by Sentinel-1 (~23.9%) and the geomorphological auxiliary metrics (~1%) for the ECOSO legends (Z > 2.53; p-value < 0.03) and CORINE (Z > 2.68; p-value < 0.01) legends (Z > 2.53; p-value < 0.03) (Figure 5).

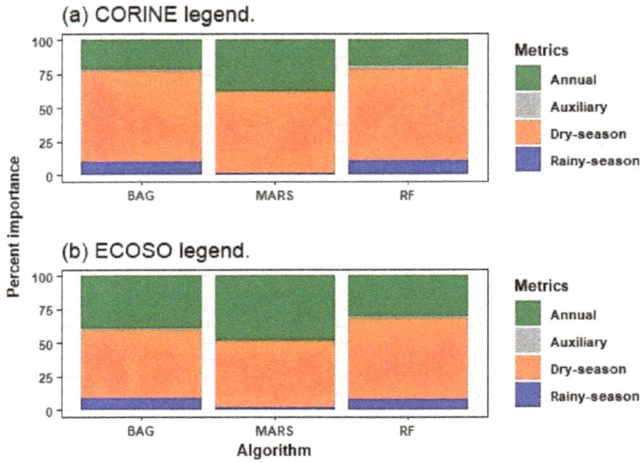

Figure 4. Comparison of the predictor importance for the land cover modeling generated by three types of temporal metrics (annual, dry-season, and rainy-season mean metrics estimated from Sentinel-1 and Sentinel-2 data) and topographic auxiliary data. Three learning algorithms were used for the predictor importance estimates: Bootstrap aggregating trees or Bagging (BAG), Random Forest (RF), and Multivariate Adaptive Regression Splines (MARS).

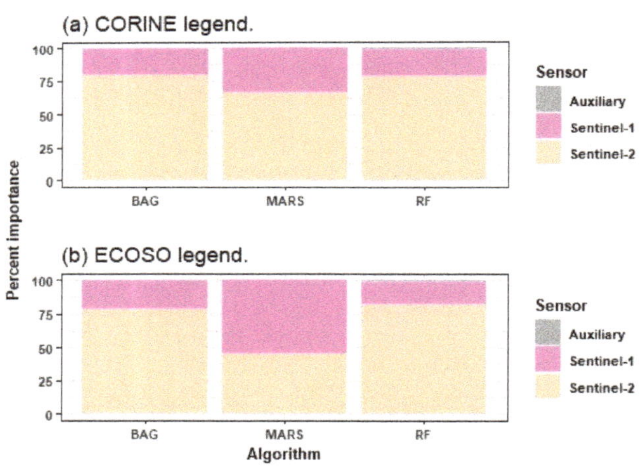

Figure 5. Comparison of the predictor importance for the land cover modeling generated by three types of remote sensing data: the SAR data of Sentinel-1, the multispectral data of Sentinel-2, and topographic auxiliary data. Three learning algorithms were used for the predictor importance estimates: Bootstrap aggregating trees or Bagging (BAG), Random Forest (RF), and Multivariate Adaptive Regression Splines (MARS).

The map for the ECOSO legend obtained higher accuracy than the map for the CORINE legend when the OA (0.81 for the ECOSO Legend and 0.73 for the CORINE legend) and kappa (0.75 for the ECOSO legend and 0.61 for the CORINE legend) were estimated using the validation data obtained from surveys in the field (Figure 6). Herbaceous, grassland, and/or shrub vegetation for the CORINE legend and herbaceous, wetland, and shrubland for the ECOSO legend produced the lowest sensitivity values (0.60 < sensitivity < 0.66), indicating that these land cover classes are the most difficult to map. All land cover of both legends generated high values of specificity (>0.95), that is, the proportion of testing data of land cover incorrectly classified as a land cover was low in general terms. Moreover, the F1 scores were relatively low for heterogeneous agricultural areas (0.53) of the CORINE legend and herbaceous vegetation and shrubland (0.59) of the ECOSO legend (Tables A3 and A4).

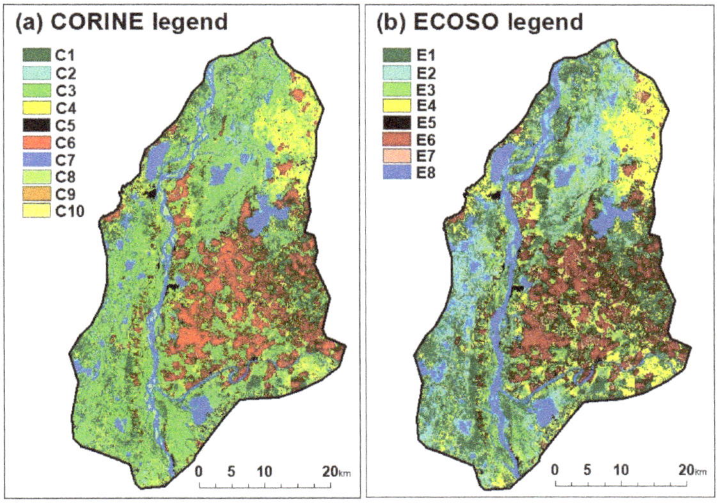

Figure 6. Land cover maps. (**a**) Map for the CORINE legend: Forest—C1, Continental humid areas—C2, Areas with herbaceous and/or shrub vegetation—C3, Grassland—C4, Urban areas—C5, Permanent crops—C6, Water—C7, Temporary crops—C8, Heterogeneous agricultural areas—C9, and Open areas with little or no vegetation—C10. (**b**) Map for the ECOSO legend: Tropical moist Forest—E1, Wetland—E2, Herbaceous and shrubland—E3, Grassland—E4, Infrastructure—E5, Mature palm plantations—E6, Young palm plantations—E7, and Water—E8.

4. Discussion

By integrating dry-season, rainy-season, and annual metrics of SAR and multispectral data with visual pixel classifications and field surveys, we obtained high accuracies for land cover modeling (kappa < 0.82) and land cover mapping (kappa < 0.76) in a tropical region with seasonal flooding at a detailed spatial resolution (10 m). We used free and open-access data and software; therefore, our methods can be adopted in other regions to construct land cover maps. We highlight that our classification analyses were based on large sample data (Big data) that were performed in a high-performance computing cluster. This could be a limitation for the development of this type of analysis; however, desktop computers can perform classifications with enough data sizes to produce land cover maps with sufficiently high accuracy.

Previous studies on the land cover modeling of tropical regions have obtained lower accuracies at coarser spatial resolutions when annual metrics are used [14,89,90], demonstrating that the inclusion of more temporal metrics can increase the accuracy of land cover mapping. On the other hand, findings using the reflectance and backscatter values of individual SAR and multispectral images as well as smaller sample sizes of the sample data set (sample data < 200) showed higher accuracies for the land cover mapping of the tropical regions of Colombia (kappa > 0.86) [56]. Theoretically, temporal metrics should increase the resulting accuracies of the land cover modeling and land cover mapping due to the inclusion of the phenological characteristics of the vegetation [40,51,53]. Although differences in the terrain's physical conditions and the landcover legends make it difficult to compare the resulting accuracies of different land cover mapping efforts, it is possible that a spatial bias or dependence between sample and validation data inflates map accuracies when only one image and low sample sizes are used [62,63]. To clarify this type of discrepancy, it is necessary to consistently evaluate the use of temporal metrics estimated from temporal stacks of imagery against the reflectance and textural metrics estimated from individual images.

The dry-season metrics estimated of SAR and multispectral data had higher importance in the land cover classifications than the rainy-season metrics. Some classes of the ECOSO and CORINE legends were built based on vegetation characteristics, and the dry season is a period of water stress for some of these vegetation types. Grassland, shrubland, and herbaceous vegetation are highly exposed to water limitations during the dry season, which significantly reduces their greenness and changes their structure (loss of leaves), while the forest and wetland vegetation are less exposed to such limitations and thus can maintain a relatively higher greenness and a higher leaf density. These phenological variations would maximize the spectral differences during the dry season among land cover types. On the other hand, during the rainy season, all vegetation types have less moisture limitation and higher photosynthesis levels and leaf density; consequently, their SAR and multispectral signatures may present similarities. These spectral similarities make it difficult to discriminate land cover classes when classification algorithms and remote sensing data are used in tropical environments [30,91,92]. Interestingly, the dry-season metrics estimated of SAR and multispectral data also had a higher importance in the land cover classifications than the annual metrics. Although annual metrics capture the spectral differences of the dry season, increasing land cover discrimination, annual metrics also capture the spectral similarities of the rainy season that reduce discrimination. This spectral ambiguity may explain why annual metrics were comparatively superior to rainy-season metrics but inferior to dry-season metrics for land cover classification.

Using the same reference dataset (sample data and predictor variables), we found that the land cover modeling and mapping for the ECOSO legend was more accurate than for the CORINE legend. The ECOSO legend included only land cover that represented the conditions of the study area while the CORINE legend contained a higher number of classes that are not well adapted to the study area, explaining its lower modeling and mapping accuracy. Some CORINE classes, such as heterogeneous agricultural areas and temporary crops, have ambiguous definitions for the study area and thus were difficult to discriminate by the learning algorithms, reducing the accuracy of the modeling and mapping. The ambiguous definition of some land cover is an issue frequently mentioned by technicians who build maps by the visual interpretation of high spatial resolution images when the CORINE legend is used in Colombia. We were able to detect the same issue in the land cover modeling of CORINE using learning algorithms.

We found lower differences in accuracy between the ECOSO and the CORINE legend for the land cover modeling (~2% for OA and ~1 for kappa values) than for the land cover mapping (8% for OA and 14 for kappa values). Previous authors have criticized the bias of land cover maps when accuracies are estimated only with partitions of the sample data (e.g., cross validations or data partitions) as we did with the accuracies of land cover modeling. These authors suggest that the non-independence of the data for estimating accuracies inflates OA and kappa due to variations of the prevalence generated by the high spatial correlation of sample data and human bias in imagery interpretation [63,93]. For those reasons, field confirmations of land cover maps prepared by accuracy estimations using surveys are essential to estimate more realistic OA and kappa values, as we did with the accuracies of the land cover mapping. The substantially lower OA and kappa for the land cover mapping of the CORINE compared to the ECOSO legend showed that the Random Forest modeling actually produced an inferior representation of the CORINE land cover.

The least accurately mapped land cover in both legends were the areas corresponding to herbaceous and shrub vegetation (C3 of the CORINE and L3 of the ECOSO legend in Table 1). Previous work has also encountered this issue from tropical to temporal environments [14,15,94,95]; these authors suggest that the performance classification of herbaceous and shrubland is relatively poor because their phenological changes produce greater intra-annual complexity and spectral variability than the other classes. We initially expected the classification of the wetland areas to be relatively poor due to their seasonal water-level changes that produce changes in their coverage areas during the same year. However, compared to previous works in other tropical regions [27,96], the wetland classifications for both legends (C2 of CORINE and L2 of the ECOSO legend in Table 1) presented higher accuracies. We demonstrated that the inclusion of seasonal predictors of SAR and multispectral data (dry-season, rainy-season, and annual metrics) and the topographic auxiliary predictor (topographic slope) reduced the errors of wetland classification; the former predictors involved the seasonal dynamic of the wetlands while the latter predictor discarded places where topography limits the wetland distribution.

After evaluating five learning algorithms with different statistical approaches to predict a response variable, we found that Random Forest was the best algorithm for our land cover modeling. Other works have also shown the better performance of Random Forest relative to other learning algorithms to predict land covers [97,98]. Although Random Forest requires some parameter tuning and predictor selection, the algorithm is relatively simple to use and is less affected by outliers, nonparametric data, and predictor correlation than other learning algorithms. Recently, deep learning algorithms (e.g., convolutional neural networks and deep learning networks) have shown competitive or superior accuracies to Random Forest for the modeling of land cover [99–101]; however, deep learning algorithms currently lack standardized methodologies to estimate the predictor importance which complicates inference related to the predictors.

The metrics generated by Sentinel-2 presented a higher importance for land cover modeling than the metrics generated by Sentinel-1 and the geomorphological auxiliary metrics in both evaluated land cover legends. A large number of Sentinel-2 predictors (42) compared with the Sentinel-1 (12) and geomorphological (1) predictors produced a higher general importance of Sentinel-2. However, the comparative importance of Sentinel-1 is not low (~23.9%), confirming that the integration of SAR (Sentinel-1) and multispectral (Sentinel-2) data is a suitable remote sensing strategy to improve land cover classifications. We also found that three Sentinel-1 predictors for the CORINE legend and one Sentinel-1 predictor for the ECOSO legend were among the most important ten predictors for the land cover classification of each legend, demonstrating the positive effect of this sensor in both land cover classifications.

5. Conclusions

We demonstrated that it is possible to improve land cover modeling by integrating the temporal metrics of SAR and multispectral data with visual pixel classifications and field surveys. The use of dry-season, rainy-season, and annual metrics of Sentinel-1 and Sentinel-2 captured the phenological and structural variation of the vegetation that comprises land cover, increasing land cover discrimination by typical learning algorithms used for modeling and mapping. We showed that these learning algorithms produced high accuracies for modeling and mapping when land cover legends were developed using the ecological and socioeconomic conditions of the study area. Conversely, accuracy estimations are lower when these learning algorithms model land cover legends developed for different contexts, as in the case of the CORINE legend adapted to Colombia. These results suggest the need to build an official land cover legend for Colombia using information on the environmental conditions of the country. Our results also confirm the importance of the independence between the sample and validation data to avoid inflating the accuracy estimation of the land cover maps. Future advances in remote sensing data and statistical methods are expected to increase the accuracy of the land cover maps generated by supervised classifications. However, the reliability of land cover legends in describing the characteristics of the regions and countries in which they are used will be a fundamental step to increasing these accuracies. For this reason, in countries such as Colombia, where the official legend has been transferred from other countries or regions, it is necessary to start thinking about the design of a national legend that brings together the country's own regional variation and facilitates future planning.

Author Contributions: Conceptualization, J.C.F. and S.R.-B.; Methodology, J.C.F., S.R.-B. and P.J.; Formal analysis, J.C.F.; Investigation, J.C.F., S.R.-B. and P.J.; Resources, J.C.F. and S.R.-B.; Writing—review and editing, J.C.F., S.R.-B. and P.J. All authors have read and agreed to the published version of the manuscript.

Funding: This research was funded by the Agencia Nacional de hidrocarburos de Colombia and the Humboldt Institute, grant number 21-117-2020. Support for J.C.F. was provided by Universidad Nacional de Colombia—Sede Bogotá; Proyecto HERMES 58648 and Proyecto HERMES 52625. Support for P.J. was provided by NASA grants 80NSSC19K0186 and 80NSCC18K0338.

Data Availability Statement: Land cover maps that support the findings of this study will be openly available at the following URL/DOI: (https://doi.org/10.5281/zenodo.7799931).

Acknowledgments: We acknowledge Instituto Humboldt, Departamento de Biología of Universidad Nacional de Colombia (Sede Bogota D.C), and School of Informatics, Computing, and Cyber Systems of Northern Arizona University for providing technical support and data. This work is one of the requirements for the forgiveness of the COLCIENCIAS-loan 529-2011 for J.C.F.

Conflicts of Interest: The authors declare no conflict of interest.

Appendix A

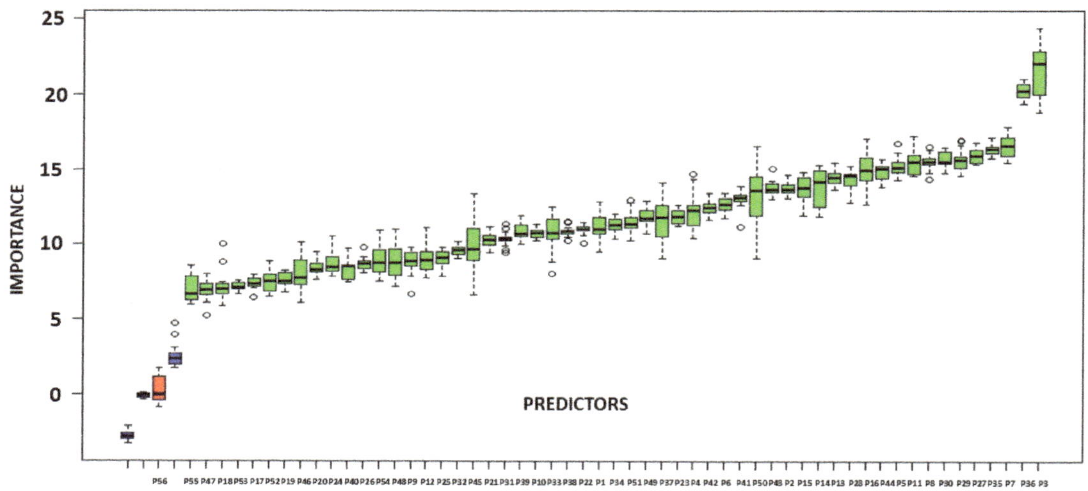

Figure A1. Importance of predictors in the Random Forest classification for the land covers of the CORINE legend. Predictors with significant importance (green boxes), predictors with no significant importance (red boxes), and the maximum, minimum, and average of the shadow predictors (blue boxes). P1 (annual VV), P2 (annual VH), P3 (annual VH/VV), P4 (annual VH-VV), P5 (dry VV), P6 (dry VH), P7 (dry VH/VV), P8 (dry VH-VV), P9 (rainy VV), P10 (rainy VH), P11 (rainy VH/VV), P12 (rainy VH-VV), P13 (annual blue), P14 (annual green), P15 (annual red), P16 (annual red edge 1), P17 (annual red edge 2), P18 (annual red edge 3), P19 (annual near infrared), P20 (annual red edge 4), P21 (annual short wave infrared 1), P22 (annual short wave infrared 2), P23 (annual NDVI), P24 (annual EVI), P25 (annual SAVI), P26 (annual RNDVI), P27 (dry blue), P28 (gry green), P29 (dry red), P30 (dry red edge 1), P31 (dry red edge 2), P32 (dry red edge 3), P33 (dry near infrared), P34 (dry red edge 4), P35 (dry short wave infrared 1), P36 (dry short wave infrared 2), P37 (dry NDVI), P38 (dry EVI), P39 (dry SAVI), P40 (dry RNDVI), P41 (rainy blue), P42 (rainy green), P43 (rainy red), P44 (rainy red edge 1), P45 (rainy red edge 2), P46 (rainy red edge 3), P47 (rainy near infrared), P48 (rainy red edge 4), P49 (rainy short wave infrared 1), P50 (rainy short wave infrared 2), P51 (rainy NDVI), P52 (rainy EVI), P53 (rainy SAVI), P54 (rainy RNDVI), P55 (slope), and P56 (TPI).

Appendix B

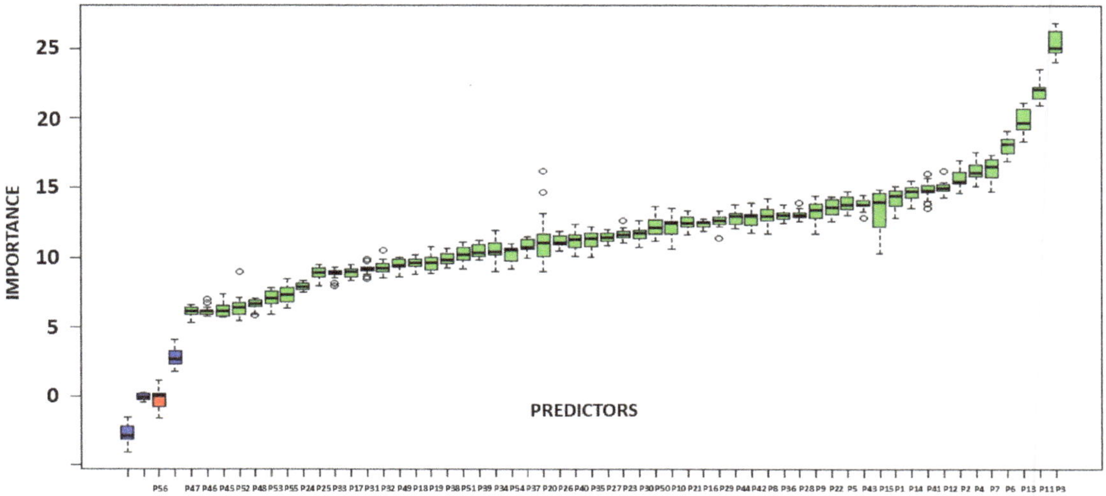

Figure A2. Importance of predictors in the Random Forest classification for the land covers of the ECOSO legend. Predictors with significant importance (green boxes), predictors with no significant importance (red boxes), and the maximum, minimum, and average of the shadow predictors (blue boxes). P1 (annual VV), P2 (annual VH), P3 (annual VH/VV), P4 (annual VH-VV), P5 (dry VV), P6 (dry VH), P7 (dry VH/VV), P8 (dry VH-VV), P9 (rainy VV), P10 (rainy VH), P11 (rainy VH/VV), P12 (rainy VH-VV), P13 (annual blue), P14 (annual green), P15 (annual red), P16 (annual red edge 1), P17 (annual red edge 2), P18 (annual red edge 3), P19 (annual near infrared), P20 (annual red edge 4), P21 (annual short wave infrared 1), P22 (annual short wave infrared 2), P23 (annual NDVI), P24 (annual EVI),P25 (annual SAVI), P26 (annual RNDVI), P27 (dry blue), P28 (dry green), P29 (dry red), P30 (dry red edge 1), P31 (dry red edge 2), P32 (dry red edge 3), P33 (dry near infrared), P34 (dry red edge 4), P35 (dry short wave infrared 1), P36 (dry short wave infrared 2), P37 (dry NDVI), P38 (dry EVI), P39 (dry SAVI), P40 (dry RNDVI), P41 (rainy blue), P42 (rainy green), P43 (rainy red), P44 (rainy red edge 1), P45 (rainy red edge 2), P46 (rainy red edge 3), P47 (rainy near infrared), P48 (rainy red edge 4), P49 (rainy short wave infrared 1), P50 (rainy short wave infrared 2), P51 (rainy NDVI), P52 (rainy EVI), P53 (rainy SAVI), P54 (rainy RNDVI), P55 (slope), and P56 (TPI).

Appendix C

Table A1. Summary of one-way ANOVAS to compare the temporal mean values of the multispectral and SAR data per land cover of the ECOSO legend for three periods of 2020: (1) Annual mean using all imagery of 2020; (2) dry season mean using the imagery of the two driest months of 2020; and (3) rain season mean using the imagery of the five rainiest months of the year. We found significant differences among these three periods that show that the temporal means per band, multispectral index, and backscatter coefficient behave differently during the evaluated periods. These differences demonstrate that the temporal mean value calculated for each of these periods can be used as a variable to discriminate land covers in the study area. The first letter and second letter in the SAR data (H or V) refer to the transmit and return signals; H stands for horizontal and V for vertical polarizations. Significant p values range; $p < 0.001$ (***), $p < 0.01$ (**), and $p < 0.05$ (*).

Satellite (Data Type)	Band, Index Name, or Backscatter Coefficient	Land Cover Class	F-Value	p-Value	Significance
	VV	Tropical moist forest	6	0.002568297	**
		Grassland	62	3.67405E-26	***
		Herbaceous and shrubland	23	1.57724E-10	***
		Infrastructure	0	0.631883799	
		Mature palm plantations	72	3.08908E-30	**
		Water	3	0.031589942	*
		Wetland	11	1.76326E-05	***
		Young palm plantations	90	3.55004E-37	***
	VH	Tropical moist forest	6	0.002387089	**
		Grassland	84	4.77421E-35	***
		Herbaceous and shrubland	26	1.27439E-11	***
		Infrastructure	1	0.343709726	
		Mature palm plantations	139	5.1437E-55	***
		Water	7	0.000870292	***
		Wetland	3	0.053333071	*
		Young palm plantations	101	3.48339E-41	***
	VVmVH	Tropical moist forest	5	0.008475596	**
		Grassland	47	3.22066E-20	***
		Herbaceous and shrubland	18	2.72475E-08	***
		Infrastructure	0	0.680842298	
		Mature palm plantations	42	1.51053E-18	***
		Shrubland	8	0.000320566	***
		Water	2	0.090936068	
		Wetland	14	1.33331E-06	***
		Young palm plantations	62	1.71158E-26	***
	VHdVV	Tropical moist forest	1	0.297645719	
		Grassland	0	0.645580222	
		Herbaceous and shrubland	0	0.846826805	
		Infrastructure	0	0.977296257	
		Mature palm plantations	9	0.000140432	***
		Water	0	0.760810131	
		Wetland	11	2.46617E-05	***
		Young palm plantations	3	0.040476133	*

Table A1. *Cont.*

Satellite (Data Type)	Band, Index Name, or Backscatter Coefficient	Land Cover Class	*F*-Value	*p*-Value	Significance
	BLUE	Tropical moist forest	2	0.153644537	
		Grassland	5	0.00609176	*
		Herbaceous and shrubland	2	0.12682676	
		Infrastructure	5	0.004748278	*
		Mature palm plantations	42	3.40328E-18	***
		Water	7	0.001516881	*
		Wetland	11	2.93761E-05	***
		Young palm plantations	29	4.75329E-13	***
	GREEN	Tropical moist forest	3	0.053247018	*
		Grassland	2	0.132665756	
		Herbaceous and shrubland	2	0.207288516	
		Infrastructure	7	0.001213031	**
		Mature palm plantations	4	0.017734227	*
		Water	1	0.225658337	
		Wetland	6	0.003982018	**
		Young palm plantations	2	0.115107982	
	RED	Tropical moist forest	5	0.010667312	*
		Grassland	17	5.21408E-08	***
		Herbaceous and shrubland	6	0.002036325	*
		Infrastructure	3	0.046986096	*
		Mature palm plantations	56	6.96644E-24	***
		Water	4	0.012729719	*
		Wetland	23	2.33559E-10	***
		Young palm plantations	44	2.82816E-19	***
	RED_E_1	Tropical moist forest	11	1.62602E-05	***
		Grassland	2	0.216948004	
		Herbaceous and shrubland	3	0.033364352	*
		Infrastructure	10	6.81317E-05	***
		Mature palm plantations	3	0.038285219	*
		Water	8	0.00024883	***
		Wetland	11	1.72503E-05	***
		Young palm plantations	2	0.164089184	
	RED_E_2	Tropical moist forest	110	1.25924E-44	***
		Grassland	113	7.53722E-46	***
		Herbaceous and shrubland	38	8.18867E-17	***
		Infrastructure	18	1.79485E-08	***
		Mature palm plantations	410	2.3864E-136	***
		Water	4	0.020459247	*
		Wetland	14	6.56859E-07	***
		Young palm plantations	341	6.1106E-118	***

Table A1. *Cont.*

Satellite (Data Type)	Band, Index Name, or Backscatter Coefficient	Land Cover Class	F-Value	p-Value	Significance
	RED_E_3	Tropical moist forest	114	6.74092E-46	***
		Grassland	113	1.40749E-45	***
		Herbaceous and shrubland	37	2.16841E-16	***
		Infrastructure	18	2.92979E-08	***
		Mature palm plantations	423	1.76E-139	***
		Water	4	0.016013878	*
		Wetland	16	1.78503E-07	***
		Young palm plantations	333	8.7789E-116	***
	NIR	Tropical moist forest	65	1.81645E-27	***
		Grassland	63	1.16778E-26	***
		Herbaceous and shrubland	21	9.85588E-10	***
		Infrastructure	9	0.000172513	***
		Mature palm plantations	291	1.5266E-103	***
		Water	3	0.056042942	*
		Wetland	10	5.68407E-05	***
		Young palm plantations	236	5.56687E-87	***
	RED_E_4	Tropical moist forest	95	4.88148E-39	***
		Grassland	77	2.82783E-32	***
		Herbaceous and shrubland	29	4.25091E-13	***
		Infrastructure	15	2.53381E-07	***
		Mature palm plantations	388	1.926E-130	***
		Water	3	0.059911938	*
		Wetland	11	1.26172E-05	***
		Young palm plantations	291	1.5251E-103	***
	SWIR_1	Tropical moist forest	13	3.99272E-06	***
		Grassland	38	1.44193E-16	***
		Herbaceous and shrubland	6	0.002912467	**
		Infrastructure	10	5.48921E-05	***
		Mature palm plantations	8	0.000514043	**
		Water	3	0.043625139	*
		Wetland	30	2.57828E-13	***
		Young palm plantations	13	3.20698E-06	***
	SWIR_2	Tropical moist forest	2	0.187005907	
		Grassland	42	1.53557E-18	***
		Herbaceous and shrubland	8	0.000266878	***
		Infrastructure	6	0.00235001	**
		Mature palm plantations	7	0.001428365	**
		Water	8	0.000312134	**
		Wetland	37	3.01375E-16	***
		Young palm plantations	26	6.6958E-12	***

Table A1. *Cont.*

Satellite (Data Type)	Band, Index Name, or Backscatter Coefficient	Land Cover Class	*F*-Value	*p*-Value	Significance
	EVI	Tropical moist forest	50	1.03023E-21	***
		Grassland	58	1.03682E-24	***
		Herbaceous and shrubland	17	3.80441E-08	***
		Infrastructure	2	0.207651799	
		Mature palm plantations	276	2.9126E-99	***
		Water	1	0.449498764	
		Wetland	11	2.70347E-05	***
		Young palm plantations	186	4.50887E-71	***
	SAVI	Tropical moist sorest	41	4.27621E-18	***
		Grassland	45	1.15582E-19	***
		Herbaceous and shrubland	14	7.42276E-07	***
		Infrastructure	1	0.384579741	
		Mature palm plantations	260	2.23105E-94	***
		Water	0	0.739025492	
		Wetland	8	0.000341478	**
		Young palm plantations	175	2.39727E-67	***
	RNDVI	Tropical moist forest	5	0.009506599	**
		Grassland	32	4.23285E-14	***
		Herbaceous and shrubland	11	1.7484E-05	***
		Infrastructure	6	0.003625327	**
		Mature palm plantations	3	0.034660805	*
		Water	3	0.043060708	*
		Wetland	13	2.06961E-06	***
		Young palm plantations	6	0.00176677	**
	NDVI	Tropical moist forest	14	7.66839E-07	***
		Grassland	25	2.5465E-11	***
		Herbaceous and shrubland	8	0.000231118	**
		Infrastructure	0	0.858098151	
		Mature palm plantations	142	6.91379E-56	***
		Water	1	0.493770717	
		Wetland	3	0.046984903	*
		Young palm plantations	92	8.82396E-38	***

Appendix D. (Long Table, It Was Added after References)

Table A2. Summary of one-way ANOVAS to compare the temporal mean values of the multispectral and SAR data per land cover of the CORINE legend for three periods of 2020: (1) Annual mean using all imagery of 2020; (2) Dry season mean using the imagery of the two driest months of 2020; and (3) Rain season mean using the imagery of the five rainiest months of the year. We found significant differences among these three periods that show that the temporal means per band, multispectral index, and backscatter coefficient behave differently during the evaluated periods. These differences demonstrate that each temporal mean value calculated for these periods can be used as a variable to discriminate land covers in the study area. The first letter and second letter in the SAR data (H or V) refer to the transmit and return signals; H stand for horizontal and V for vertical polarizations. Significant p values range; $p < 0.001$ (***), $p < 0.01$ (**), and $p < 0.05$ (*).

Satellite (Data Type)	Band, Index Name, or Backscatter Coefficient	Land Cover Class	*F*-Value	*p*-Value	Significance
Sentinel-1 (SAR)	VV	Urban areas	3.8	0.022657	*
		Temporary crops	77.9	1.57E-32	***
		Permanent crops	34.5	2.61E-15	***
		Grassland	33.0	1.1E-14	***
		Heterogeneous agricultural areas	149.5	1.1E-58	***
		Forest	9.8	5.93E-05	***
		Areas with herbaceous and/or shrub vegetation	7.0	0.000905	***
		Open areas with little or no vegetation	1.7	0.182592	
		Continental humid areas	41.8	2.82E-18	***
		Water	1.9	0.151496	
	VH	Urban areas	2.5	0.080392	
		Temporary crops	157.8	1.55E-61	***
		Permanent crops	62.9	1.08E-26	***
		Grassland	51.8	2.66E-22	***
		Heterogeneous agricultural areas	386.7	2.9E-130	***
		Forest	8.3	0.000261	***
		Areas with herbaceous and/or shrub vegetation	7.4	0.000615	***
		Open areas with little or no vegetation	22.7	2.11E-10	***
		Continental humid areas	9.5	8E-05	***
		Water	4.3	0.014194	**
	VVmVH	Urban areas	3.8	0.023579	*
		Temporary crops	57.3	1.77E-24	***
		Permanent crops	19.3	5.67E-09	***
		Grassland	23.2	1.26E-10	***
		Heterogeneous agricultural areas	71.5	4.72E-30	***
		Forest	8.6	0.000187	***
		Areas with herbaceous and/or shrub vegetation	5.7	0.003311	*
		Open areas with little or no vegetation	0.0	0.98551	
		Continental humid areas	50.6	7.73E-22	***
		Water	1.3	0.279511	

Table A2. *Cont.*

Satellite (Data Type)	Band, Index Name, or Backscatter Coefficient	Land Cover Class	*F*-Value	*p*-Value	Significance
	VHdVV	Urban areas	2.9	0.058313	
		Temporary crops	0.4	0.694037	
		Permanent crops	3.4	0.032289	*
		Grassland	3.2	0.041195	*
		Heterogeneous agricultural areas	6.8	0.001139	*
		Forest	2.5	0.07944	
		Areas with herbaceous and/or shrub vegetation	1.5	0.221896	
		Open areas with little or no vegetation	188.8	5.28E-72	***
		Continental humid areas	15.3	2.66E-07	***
		Water	0.7	0.514729	
Sentinel-2 (Multispectral)	BLUE	Urban areas	1.0	0.352686	
		Temporary crops	1.0	0.351463	
		Permanent crops	21.2	9.04E-10	***
		Grassland	10.9	2.06E-05	***
		Heterogeneous agricultural areas	4.0	0.018951	*
		Forest	10.1	4.43E-05	***
		Areas with herbaceous and/or shrub vegetation	1.5	0.232523	
		Open areas with little or no vegetation	281.4	7E-101	***
		Continental humid areas	24.2	5.18E-11	***
		Water	14.8	4.53E-07	***
	GREEN	Urban areas	1.1	0.323845	
		Temporary crops	4.0	0.018853	*
		Permanent crops	1.1	0.342954	
		Grassland	3.3	0.036229	*
		Heterogeneous agricultural areas	43.3	6.89E-19	***
		Forest	0.3	0.722477	
		Areas with herbaceous and/or shrub vegetation	2.2	0.110736	
		Open areas with little or no vegetation	278.3	5.7E-100	***
		Continental humid areas	6.7	0.001316	*
		Water	14.8	4.31E-07	***
	RED	Urban areas	0.2	0.855469	
		Temporary crops	6.3	0.001832	**
		Permanent crops	17.2	4.15E-08	***
		Grassland	34.5	2.75E-15	***
		Heterogeneous agricultural areas	19.8	3.45E-09	***
		Forest	12.1	6.19E-06	***
		Areas with herbaceous and/or shrub vegetation	5.2	0.005423	*
		Open areas with little or no vegetation	213.5	5.11E-80	***
		Continental humid areas	73.6	7.3E-31	***
		Water	36.6	3.91E-16	***

Table A2. *Cont.*

Satellite (Data Type)	Band, Index Name, or Backscatter Coefficient	Land Cover Class	*F*-Value	*p*-Value	Significance
	RED_E_1	Urban areas	0.6	0.568254	
		Temporary crops	8.8	0.000166	***
		Permanent crops	13.9	1.07E-06	***
		Grassland	5.0	0.006749	*
		Heterogeneous agricultural areas	50.8	6.9E-22	***
		Forest	5.0	0.006819	*
		Areas with herbaceous and/or shrub vegetation	6.1	0.002329	*
		Open areas with little or no vegetation	204.9	3.01E-77	***
		Continental humid areas	16.7	7.11E-08	***
		Water	55.4	9.94E-24	***
	RED_E_2	Urban areas	28.0	2.16E-12	***
		Temporary crops	238.3	7.46E-88	***
		Permanent crops	166.7	1.38E-64	***
		Grassland	135.6	8.32E-54	***
		Heterogeneous agricultural areas	166.1	2.22E-64	***
		Forest	66.2	5.19E-28	***
		Areas with herbaceous and/or shrub vegetation	24.3	4.69E-11	***
		Open areas with little or no vegetation	33.0	1.16E-14	***
		Continental humid areas	34.0	4.49E-15	***
		Water	8.9	0.00014	*
	RED_E_3	Urban areas	23.3	1.7E-10	***
		Temporary crops	285.3	4.7E-102	***
		Permanent crops	143.5	1.36E-56	***
		Grassland	137.9	1.25E-54	***
		Heterogeneous agricultural areas	189.4	3.55E-72	***
		Forest	75.5	1.34E-31	***
		Areas with herbaceous and/or shrub vegetation	24.6	3.42E-11	***
		Open areas with little or no vegetation	21.5	6.91E-10	***
		Continental humid areas	40.7	7.86E-18	***
		Water	7.6	0.000514	**
	NIR	Urban areas	6.1	0.00236595	*
		Temporary crops	211.6	2.0775E-79	***
		Permanent crops	70.3	1.348E-29	***
		Grassland	74.0	4.8769E-31	***
		Heterogeneous agricultural areas	107.5	1.1997E-43	***
		Forest	46.1	4.9883E-20	***
		Areas with herbaceous and/or shrub vegetation	14.6	5.2772E-07	***
		Open areas with little or no vegetation	23.9	6.8998E-11	***
		Continental humid areas	22.2	3.4788E-10	***
		Water	5.4	0.00482082	*

Table A2. *Cont.*

Satellite (Data Type)	Band, Index Name, or Backscatter Coefficient	Land Cover Class	*F*-Value	*p*-Value	Significance
	RED_E_4	Urban areas	14.0	1.13E-06	***
		Temporary crops	283.1	2.1E-101	***
		Permanent crops	120.8	1.62E-48	***
		Grassland	101.1	2.77E-41	***
		Heterogeneous agricultural areas	147.3	6.28E-58	***
		Forest	63.1	8.5E-27	***
		Areas with herbaceous and/or shrub vegetation	17.6	2.98E-08	***
		Open areas with little or no vegetation	24.0	6.21E-11	***
		Continental humid areas	25.0	2.29E-11	***
		Water	5.8	0.003018	*
	SWIR_1	Urban areas	15.2	3.47E-07	***
		Temporary crops	4.0	0.017824	*
		Permanent crops	13.7	1.27E-06	***
		Grassland	46.3	4.21E-20	***
		Heterogeneous agricultural areas	4.6	0.01007	*
		Forest	7.7	0.000479	**
		Areas with herbaceous and/or shrub vegetation	7.0	0.000932	**
		Open areas with little or no vegetation	253.7	1.41E-92	***
		Continental humid areas	32.4	1.94E-14	***
		Water	0.2	0.822389	
	SWIR_2	Urban areas	12.9	3.21E-06	***
		Temporary crops	4.9	0.007743	*
		Permanent crops	3.4	0.032219	*
		Grassland	60.9	6.76E-26	***
		Heterogeneous agricultural areas	16.9	5.65E-08	***
		Forest	1.7	0.182995	
		Areas with herbaceous and/or shrub vegetation	8.0	0.000366	*
		Open areas with little or no vegetation	384.5	1.1E-129	***
		Continental humid areas	50.8	6.87E-22	***
		Water	0.6	0.568736	
	EVI	Urban areas	1.4	0.236922	
		Temporary crops	122.9	2.92E-49	***
		Permanent crops	56.1	4.96E-24	***
		Grassland	81.5	6.47E-34	***
		Heterogeneous agricultural areas	99.1	1.51E-40	***
		Forest	39.0	3.9E-17	***
		Areas with herbaceous and/or shrub vegetation	10.5	2.97E-05	***
		Open areas with little or no vegetation	1.4	0.241734	
		Continental humid areas	39.2	3.24E-17	***
		Water	0.1	0.877852	

Table A2. *Cont.*

Satellite (Data Type)	Band, Index Name, or Backscatter Coefficient	Land Cover Class	*F*-Value	*p*-Value	Significance
	SAVI	Urban areas	0.9	0.400928	
		Temporary crops	95.4	3.73E-39	***
		Permanent crops	55.1	1.3E-23	***
		Grassland	66.6	3.65E-28	***
		Heterogeneous agricultural areas	70.6	1.02E-29	***
		Forest	35.1	1.53E-15	***
		Areas with herbaceous and/or shrub vegetation	7.3	0.000692	**
		Open areas with little or no vegetation	1.1	0.347356	
		Continental humid areas	28.3	9.89E-13	***
		Water	0.1	0.914428	
		Urban areas	0.9	0.400928	
	RNDVI	Urban areas	3.6	0.027078146	*
		Temporary crops	2.2	0.108978781	
		Permanent crops	17.3	4.06165E-08	***
		Grassland	17.9	2.11557E-08	***
		Heterogeneous agricultural areas	2.1	0.118146032	
		Forest	6.7	0.001281665	**
		Areas with herbaceous and/or shrub vegetation	10.8	2.19194E-05	***
		Open areas with little or no vegetation	8.4	0.000228264	*
		Continental humid areas	8.7	0.000170631	**
		Water	5.8	0.003089327	**
	NDVI	Urban areas	0.5	0.579018	
		Temporary crops	30.7	1.03E-13	***
		Permanent crops	31.2	6.18E-14	***
		Grassland	40.5	9.95E-18	***
		Heterogeneous agricultural areas	31.5	4.58E-14	***
		Forest	11.4	1.27E-05	***
		Areas with herbaceous and/or shrub vegetation	1.7	0.191762	
		Open areas with little or no vegetation	0.4	0.692677	
		Continental humid areas	15.4	2.47E-07	***
		Water	1.1	0.335542	

Appendix E

Table A3. Estimates of sensitivity (true positive rate) and specificity (true negative rate) for the CORINE landcover legend. Sensitivity and specificity were estimated by performing data partitions of the 49,500 sample data (training 70% and testing 30%).

Land Cover	Sensitivity	Specificity	F1 Score	Prevalence
Forest (C1)	0.72	0.96	0.68	0.09
Continental humid areas(C2)	0.74	0.97	0.60	0.04
Areas with herbaceous and/or shrub vegetation (C3)	0.62	0.95	0.69	0.22
Grassland (C4)	0.66	0.97	0.61	0.05
Urban areas (C5)	0.90	1.00	0.87	0.02
Permanent crops (C6)	0.83	0.98	0.82	0.09
Water (C7)	0.95	0.98	0.96	0.43
Temporary crops (C8)	0.93	1.00	0.71	0.00
Heterogeneous agricultural areas (C9)	0.82	1.00	0.53	0.00
Open areas with little or no vegetation (C10)	0.88	0.98	0.84	0.07

Appendix F

Table A4. Estimates of sensitivity (true positive rate) and specificity (true negative rate) for the ECOSO land cover legend. Sensitivity and specificity were estimated by performing data partitions of the 49,500 sample data (training 70% and testing 30%).

Land Cover	Sensitivity	Specificity	F1 Score	Prevalence
Tropical moist forest (L1):	0.75	0.97	0.78	0.15
Wetland (L2):	0.66	0.97	0.70	0.12
Herbaceous and shrubland (L3):	0.60	0.95	0.59	0.09
Grassland (L4):	0.72	0.98	0.66	0.05
Infrastructure (L5):	1.00	1.00	0.78	0.01
Mature palm plantations (L6):	0.83	0.98	0.78	0.05
Young palm plantations (L7):	0.77	0.99	0.71	0.03
Water (L8):	0.97	0.97	0.97	0.51

References

1. Winkler, K.; Fuchs, R.; Rounsevell, M.; Herold, M. Global land use changes are four times greater than previously estimated. *Nat. Commun.* **2021**, *12*, 2501. [CrossRef] [PubMed]
2. Rudel, T.K.; Coomes, O.T.; Moran, E.; Achard, F.; Angelsen, A.; Xu, J.C.; Lambin, E. Forest transitions: Towards a global understanding of land use change. *Glob. Environ. Chang. Policy Dimens.* **2005**, *15*, 23–31. [CrossRef]
3. Turner, B.L.; Lambin, E.F.; Reenberg, A. The emergence of land change science for global environmental change and sustainability. *Proc. Natl. Acad. Sci. USA* **2007**, *104*, 20666–20671. [CrossRef]
4. Song, X.P.; Hansen, M.C.; Stehman, S.V.; Potapov, P.V.; Tyukavina, A.; Vermote, E.F.; Townshend, J.R. Global land change from 1982 to 2016. *Nature* **2018**, *560*, 639. [CrossRef] [PubMed]
5. Friedlingstein, P.; Jones, M.W.; O'Sullivan, M.; Andrew, R.M.; Bakker, D.C.E.; Hauck, J.; Le Quéré, C.; Peters, G.P.; Peters, W.; Pongratz, J.; et al. Global Carbon Budget 2021. *Earth Syst. Sci. Data Discuss.* **2021**, *2021*, 1–191. [CrossRef]
6. Powers, R.P.; Jetz, W. Global habitat loss and extinction risk of terrestrial vertebrates under future land-use-change scenarios. *Nat. Clim. Chang.* **2019**, *9*, 323–329. [CrossRef]
7. Venter, Z.S.; Sydenham, M.A.K. Continental-Scale Land Cover Mapping at 10 m Resolution Over Europe (ELC10). *Remote Sens.* **2021**, *13*, 2301. [CrossRef]
8. Tulbure, M.G.; Hostert, P.; Kuemmerle, T.; Broich, M. Regional matters: On the usefulness of regional land-cover datasets in times of global change. *Remote Sens. Ecol. Conserv.* **2022**, *8*, 272–283. [CrossRef]
9. Cihlar, J. Land cover mapping of large areas from satellites: Status and research priorities. *Int. J. Remote Sens.* **2000**, *21*, 1093–1114. [CrossRef]
10. Congalton, R.G.; Gu, J.; Yadav, K.; Thenkabail, P.S.; Ozdogan, M. Global land cover mapping: A review and uncertainty analysis. *Remote Sens.* **2014**, *6*, 12070–12093. [CrossRef]
11. Vancutsem, C.; Achard, F.; Pekel, J.F.; Vieilledent, G.; Carboni, S.; Simonetti, D.; Gallego, J.; Aragão, L.E.O.C.; Nasi, R. Long-term (1990–2019) monitoring of forest cover changes in the humid tropics. *Sci. Adv.* **2022**, *7*, eabe1603. [CrossRef]
12. NASA. Land-Cover and Land-Use Change (LCLUC) Program. Available online: https://lcluc.umd.edu/ (accessed on 23 February 2023).
13. ESA. Land Cover Project. Available online: https://climate.esa.int/en/projects/land-cover/ (accessed on 23 February 2023).
14. Buchhorn, M.; Lesiv, M.; Tsendbazar, N.-E.; Herold, M.; Bertels, L.; Smets, B. Copernicus Global Land Cover Layers—Collection 2. *Remote Sens.* **2020**, *12*, 1044. [CrossRef]
15. Tsendbazar, N.E.; de Bruin, S.; Mora, B.; Schouten, L.; Herold, M. Comparative assessment of thematic accuracy of GLC maps for specific applications using existing reference data. *Int. J. Appl. Earth Obs. Geoinf.* **2016**, *44*, 124–135. [CrossRef]
16. Mushtaq, F.; Henry, M.; O'Brien, C.D.; Di Gregorio, A.; Jalal, R.; Latham, J.; Muchoney, D.; Hill, C.T.; Mosca, N.; Tefera, M.G.; et al. An International Library for Land Cover Legends: The Land Cover Legend Registry. *Land* **2022**, *11*, 1083. [CrossRef]
17. Tsendbazar, N.E.; de Bruin, S.; Herold, M. Assessing global land cover reference datasets for different user communities. *ISPRS J. Photogramm. Remote Sens.* **2015**, *103*, 93–114. [CrossRef]
18. Herold, M.; Mayaux, P.; Woodcock, C.E.; Baccini, A.; Schmullius, C. Some challenges in global land cover mapping: An assessment of agreement and accuracy in existing 1 km datasets. *Remote Sens. Environ.* **2008**, *112*, 2538–2556. [CrossRef]
19. ESA. CORINE land cover. Available online: https://land.copernicus.eu/pan-european/corine-land-cover (accessed on 23 February 2023).
20. Büttner, G. CORINE land cover and land cover change products. In *Land Use and Land Cover Mapping in Europe*; Manakos, I., Braun, M., Eds.; Springer: Dordrecht, The Netherlands, 2014; ISBN 978-94-007-7969-3.
21. Bielecka, E.; Jenerowicz, A. Intellectual Structure of CORINE Land Cover Research Applications in Web of Science: A Europe-Wide Review. *Remote Sens.* **2019**, *11*, 2017. [CrossRef]

22. Cruickshank, M.M.; Tomlinson, R.W. Application of CORINE Land Cover Methodology to the U.K.-Some Issues Raised from Northern Ireland. *Glob. Ecol. Biogeogr. Lett.* **1996**, *5*, 235–248. [CrossRef]
23. Diaz-Pacheco, J.; Gutiérrez, J. Exploring the limitations of CORINE Land Cover for monitoring urban land-use dynamics in metropolitan areas. *J. Land Use Sci.* **2014**, *9*, 243–259. [CrossRef]
24. Di Sabatino, A.; Coscieme, L.; Vignini, P.; Cicolani, B. Scale and ecological dependence of ecosystem services evaluation: Spatial extension and economic value of freshwater ecosystems in Italy. *Ecol. Indic.* **2013**, *32*, 259–263. [CrossRef]
25. IDEAM. *Leyenda Nacional de Coberturas de la Tierra. Metodología CORINE Land Cover Adaptada para Colombia Escala 1:100.000*; IDEAM: Bogotá D.C., Colombia, 2010; ISBN 978-958-806729-2.
26. Foody, G.M. Approaches for the production and evaluation of fuzzy land cover classifications from remotely-sensed data. *Int. J. Remote Sens.* **1996**, *17*, 1317–1340. [CrossRef]
27. Gumbricht, T.; Roman-Cuesta, R.M.; Verchot, L.; Herold, M.; Wittmann, F.; Householder, E.; Herold, N.; Murdiyarso, D. An expert system model for mapping tropical wetlands and peatlands reveals South America as the largest contributor. *Glob. Chang. Biol.* **2017**, *23*, 3581–3599. [CrossRef]
28. Doyle, C.; Beach, T.; Luzzadder-Beach, S. Tropical Forest and Wetland Losses and the Role of Protected Areas in Northwestern Belize, Revealed from Landsat and Machine Learning. *Remote Sens.* **2021**, *13*, 379. [CrossRef]
29. Mizuochi, H.; Nishiyama, C.; Ridwansyah, I.; Nasahara, K.N. Monitoring of an Indonesian Tropical Wetland by Machine Learning-Based Data Fusion of Passive and Active Microwave Sensors. *Remote Sens.* **2018**, *10*, 1235. [CrossRef]
30. Hansen, M.C.; Potapov, P.V.; Moore, R.; Hancher, M.; Turubanova, S.A.; Tyukavina, A.; Thau, D.; Stehman, S.V.; Goetz, S.J.; Loveland, T.R.; et al. High-Resolution Global Maps of 21st-Century Forest Cover Change. *Science* **2013**, *342*, 850–853. [CrossRef]
31. Jiang, Y.; Wang, G.; Liu, W.; Erfanian, A.; Peng, Q.; Fu, R. Modeled Response of South American Climate to Three Decades of Deforestation. *J. Clim.* **2021**, *34*, 2189–2203. [CrossRef]
32. Hansen, M.C.; Potapov, P.V.; Pickens, A.H.; Tyukavina, A.; Hernandez-Serna, A.; Zalles, V.; Turubanova, S.; Kommareddy, I.; Stehman, S.V.; Song, X.-P.; et al. Global land use extent and dispersion within natural land cover using Landsat data. *Environ. Res. Lett.* **2022**, *17*, 34050. [CrossRef]
33. Salazar, A.; Salazar, J.F.; Sánchez-Pacheco, S.J.; Sanchez, A.; Lasso, E.; Villegas, J.C.; Arias, P.A.; Poveda, G.; Rendón, Á.M.; Uribe, M.R.; et al. Undermining Colombia's peace and environment. *Science* **2021**, *373*, 289 LP–290 LP.
34. Murillo-Sandoval, P.J.; Gjerdseth, E.; Correa-Ayram, C.; Wrathall, D.; Van Den Hoek, J.; Dávalos, L.M.; Kennedy, R. No peace for the forest: Rapid, widespread land changes in the Andes-Amazon region following the Colombian civil war. *Glob. Environ. Chang.* **2021**, *69*, 102283. [CrossRef]
35. Armenteras, D.; Schneider, L.; Davalos, L.M. Fires in protected areas reveal unforeseen costs of Colombian peace. *Nat. Ecol. Evol.* **2019**, *3*, 20–23. [CrossRef] [PubMed]
36. Clerici, N.; Armenteras, D.; Kareiva, P.; Botero, R.; Ramírez-Delgado, J.P.; Forero-Medina, G.; Ochoa, J.; Pedraza, C.; Schneider, L.; Lora, C.; et al. Deforestation in Colombian protected areas increased during post-conflict periods. *Sci. Rep.* **2020**, *10*, 4971. [CrossRef]
37. Gorelick, N.; Hancher, M.; Dixon, M.; Ilyushchenko, S.; Thau, D.; Moore, R. Google Earth Engine: Planetary-scale geospatial analysis for everyone. *Remote Sens. Environ.* **2017**, *202*, 18–27. [CrossRef]
38. Potapov, P.; Tyukavina, A.; Turubanova, S.; Talero, Y.; Hernandez-Serna, A.; Hansen, M.C.; Saah, D.; Tenneson, K.; Poortinga, A.; Aekakkararungroj, A.; et al. Annual continuous fields of woody vegetation structure in the Lower Mekong region from 2000-2017 Landsat time-series. *Remote Sens. Environ.* **2019**, *232*, 111278. [CrossRef]
39. Potapov, P.; Hansen, M.C.; Kommareddy, I.; Kommareddy, A.; Turubanova, S.; Pickens, A.; Adusei, B.; Tyukavina, A.; Ying, Q. Landsat Analysis Ready Data for Global Land Cover and Land Cover Change Mapping. *Remote Sens.* **2020**, *12*, 426. [CrossRef]
40. Fagua, J.C.; Ramsey, R.D. Geospatial modeling of land cover change in the Chocó-Darien global ecoregion of South America; One of most biodiverse and rainy areas in the world. *PLoS ONE* **2019**, *14*, e0211324. [CrossRef] [PubMed]
41. Didan, K. MYD13Q1 MODIS/Aqua Vegetation Indices 16-Day L3 Global 250m SIN Grid V006 [Data Set]. Available online: https://doi.org/10.5067/MODIS/MYD13Q1.006 (accessed on 23 February 2023). [CrossRef]
42. Xian, G.; Homer, C.; Rigge, M.; Shi, H.; Meyer, D. Characterization of shrubland ecosystem components as continuous fields in the northwest United States. *Remote Sens. Environ.* **2015**, *168*, 286–300. [CrossRef]
43. Homer, C.; Dewitz, J.; Yang, L.; Jin, S.; Danielson, P.; Xian, G.; Coulston, J.; Herold, N.; Wickham, J.; Megown, K. Completion of the 2011 National Land Cover Database for the Conterminous United States—Representing a Decade of Land Cover Change Information. *Photogramm. Eng. Remote Sens.* **2015**, *81*, 345–354.
44. Potapov, P.; Hansen, M.C.; Pickens, A.; Hernandez-Serna, A.; Tyukavina, A.; Turubanova, S.; Zalles, V.; Li, X.; Khan, A.; Stolle, F.; et al. The Global 2000-2020 Land Cover and Land Use Change Dataset Derived From the Landsat Archive: First Results. *Front. Remote Sens.* **2022**, *3*, 856903. [CrossRef]
45. Zhu, Z.; Woodcock, C.E. Continuous change detection and classification of land cover using all available Landsat data. *Remote Sens. Environ.* **2014**, *144*, 152–171. [CrossRef]
46. Kennedy, R.E.; Yang, Z.; Cohen, W.B. Detecting trends in forest disturbance and recovery using yearly Landsat time series: 1. LandTrendr—Temporal segmentation algorithms. *Remote Sens. Environ.* **2010**, *114*, 2897–2910. [CrossRef]

47. Huang, C.; Goward, S.N.; Masek, J.G.; Thomas, N.; Zhu, Z.; Vogelmann, J.E. An automated approach for reconstructing recent forest disturbance history using dense Landsat time series stacks. *Remote Sens. Environ.* **2010**, *114*, 183–198. [CrossRef]
48. Maus, V.; Câmara, G.; Cartaxo, R.; Sanchez, A.; Ramos, F.M.; Queiroz, G.R. A Time-Weighted Dynamic Time Warping Method for Land-Use and Land-Cover Mapping. *IEEE J. Sel. Top. Appl. Earth Obs. Remote Sens.* **2016**, *9*, 3729–3739. [CrossRef]
49. Verbesselt, J.; Hyndman, R.; Newnham, G.; Culvenor, D. Detecting trend and seasonal changes in satellite image time series. *Remote Sens. Environ.* **2010**, *114*, 106–115. [CrossRef]
50. Arevalo, P.; Olofsson, P.; Woodcock, C.E. Continuous monitoring of land change activities and post-disturbance dynamics from Landsat time series: A test methodology for REDD plus reporting. *Remote Sens. Environ.* **2020**, *238*, 111051. [CrossRef]
51. Potapov, P.; Li, X.; Hernandez-Serna, A.; Tyukavina, A.; Hansen, M.C.; Kommareddy, A.; Pickens, A.; Turubanova, S.; Tang, H.; Silva, C.E.; et al. Mapping global forest canopy height through integration of GEDI and Landsat data. *Remote Sens. Environ.* **2020**, *253*, 112165. [CrossRef]
52. Fagua, J.C.; Jantz, P.; Burns, P.; Massey, R.; Buitrago, J.Y.; Saatchi, S.; Hakkenberg, C.; Goetz, S.J. Mapping tree diversity in the tropical forest region of Chocó-Colombia. *Environ. Res. Lett.* **2021**, *16*, 54024. [CrossRef]
53. Lang, N.; Jetz, W.; Schindler, K.; Wegner, J.D. A high-resolution canopy height model of the Earth 2022. *arXiv* **2022**, arXiv:2204.08322.
54. Prasad, P.; Loveson, V.J.; Chandra, P.; Kotha, M. Evaluation and comparison of the earth observing sensors in land cover/land use studies using machine learning algorithms. *Ecol. Inform.* **2022**, *68*, 101522. [CrossRef]
55. De Luca, G.; Silva, J.M.N.; Di Fazio, S.; Modica, G. Integrated use of Sentinel-1 and Sentinel-2 data and open-source machine learning algorithms for land cover mapping in a Mediterranean region. *Eur. J. Remote Sens.* **2022**, *55*, 52–70. [CrossRef]
56. Clerici, N.; Valbuena Calderón, C.A.; Posada, J.M. Fusion of Sentinel-1A and Sentinel-2A data for land cover mapping: A case study in the lower Magdalena region, Colombia. *J. Maps* **2017**, *13*, 718–726. [CrossRef]
57. Fagua, J.C.; Jantz, P.; Rodriguez-Buritica, S.; Laura, D.; Goetz, S.J. Integrating LiDAR, Multispectral and SAR Data to Estimate and Map Canopy Height in Tropical Forests. *Remote Sens.* **2019**, *11*, 20. [CrossRef]
58. Kacic, P.; Hirner, A.; Da Ponte, E. Fusing Sentinel-1 and -2 to Model GEDI-Derived Vegetation Structure Characteristics in GEE for the Paraguayan Chaco. *Remote Sens.* **2021**, *13*, 5105. [CrossRef]
59. Foody, G.M.; Arora, M.K. An evaluation of some factors affecting the accuracy of classification by an artificial neural network. *Int. J. Remote Sens.* **1997**, *18*, 799–810. [CrossRef]
60. Hermosilla, T.; Wulder, M.A.; White, J.C.; Coops, N.C. Land cover classification in an era of big and open data: Optimizing localized implementation and training data selection to improve mapping outcomes. *Remote Sens. Environ.* **2022**, *268*, 112780. [CrossRef]
61. Foody, G.M.; Pal, M.; Rocchini, D.; Garzon-Lopez, C.X.; Bastin, L. The Sensitivity of Mapping Methods to Reference Data Quality: Training Supervised Image Classifications with Imperfect Reference Data. *ISPRS Int. J. Geo-Information* **2016**, *5*, 199. [CrossRef]
62. Hastie, T.; Tibshirani, R.; Friedman, J. *The Elements of Statistical Learning: Data Mining, Inference, and Prediction*, 2nd ed.; Bickel, P., Diggle, P., Fienberg, S., Gather, U., Olkin, I., Zeger, S., Eds.; Springer Science+Business Media: New York, NY, USA, 2017.
63. Olofsson, P.; Foody, G.M.; Herold, M.; Stehman, S.V.; Woodcock, C.E.; Wulder, M.A. Good practices for estimating area and assessing accuracy of land change. *Remote Sens. Environ.* **2014**, *148*, 42–57. [CrossRef]
64. Copass, C.; Antonova, N.; Kennedy, R. Comparison of Office and Field Techniques for Validating Landscape Change Classification in Pacific Northwest National Parks. *Remote Sens.* **2019**, *11*, 3. [CrossRef]
65. Lu, N.; Hernandez, A.J.; Ramsey, R.D. Land cover dynamics monitoring with Landsat data in Kunming, China: A cost-effective sampling and modelling scheme using Google Earth imagery and random forests. *Geocarto Int.* **2015**, *30*, 186–201. [CrossRef]
66. Toure, S.I.; Stow, D.A.; Shih, H.; Weeks, J.; Lopez-Carr, D. Land cover and land use change analysis using multi-spatial resolution data and object-based image analysis. *Remote Sens. Environ.* **2018**, *210*, 259–268. [CrossRef]
67. Ling, F.; Li, X.D.; Du, Y.; Xiao, F. Super-Resolution Land Cover Mapping with Spatial-Temporal Dependence by Integrating a Former Fine Resolution Map. *IEEE J. Sel. Top. Appl. Earth Obs. Remote Sens.* **2014**, *7*, 1816–1825. [CrossRef]
68. Momeni, R.; Aplin, P.; Boyd, D.S. Mapping Complex Urban Land Cover from Spaceborne Imagery: The Influence of Spatial Resolution, Spectral Band Set and Classification Approach. *Remote Sens.* **2016**, *8*, 88. [CrossRef]
69. Olson, D.M.; Dinerstein, E.; Wikramanayake, E.D.; Burgess, N.D.; Powell, G.V.N.; Underwood, E.C.; D'Amico, J.A.; Itoua, I.; Strand, H.E.; Morrison, J.C.; et al. Terrestrial ecoregions of the worlds: A new map of life on Earth. *Bioscience* **2001**, *51*, 933–938. [CrossRef]
70. IAvH. *Informe del Estado del Ambiente y de los Recursos Naturales Renovables*; Casallas, I.M., del Guardiola, L.P., Moreno, Y.A., Eds.; IDEAM: Bogota D.C., Colombia, 2016; ISBN 978-958-8067-87-2.
71. IDEAM. Marco nacional de Servicios Climáticos. Available online: http://www.ideam.gov.co/web/tiempo-y-clima/marco-nacional-de-servicios-climaticos (accessed on 23 February 2023).
72. Etter, A.; McAlpine, C.; Possingham, H. Historical Patterns and Drivers of Landscape Change in Colombia Since 1500: A Regionalized Spatial Approach. *Ann. Assoc. Am. Geogr.* **2008**, *98*, 2–23. [CrossRef]
73. Primack, R.B.; Corlett, R.T. *Tropical Rain Forests: An Ecological and Biogeographical Comparison*; Blackwell Publishing: Oxford, UK, 2009; ISBN 978-1-4051-4109-3.

74. One Earth Magdalena-Urabá Moist Forests. Available online: https://www.oneearth.org/ecoregions/magdalena-uraba-moist-forests/ (accessed on 23 February 2023).
75. García, Y.; Rangel, O.; Rodríguez, P. Cambios en la vegetación y el clima durante los últimos 200 años en ciénagas del sur del departamento del cesar. In *Complejo Cenagoso Zapatosa y Ciénaga del sur del Cesar*; Rangel, O., Ed.; Universidad Nacional de Colombia: Bogota D.C., Colombia, 2012; pp. 133–163, ISBN 9787587615067.
76. Fedepalma | Federación Nacional de Cultivadores de Palma de Aceite CID Palmero. Available online: http://cidpalmero.fedepalma.org/recursos-internet (accessed on 23 February 2023).
77. ESA. Sentinel-1 SAR GRD: C-Band Synthetic Aperture Radar Ground Range Detected, Log Scaling. Available online: https://developers.google.com/earth-engine/datasets/catalog/COPERNICUS_S1_GRD (accessed on 23 February 2023).
78. Vollrath, A.; Mullissa, A.; Reiche, J. Angular-Based Radiometric Slope Correction for Sentinel-1 on Google Earth Engine. *Remote Sens.* **2020**, *12*, 1867. [CrossRef]
79. ESA. Sentinel-2 MSI: MultiSpectral Instrument, Level-2A. Available online: https://developers.google.com/earth-engine/datasets/catalog/COPERNICUS_S2_SR (accessed on 23 February 2023).
80. ESA. Sentinel-2: Cloud Probability. Available online: https://developers.google.com/earth-engine/datasets/catalog/COPERNICUS_S2_CLOUD_PROBABILITY (accessed on 23 February 2023).
81. Braaten, J. Sentinel-2 Cloud Masking with s2cloudless. Available online: https://developers.google.com/earth-engine/tutorials/community/sentinel-2-s2cloudless (accessed on 23 February 2023).
82. Jenness, J. Topographic Position Index (TPI) v. 1.2. Available online: http://www.jennessent.com/downloads/tpi_documentation_online.pdf (accessed on 23 February 2023).
83. Čučković, Z. Terrain Position Index for QGIS. Available online: https://landscapearchaeology.org/2019/tpi/ (accessed on 23 February 2023).
84. Boehmke, B.; Greenwell, B.M. *Hands-On Machine Learning with R*; CRC Press: Boca Raton, FL, USA, 2019.
85. Kuhn, M.; Wing, J.; Weston, S.; Williams, A.; Keefer, C.; Engelhardt, A.; Cooper, T.; Mayer, Z.; Kenkel, B.; Benesty, M.; et al. Package 'caret': Classification and Regression Training. Available online: https://github.com/topepo/caret/ (accessed on 23 February 2023).
86. Liaw, A. Package 'randomForest': Breiman and Cutler's Random Forests for Classification and Regression; Berkeley, CA, USA. 2018. Available online: https://cran.r-project.org/web/packages/randomForest/index.html (accessed on 23 February 2023).
87. Milborrow, S.; Tibshirani, R. Package 'earth': Multivariate Adaptive Regression Splines; 2019. Available online: https://cran.r-project.org/web/packages/earth/index.html (accessed on 23 February 2023).
88. Kursa, M.B.; Rudnicki, W.R. Feature Selection with the Boruta Package. *J. Stat. Softw.* **2010**, *36*, 1–13. [CrossRef]
89. Tsendbazar, N.-E.; Herold, M.; de Bruin, S.; Lesiv, M.; Fritz, S.; Van De Kerchove, R.; Buchhorn, M.; Duerauer, M.; Szantoi, Z.; Pekel, J.-F. Developing and applying a multi-purpose land cover validation dataset for Africa. *Remote Sens. Environ.* **2018**, *219*, 298–309. [CrossRef]
90. Xu, Y.; Yu, L.; Feng, D.; Peng, D.; Li, C.; Huang, X.; Lu, H.; Gong, P. Comparisons of three recent moderate resolution African land cover datasets: CGLS-LC100, ESA-S2-LC20, and FROM-GLC-Africa30. *Int. J. Remote Sens.* **2019**, *40*, 6185–6202. [CrossRef]
91. Fagua, J.C.; Ramsey, R.D. Comparing the accuracy of MODIS data products for vegetation detection between two environmentally dissimilar ecoregions: The Chocó-Darien of South America and the Great Basin of North America. *GISci. Remote Sens.* **2019**, *56*, 1046–1064. [CrossRef]
92. Helmer, E.H.; Ruzycki, T.S.; Benner, J.; Voggesser, S.M.; Scobie, B.P.; Park, C.; Fanning, D.W.; Ramnarine, S. Detailed maps of tropical forest types are within reach: Forest tree communities for Trinidad and Tobago mapped with multiseason Landsat and multiseason fine-resolution imagery. *For. Ecol. Manag.* **2012**, *279*, 147–166. [CrossRef]
93. Foody, G.M. Explaining the unsuitability of the kappa coefficient in the assessment and comparison of the accuracy of thematic maps obtained by image classification. *Remote Sens. Environ.* **2020**, *239*, 111630. [CrossRef]
94. Zhang, X.; Liu, L.; Chen, X.; Xie, S.; Gao, Y. Fine Land-Cover Mapping in China Using Landsat Datacube and an Operational SPECLib-Based Approach. *Remote Sens.* **2019**, *11*, 1056. [CrossRef]
95. Ghulam, A.; Ghulam, O.; Maimaitijiang, M.; Freeman, K.; Porton, I.; Maimaitiyiming, M. Remote Sensing Based Spatial Statistics to Document Tropical Rainforest Transition Pathways. *Remote Sens.* **2015**, *7*, 6257–6279. [CrossRef]
96. Wang, H.; Hu, Y.; Feng, Z. Fusion and Analysis of Land Use/Cover Datasets Based on Bayesian-Fuzzy Probability Prediction: A Case Study of the Indochina Peninsula. *Remote Sens.* **2022**, *14*, 5786. [CrossRef]
97. Talukdar, S.; Singha, P.; Mahato, S.; Shahfahad; Pal, S.; Liou, Y.-A.; Rahman, A. Land-Use Land-Cover Classification by Machine Learning Classifiers for Satellite Observations—A Review. *Remote Sens.* **2020**, *12*, 1135. [CrossRef]
98. Alabi, T.R.; Adewopo, J.; Duke, O.P.; Kumar, P.L. Banana Mapping in Heterogenous Smallholder Farming Systems Using High-Resolution Remote Sensing Imagery and Machine Learning Models with Implications for Banana Bunchy Top Disease Surveillance. *Remote Sens.* **2022**, *14*, 5206. [CrossRef]
99. Liu, T.; Abd-Elrahman, A.; Morton, J.; Wilhelm, V.L. Comparing fully convolutional networks, random forest, support vector machine, and patch-based deep convolutional neural networks for object-based wetland mapping using images from small unmanned aircraft system. *GISCIENCE Remote Sens.* **2018**, *55*, 243–264. [CrossRef]

100. Mazzia, V.; Khaliq, A.; Chiaberge, M. Improvement in Land Cover and Crop Classification based on Temporal Features Learning from Sentinel-2 Data Using Recurrent-Convolutional Neural Network (R-CNN). *Appl. Sci.* **2020**, *10*, 238. [CrossRef]
101. Boston, T.; Van Dijk, A.; Larraondo, P.R.; Thackway, R. Comparing CNNs and Random Forests for Landsat Image Segmentation Trained on a Large Proxy Land Cover Dataset. *Remote Sens.* **2022**, *14*, 3396. [CrossRef]

remote sensing

MDPI

Article

Spatial–Temporal Changes in Land Use and Their Driving Forces in the Circum-Bohai Coastal Zone of China from 2000 to 2020

Jian Cui [1], Wenxin Ji [1], Peng Wang [2], Mingshui Zhu [2] and Yaohui Liu [1,*]

1 School of Surveying and Geo-Informatics, Shandong Jianzhu University, Jinan 250101, China
2 Ji'nan Institute of Survey and Investigation, Jinan 250101, China
* Correspondence: liuyaohui20@sdjzu.edu.cn

Abstract: Over the past two decades, the location and morphology of the coastline, as well as the land use/land cover (LULC) in the Circum-Bohai region in China, have undergone significant changes due to rapid industrialization and urbanization. Analyzing the temporal and spatial variation in coastal lines and LULC can provide a meaningful basis for the rational allocation of land resources. Using Landsat TM/OLI series dates from the Google Earth Engine (GEE) platform, this study applied the Linear Superposition Water Index (LSWI) and the Otsu threshold method (OTSU) algorithm to extract and analyze the coastline of the Circum-Bohai region. Additionally, the Random Forests (RF) method was employed to extract LULC information in the coastal zone. Using the geographical detector, we further explored the influence of social and economic factors, as well as natural factors, on spatial differentiation mechanisms of LULC change in the Circum-Bohai. Our results show that between 2000 and 2020, the Circum-Bohai coastline generally expanded towards the ocean by a total of 1062.99 km. The highest rate of change occurred during 2010 to 2015, and human activities were the primary cause of most of the changes, with the exception of the Yellow River Delta, where natural factors were dominant. The main types of LULC in the study area from 2000 to 2020 were farmland and construction land. The area of farmland proportion decreased by 1.75%, while the area of construction land proportion increased from 16.73% to 29.54%. Our findings indicate that the degree of land use in the Circum-Bohai is deepening. Based on our factor detection analysis, the added value of the secondary industry was the most critical influencing factor on LULC. Furthermore, the combined effect of the added value of the secondary industry and gross domestic product (GDP) has a significant driving impact on LULC. These findings can provide reference and data support for the sustainable development and comprehensive management of land resources. The relevant departments can use these results to prompt corresponding policies for the rational allocation of land resources.

Keywords: Google Earth Engine (GEE); Circum-Bohai coastal zone; Linear Superposition Water Index (LSWI); Random Forests (RF); geographical detector; driving force

1. Introduction

Affected by the rapid improvement of the world's economy and the extensive comprehensive impact of human and marine activities, global coastlines have undergone varying degrees of change, and the land use degree (LUD) in the coastal zones has also continued to deepen. Countries worldwide have increasingly prioritized the development, utilization, and conservation of their marine resources [1]. Monitoring the spatial and temporal changes in coastline morphology and coastal zone land use/land cover (LULC) is crucial for regional social and economic development, resource environment balance, and promoting sustainable economic development in coastal areas [2].

With the increase in the number of remote sensing satellites and the improvement of image resolution [3–5], advanced remote sensing technology and multi-spatial resolution remote sensing image data have gradually provided reliable technical means and data support for monitoring coastline change and coastal land use/land cover change (LUCC). The selection and application of the coastline extraction method is the key aspect in the process of monitoring coastline change using remote sensing technology [6]. Relevant studies have employed various methods to extract and analyze different types of coastlines at various scales in different regions based on multi-source multi-spatial resolution remote sensing image data, including visual interpretation [7], Normalized Difference Water Index (NDWI) [8], Tasseled Cap Transformation (TCT) [9], Automated Water Extraction Index (AWEI) [10], automated edge detection algorithms [11], Support Vector Machine (SVM), Maximum Likelihood (ML) [12], Modified Normalized Difference Water Index (MNDWI) [13], active connection matrix (ACM) [14], convolutional neural network (CNN) [15], and others. Particularly, the water index method is usually employed for research on large-scale and long-term monitoring of coastal changes. The NDWI, TCT, and MNDWI are the most commonly used methods [16]. Since the Otsu threshold method (OTSU) [17] is very suitable for cases where the gray values of the segmentation target and the segmentation background are significantly different [18], it is often integrated with NDWI, TCT, MNDWI, and other water index algorithms to extract large-scale coastlines.

In previous studies on LUCC monitoring in coastal zones, the acquisition of LULC information can be categorized into two general types. The first method is mainly to obtain LULC information in the study area through relevant datasets [19]. Based on different LULC classification datasets, researchers have analyzed and studied the land use of coastal zones [20,21]. The second method involves using supervised or semi-supervised classification methods to classify coastal zone LULC and then obtain coastal LULC information. Due to the superior classification effect and experimental accuracy, the Random Forest (RF) [22] algorithm is often used in the LULC information classification and extraction task [23]. Researchers usually employ the RF classification model to classify the LULC information [24], and other methods are combined to investigate and analyze its spatial and temporal evolution [25,26]. The selection of LULC research areas also has its own characteristics, including inland areas [23], river coasts [27], and coastal areas [26,28]. Exploring the driving factors that affect LUCC is helpful to further analyze the LUCC model in the Circum-Bohai region, which is of great significance for natural resources management. At present, qualitative or quantitative methods are usually utilized to determine the driving factors affecting a specific change [27]. Among them, qualitative analysis can only characterize the influence of various impact factors and cannot quantitatively analyze their effect. Moreover, it often ignores the spatial relationship between the change and the influencing factors. Geodetector is a method that can detect a certain geographical phenomenon or a certain spatial variation and reveal the driving factors behind it [23,27]. It has been widely used in vegetation coverage change [29], LUCC [23,27], ecological environment vulnerability [30], and so on.

In remote-sensing-technology-based research and remote sensing image analysis, the screening and processing of massive remote sensing data are crucial. However, in previous related research, difficulties in data acquisition and processing and low interpretation efficiency have often been observed [31]. To solve these problems, the emergence of Google Earth Engine (GEE) [27] provides new ideas and methods for massive data processing in large-scale and long-term research projects [27]. The GEE platform has the capability to process data in batches rapidly, regardless of temporal and spatial limitations. [31]. Using the GEE platform, relevant studies have realized the extraction of crop area [32], surface water changes [33], ecological environment quality evaluation [34,35], and LUCC [36,37]. These studies have fully verified the feasibility and effectiveness of the GEE platform in downloading, processing, and computing data. Therefore, we can efficiently research coastline spatial–temporal variation and coastal LUCC monitoring using the GEE platform. However, in related research, coastlines are often extracted using a single method, and coastal LUCC information extraction is typically focused on large-scale areas. Furthermore, there are computational complexities in exploring the driving factors behind coastal LUCC.

As one of the areas most affected by human activities in China, the Bohai Sea occupies an important position in the development of national economic strategy [38]. However, with the human activities and deepening of land development and utilization in the Circum-Bohai coastal area, artificial breeding and port construction have caused spatial differences in LULC in the coastline and coastal zone over time, which pose challenges to the protection and development of the marine and land ecological environment in the Circum-Bohai area. Therefore, it is essential to explore the changes in the Circum-Bohai coastlines, as well as to explore the spatial pattern of LULC and the driving factors behind LUCC in the Circum-Bohai coastal area. Hence, investigating alterations in the Bohai Sea coastline and analyzing variations in the LULC spatial model are imperative to comprehend the key driving factors.

In this study, we aim to explore the process of coastline changes and coastal LUCC in the Circum-Bohai coastal area from 2000 to 2020 using the Landsat TM/OLI series data on the GEE platform. First, the coastline of the Circum-Bohai area was extracted through the linear fusion of TCT and MNDWI combined with the OTSU method. Then, we explored the spatial and temporal pattern change in LULC information in the Circum-Bohai coastal area based on the RF classification method. Lastly, the transfer matrix and driving factors behind the changes in the LUD are discussed using the Geodetector model.

The remainder of this study is organized as follows: Section 2 presents the study area and data. Section 3 describes the methods used in this study. Section 4 presents the experimental results. The discussion and conclusion are presented in Sections 5 and 6, respectively.

2. Study Area and Data

2.1. Study Area

The study area is situated in the eastern part of China, stretching from Dalian City in the Liaodong Peninsula to Yantai City in the Shandong Peninsula, spanning from 117°26′58″E to 122°23′47″E and 36°59′22″N to 41°0′34″N. Additionally, the Bohai Sea is the only inner sea in China [28]. The study area covers 13 cities (presented in Figure 1). Within the Bohai Bay region, there are three predominant bays: Bohai Bay, Liaodong Bay, and Laizhou Bay. Among them, Pulandian Bay area (A) in Dalian is one of the key areas of coastline change in the Circum-Bohai. Liaodong Bay Area and Laizhou Bay area contain two large estuary deltas, Liaohe River Delta (B) and Yellow River Delta (D), respectively, which are two important wetland nature reserves in China [39]. Additionally, Tangshan Port Area (C) in Tangshan City, Bohai Bay area, features significant artificial land [40].

Due to the uniqueness of the geographical conditions and the complexity of economic development in the Circum-Bohai, this study utilized the land-side buffers along the coastline to define the regional scope of LUCC research in the area to more accurately and intuitively express the LUCC in the Circum-Bohai. Specifically, the coastline was extracted every five years from 2000 to 2020 in the Circum-Bohai region, and a unilateral buffer with a radius of 10 km was created on the land side along the coastline [41]. Five buffers were then obtained and synthesized to derive the LUCC results in the Circum-Bohai coastal zone.

2.2. Data

The study employed Landsat TM/OLI series data from 2000 to 2020 obtained from the GEE platform. A total of 1533 remote sensing image data were used in the coastline extraction experiment on the Circum-Bohai, including data from 2000, 2005, 2010, 2015, and 2020. For the 2000, 2010, and 2020 coastal land use and land cover (LULC) classification, a total of 1130 scene images data were utilized. The remote sensing image data underwent several preprocessing stages such as filtering, cloud removal, splicing, cropping, masking, and more. The data sources are provided in Table 1.

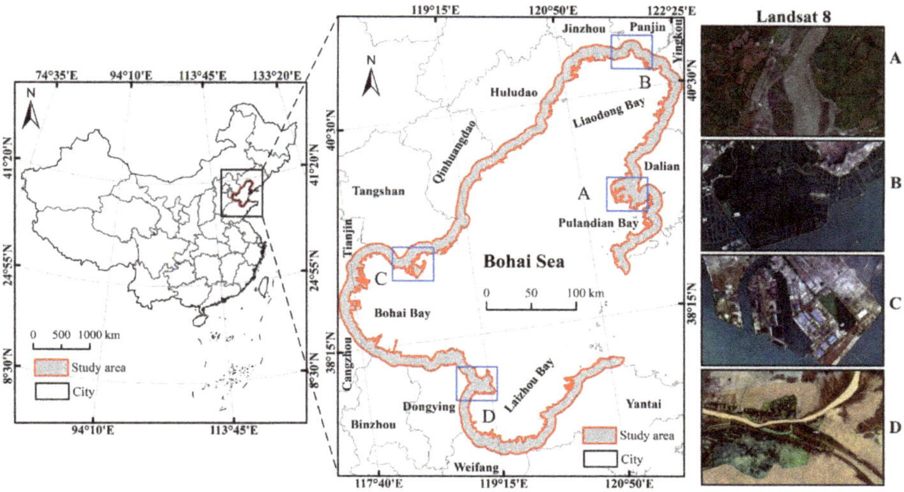

Figure 1. Study area: (**A**) Pulandian Bay area; (**B**) Liaohe River Delta; (**C**) Tangshan Port; (**D**) Yellow River Delta.

Table 1. Data sources.

Sensor	Spatial Resolution(m)	Year	Band Used
Landsat-5 TM	30	2000, 2005, 2010	Blue, Green, Red
Landsat-8 OLI	30	2015, 2020	NIR, SWIR

Considering the unique geographical conditions of the Circum-Bohai coastal zone, such as the frequent occurrence of two estuary deltas and reclamation, we divided the LULC types into eight categories according to the actual situation of the study area, namely, farmland, grassland, forest land, construction land (urban, residential area, traffic land, and industrial land), waterbody (seawater and inland freshwater), unused land (sand and bare land), natural wetland (beach and wetland), and artificial wetland (saltern, coastal reclamation, and artificial breeding). The RF classification method was then employed to classify the LULC of the Circum-Bohai coastal zone [42].

To better distinguish the various categories in the study area, this study comprehensively considered and calculated the normalized difference vegetation index (NDVI), normalized difference build and soil index (NDBI), enhanced vegetation index (EVI), NDWI, and MNDWI, and constructed a classification feature set [27]. To construct a classification sample set, this study utilized high-resolution historical data from Google Earth Pro and performed visual interpretation. The sample points for the Circum-Bohai coastal zone in 2000, 2010, and 2020 were 1813, 1821, and 1828, respectively. To ensure the accuracy and reliability of the experiment, 70% of the samples from each period were used to establish the training sample, while 30% of the samples were utilized to establish the verification sample [43].

The geographic detector model was utilized to explore the internal driving mechanisms of LUCC in the study area by analyzing both socioeconomic and natural aspects [25]. Relevant information on some of the impact factors was obtained from the Chinese Academy of Sciences (https://www.resdc.cn/ (accessed on 10 February 2023)) [27], including population density (X1), gross domestic product (GDP) (X2), annual average precipitation (X8), annual average temperature (X9), slope (X10), aspect (X11), and elevation (X12). At the same time, the other part used the statistical yearbooks of the provinces and cities in the study area to collect data on the secondary industry added value (X3), tertiary industry added value (X4), urbanization rate (X5), the total value of fishery production (X6), and port throughput (X7) [27]. The information on influencing factors is shown in Table 2.

Table 2. Influencing factors in this study.

Type	Influencing Factors	Index	Unit
Socioeconomic factor	X1	population density	Ten thousand people/km^2
	X2	gross domestic product	billion CNY
	X3	secondary industry added value	million CNY
	X4	tertiary industry added value	million CNY
	X5	urbanization rate	%
	X6	total fishery output	million tons
	X7	port throughput	million tons
Physical factor	X8	annual average precipitation	mm
	X9	annual average temperature	°C
	X10	slope	/
	X11	aspect	°
	X12	elevation	m

3. Methodology

The Landsat TM/OLI series data after preprocessing in the experimental year were selected from the GEE platform, and utilized the linear fusion TCT, MNDWI method, and OTSU algorithm to extract and analyze the changes in the Circum-Bohai coastline. The RF classification was employed to classify the LULC types in the study area, while the land use transfer matrix was utilized to explore LULC type transfers. Additionally, the land use degree index (LUDI) was calculated to evaluate the change in land use. To explore the mechanisms of socioeconomic and natural factors influencing LUCC, factor detection and interaction detector in Geodetector were used for analysis. The experimental workflow is depicted in Figure 2.

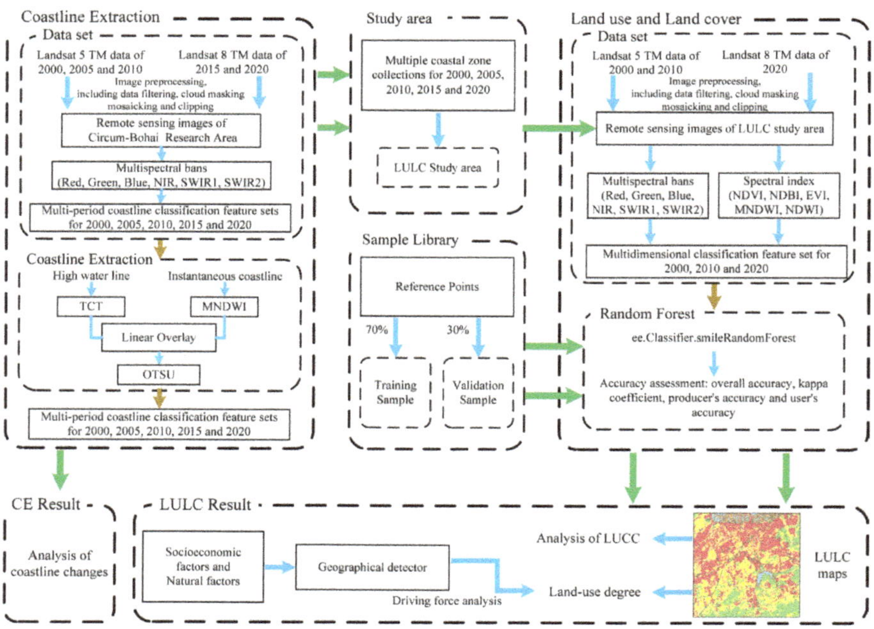

Figure 2. The flowchart of this study.

3.1. Long Time-Series Coastline Extraction Method

3.1.1. Water Index Method

Landsat TM/OLI series data in the GEE platform were utilized to accurately differentiate water and land in the Circum-Bohai by linearly fusing TCT [9] and MNDWI [13]. Specifically, TCT and MNDWI were used to calculate the remote sensing data for five periods from 2000 to 2020. Two gray image data of each period were obtained, and each pixel was linearly superimposed based on its gray value to synthesize gray image data. Finally, five years of gray images with varying gray values were generated to reflect water and land. This method is named the Linear Superposition Water Index (LSWI) for ease of reference. The mathematical expression for TCT is as follows:

$$\mu = R_X^T + r \tag{1}$$

where R is the tasseled cap transformation coefficient; X is the pixel value of different bands; r is the constant offset; and μ is the gray value of each pixel in different bands after tasseled cap transformation [44].

MNDWI mathematical is presented as follows:

$$MNDWI = (LG - LS)/(LG + LS) \tag{2}$$

where LG represents Green in Landsat TM/OLI series and LS represents SWIR/SWIR1 in Landsat TM/OLI series. $MNDWI$ is the gray value calculated for each pixel.

3.1.2. Otsu Threshold Method (OTSU)

The OTSU is an adaptive image segmentation algorithm based on the least square method [17]. Its principle is to calculate the segmentation threshold of the gray image by using the maximum inter-class variance and is therefore also called the maximum inter-class difference method [45]. OTSU possesses the characteristics of being unaffected by image brightness and contrast, along with fast segmentation speed and simple calculation.

The OTSU mathematical expression is as follows:

$$\sigma_B^2(t) = F_0(t)F_1(t)(\rho_0(t) - \rho_1(t))^2 \tag{3}$$

$$T = \arg \max_{0 \le t \le L-1} \sigma_B^2(t) \tag{4}$$

where $F_0(t)$ and $F_1(t)$ are the ratios of the pixels in the two regions; $\rho_0(t)$ and $\rho_1(t)$ are the pixel mean of the two regions, respectively. L represents the total gray level of the input image; T is the segmentation threshold obtained when $\sigma_B^2(t)$ reaches the maximum value [46].

3.2. Coastal Land Use Classification

3.2.1. Coastal LULC Classification and Its Accuracy Evaluation

The RF classification method commonly utilized in LULC information classification is a classification method based on the combination of classification decision trees [27]. Its basic units and principles are sets of decision trees and decision tree classifiers, respectively [47]. By utilizing each decision tree to perform multi-decision voting on the given samples for training and prediction, a classification selection can be obtained. This effectively improves the overfitting or underfitting problems that commonly arise in single decision trees [48]. The RF method possesses strong anti-interference performance, high classification accuracy, and classification efficiency [49]. In this study, the LULC is classified using the ee.Classifier.smileRandomForest (number of trees) function on the GEE platform. Experimental verification shows that when the number of decision trees is 500, the classification effect is comparatively precise, so 500 trees were then employed for RF classification [27].

The confusion matrix is a common method used to evaluate the accuracy of remote sensing classification results [50]. The accuracy of classification results is quantitatively analyzed by calculating user accuracy (UA), producer accuracy (PA), kappa coefficient (kappa), and overall accuracy (OA) based on the confusion matrix [23].

3.2.2. Land Use Degree Index

In this study, the LUDI is an indicator that reflects the degree of actual land use and development by humans. It is often utilized to explain the LUD and development [51]. The higher the value calculated by the grading index, the greater the level of land development and utilization in the region [23], and therefore the higher the degree of social and economic development in the area [52]. In order to analyze the LUD in the Circum-Bohai, the level of land development and utilization was evaluated using the LUDI. The calculation formula for l_a of the LUD is as follows:

$$l_\alpha = 100 \times \sum_{j=1}^{m} I_j \times P_j \tag{5}$$

where l_a is the calculated value representing the degree of land use, and its range is 100–400. I_j is a value representing the classification of different LUDI, P_j is the percentage of the j-th LUD classification area in the total area of the study area. According to the types of LULC and related research [25], this study divided the LULC into four different grades based on the land resource utilization degree and gives different classification indexes to each grade (presented in Table 3).

Table 3. Types and classification of LULC in the Circum-Bohai coastal zone.

Type	Uncultivated Land	Ecological Land	Agricultural Land	Construction Land
LULC types	Unused land (sand and bare land) Natural wetlands (beach, wetland)	Grassland Forest land Waterbody (seawater, inland freshwater)	Farmland Artificial wetland (saltern, coastal reclamation, artificial breeding)	Construction land (urban, residential area, traffic land, and industrial land)
Index of Classification	1	2	3	4

3.3. Geographic Detector

Geodetector [53] is a statistical technique for assessing the impact of selected variables on specific changes and identifying spatial changes. Geodetector includes factor detection, interactive detection, ecological detection, and risk detection [54]. This study mainly used factor detection and interactive detection in geographic detectors, which can explore the spatial divergence interpretation of each influencing factor alone or after interaction [27]. Specifically, LUDI is used as Y, the impact factor is used as X, and the decisive index q is introduced to evaluate each impact factor. The geographic detector model is presented as follows:

$$q = 1 - \frac{\sum_{h=1}^{L} N_h \sigma_h^2}{N\sigma^2} \tag{6}$$

$$W_h = \sum_{h=1}^{L} N_h \sigma_h^2 \tag{7}$$

$$T_h = N\sigma^2 \tag{8}$$

where $h = 1, \ldots, L$ is the strata of variable Y or factor X. N_h are the number of elements in the layer and N is the whole region. σ_h^2 and σ^2 are the variances in the Y values in the layer h and the whole region, respectively. W_h and T_h are within the sum of squares and the total sum of squares, respectively. The range of q is [0, 1]; the larger the value is, the more obvious the spatial differentiation of Y is.

4. Results

4.1. Coastline Change Analysis

4.1.1. Validation of Coastline Extraction Method

To validate the accuracy of coastline extraction, we utilized four methods, namely NDWI, TCT, MNDWI, and LSWI, to distinguish between sea and land in three different areas: A, B, and C, in the year 2010. These areas represent different shoreline types, including natural (A), artificial (B), and beach (C), which are affected by diverse factors such as human activities and natural processes. The experimental results of coastline extraction by the four methods are shown in Figure 3.

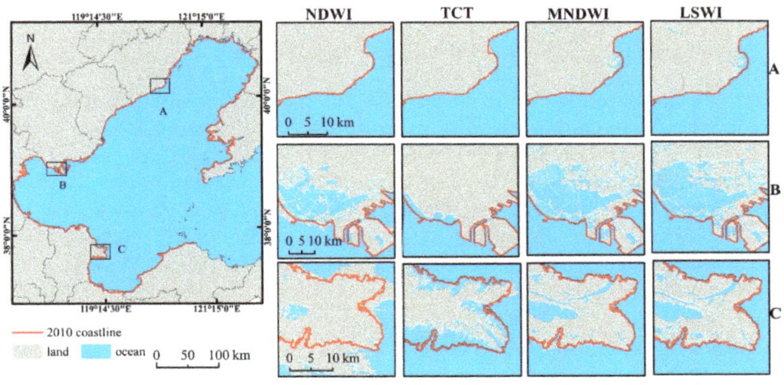

Figure 3. Three typical areas (**A**–**C**) and their classification results: (**A**) is part of the coastline of Huludao City; (**B**) is Tangshan Port; (**C**) is the Yellow River Delta.

The four methods were evaluated based on their effectiveness in segmenting water and land in each area. The results show that the segmentation effects of all four methods are good in the natural shoreline area (A). In the artificial shoreline area (B), only the MNDWI and LSWI methods produced better results. In the beach shoreline area (C), the LSWI method outperformed the other methods. Therefore, the LSWI water extraction method was selected for the segmentation of waterbodies in the Circum-Bohai coastal in this study, as it provides a better sea–land segmentation effect.

In summary, the LSWI method was found to be the most effective for distinguishing between sea and land, particularly in areas that are strongly influenced by natural or human factors. This method was thus selected to extract waterbodies and segment the ocean and land, ensuring the accuracy of the study's results.

4.1.2. Coastline Change Analysis

Figure 4 displays the extraction results of coastline length change in the Circum-Bohai from 2000 to 2020. It demonstrates that the coastline length fluctuation in the Circum-Bohai region has generally increased over the past 20 years, with a total increase of 1062.99 km. Specifically, the coastline of the Circum-Bohai region exhibited a slight upward trend from 2000 to 2005, increasing by 194.1 km. From 2005 to 2015, the coastline length of the Circum-Bohai gradually increased, and the rate of change accelerated, with a total increase of 1087.9 km during this period. Finally, from 2015 to 2020, the coastline length decreased by 24.91 km.

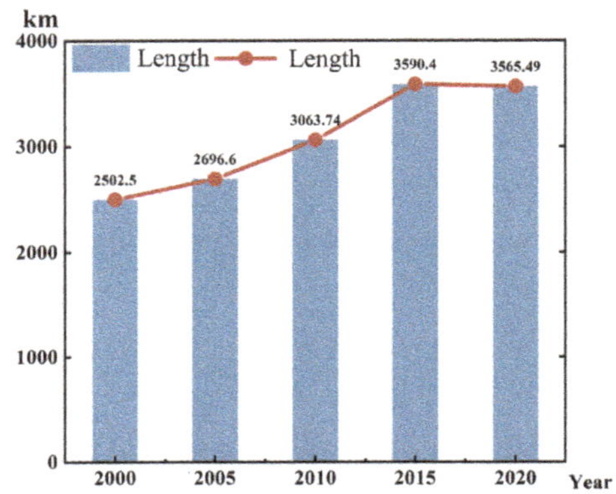

Figure 4. Length changes of Circum-Bohai coastline from 2000 to 2020.

Comparing the coastline change results obtained by the LSWI method with the classified images, it can be seen that the primary change areas of the Circum-Bohai coastline are Laizhou Bay, Bohai Bay, and Liaodong Bay. Among them, the artificial aquaculture and salt field construction in Liaodong Bay and Laizhou Bay are more prominent, and the changes in the Circum-Bohai from 2000 to 2020 are more significant. The most significant changes in the region are mainly distributed in the Bohai Bay artificial reclamation, wharf construction, and other related areas, particularly in Tianjin and Tangshan. These changes are also related to urban development policy directives and the rapid development trend of the maritime economy. Unlike these reasons, the changes in the Yellow River Delta's coastline are due to natural factors, as the coastline type of the Yellow River Delta is mainly a tidal flat coast formed by sediment accumulation brought by the Yellow River. From 2000 to 2020, the Yellow River Delta's location changed, leading to significant changes in the coastline's position in the region. In addition to the above areas, the coastline changes in other areas are relatively small because the coastlines of these regions are mostly natural and have not been extensively developed and utilized.

Figure 5 illustrates the overall changes in the coastline, with typical change areas a and b representing the artificial coastline, while c represents the natural coastline. Among them, a is the artificial breeding area, b is the Huanghua Port, and c is the Yellow River Delta. The analysis of the Circum-Bohai coastline changes revealed that the majority of changes occurred in regions where human activities, such as artificial aquaculture, artificial land, and wharves, were prevalent, while changes in the Yellow River Delta are due to natural factors. Limited coastline changes in other areas are attributed to their relatively natural state. This study provides vital insights into the complex interplay between natural and human factors contributing to changes in coastal environments.

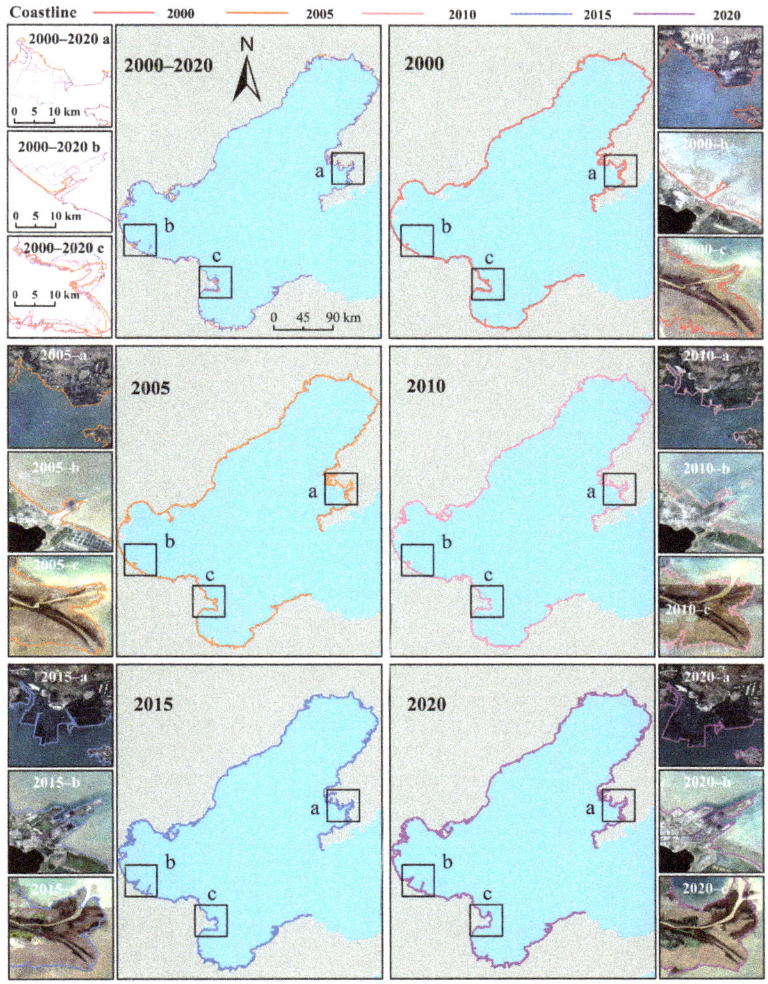

Figure 5. The changes of coastlines in the Circum-Bohai region from 2000 to 2020, where a and b are artificial coastlines and c is the natural coastline. (**a**) Pulandian Bay area; (**b**) Huanghua Port; (**c**) Yellow River Delta.

4.2. Long Time-Series LUCC Analysis of Coastal Zone

4.2.1. Accuracy Evaluation

The accuracy of LULC information classification in LUCC analysis is essential for further research. In this study, the confusion matrix was employed to evaluate the accuracy of LULC classification results in the Circum-Bohai region in 2000, 2010, and 2020. The results of the LULC classification accuracy evaluation are presented in Table 4. The OA for 2000, 2010, and 2020 was 91.15%, 88.54%, and 92.08%, respectively. Meanwhile, the kappa was 0.89, 0.86, and 0.90, respectively. The classification results of OA and kappa for each period were above 85%, which indicates high classification accuracy for different LULC types. Therefore, the classification achieved an acceptable level of accuracy, indicating that the results are both reasonable and reliable.

Table 4. Accuracy evaluation.

LUCC Types	2000		2010		2020	
	P_{UA} (%)	P_{PA} (%)	P_{UA} (%)	P_{PA} (%)	P_{UA} (%)	P_{PA} (%)
Farmland	88.42	94.28	87.76	93.93	89.57	94.5
Forest land	90.63	93.55	92.3	93.98	91.72	92.36
Grassland	100	50	100	72.73	100	25
Waterbody	96.95	92.03	86.25	92.62	87.5	90.51
Construction land	90.77	91.47	89.34	90.76	94.72	96.34
Unused land	85.71	88.24	90	46.15	100	60.71
Natural wetlands	90.09	86.96	71.19	79.25	91.04	88.41
Artificial wetland	94.06	90.91	90.86	82.87	93.13	91.35
P_{OA} (%)	91.15		88.54		92.08	
Kappa coefficient	0.89		0.86		0.90	

4.2.2. LULC Change Analysis

In the long-term LUCC analysis of the coastal zone between 2000 and 2020, the proportion and distribution of LULC in the 10 km buffer of the coastal zone of the Circum-Bohai have changed significantly. Figure 6 shows that this area's dominant LULC types are primarily farmland and construction land. The area of farmland decreased slightly from 31.23% to 29.48% from 2000 to 2020. In contrast, construction land improved from 16.73% to 29.54%, with an overall increase of 2289.84 km^2. Constructed wetland areas accounted for approximately 22% during 2000 to 2020, and the overall change was not significant. However, natural wetlands experienced a decreasing tendency, in 2020 the area accounted for around 10%. The proportion of forest land and waterbodies showed a decreasing trend. The main reason for this is that the continuous expansion of construction land between 2000 and 2020 encroached on the distribution areas of other LULC categories.

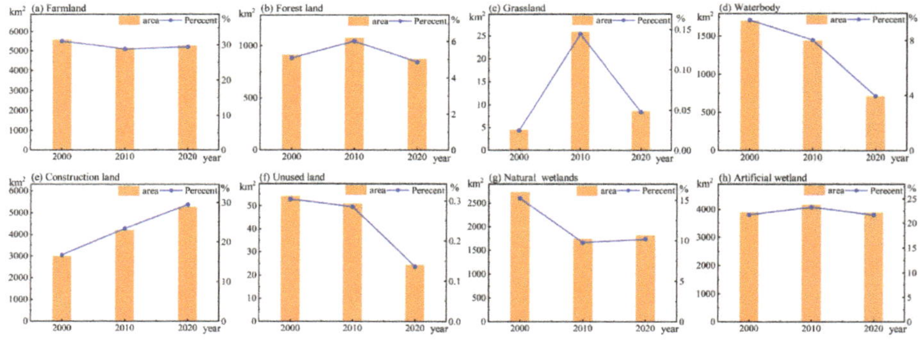

Figure 6. The proportion of LULC types changed from 2000 to 2020.

The LULC distribution of each category in the study area is shown in Figure 7; farmland is primarily found in the coastal zones of Yantai City in Shandong Province, Qinhuangdao City in Hebei Province, Huludao City, and Dalian City in Liaoning Province, which the primary industry economy has a relatively high proportion. The distribution of construction land is more dispersed, with concentrations in the economically developed and mature marine transport industry of the Tianjin and Tangshan coastal zones, e.g., the b in Figure 7 is Tianjin Port. Artificial wetlands are mainly distributed in the northeast of Laizhou Bay, Bohai Bay, and Liaodong Bay, where artificial breeding and reclamation are vigorously developed. Water is mainly distributed in rivers, lakes, and buffer zone at both ends of the sea. Natural wetlands are primarily located in the Liaohe River Delta (a) and Yellow River Delta (c).

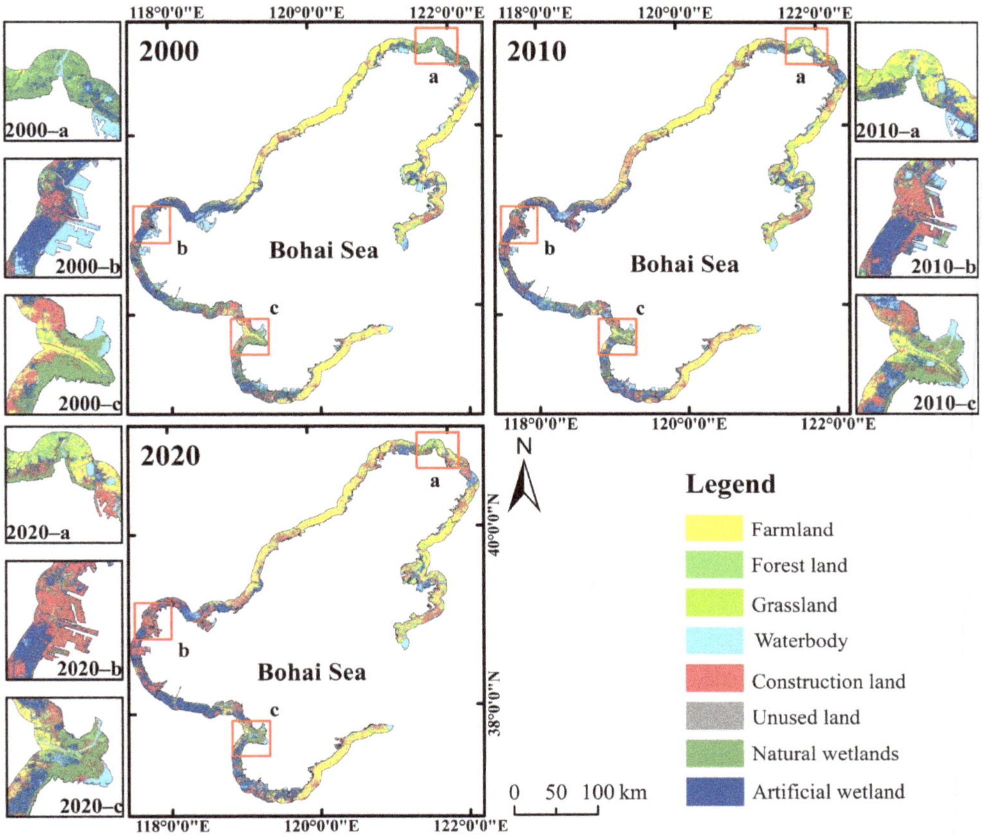

Figure 7. Spatial distribution of LULC in the Circum-Bohai from 2000 to 2020: (**a**) Liaohe River Delta; (**b**) Tianjin Port; (**c**) Yellow River Delta.

4.2.3. LULC Spatiotemporal Change Analysis

To provide a clear representation of the transfer relationship between the LULC types, we employed the transfer matrix to further analyze the transformation of LULC types from 2000 to 2010, 2010 to 2020, and 2000 to 2020, as shown in Figure 8. Overall, the construction land area has increased significantly, while the areas of waterbodies and natural wetlands have decreased significantly. The farmland area has decreased slightly, while the areas of forest land and artificial wetlands remained unchanged. Figure 8a shows that from 2000 to 2010, the construction land area increased rapidly, and the natural wetlands area showed a downward trend, mainly converted into artificial wetlands. Forests increased slightly, mainly from cropland. Figure 8b shows that from 2010 to 2020, the area of construction land increased, mainly at the expense of constructed wetlands. The water area decreased, mainly being transferred to artificial wetlands, and the forest land area was reduced due to the conversion of a small amount of forest land into farmland. Figure 8c illustrates the main transformation of LULC types from 2000 to 2020, in which the construction land area is increasing, mainly from farmland, constructed wetlands, waterbodies, and natural wetlands. The areas of waterbodies and natural wetlands are decreasing. By contrast, the overall change in farmland, forest land, and artificial wetland areas is small.

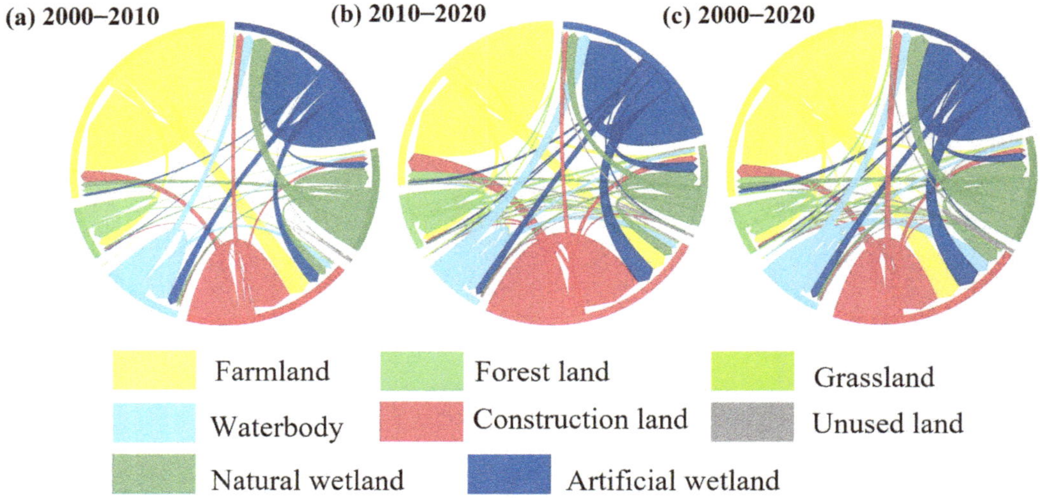

Figure 8. LULC type transfer of the Circum-Bohai coastal zone: (**a**) from 2000 to 2010; (**b**) from 2010 to 2020; (**c**) from 2000 to 2020.

4.2.4. Land Use Degree Index

The study area's comprehensive land use degree index was calculated based on the LULC classification results and used to classify different land types. The LUD spatial distribution map of the study area is displayed in Figure 9, utilizing a 1 km × 1 km grid for display. The findings reveal that LUD exhibits significant spatial differentiation, with the Bohai Bay area, which is significantly affected by human activities and economic development, having a much higher LUD than the Liaodong Bay and Laizhou Bay areas. This is likely due to the relatively developed urban economy in the Bohai Bay region, with construction land and artificial wetland being the dominant land types classified at a higher level of land use. In contrast, the Liaohe River Delta, a, and the Yellow River Delta, c, in Liaodong Bay and Laizhou Bay are dominated by natural wetlands, which experience fewer human disturbance activities leading to lower LUD. Additionally, there was an increasing trend observed in the degree of land use in coastal areas from 2000 to 2020, especially in region b in Figure 9, where LUD increased significantly.

4.3. *Analysis of LUCC Driving Factors*

Factor detection and interactive detection methods are commonly used to explore the impact of influencing factors on the spatial differentiation of LUD. Table 5 presents the p-values for the factor detection results from 2000 to 2020, which are all 0, indicating that the selected detection factors have a significant impact on the spatial heterogeneity of LUD. The q value in Table 5 represents the degree of spatial differentiation of the influencing factors on LUD in the study area. A higher value indicates a greater influence of the factor, or a stronger explanatory power of spatial differentiation.

As shown in Table 5, the added value of the secondary industry, urbanization rate, and population density between 2000 and 2020 are the key factors affecting LUD. However, over time, the q values of most influencing factors showed a downward trend, indicating that the driving effect of these factors gradually weakened.

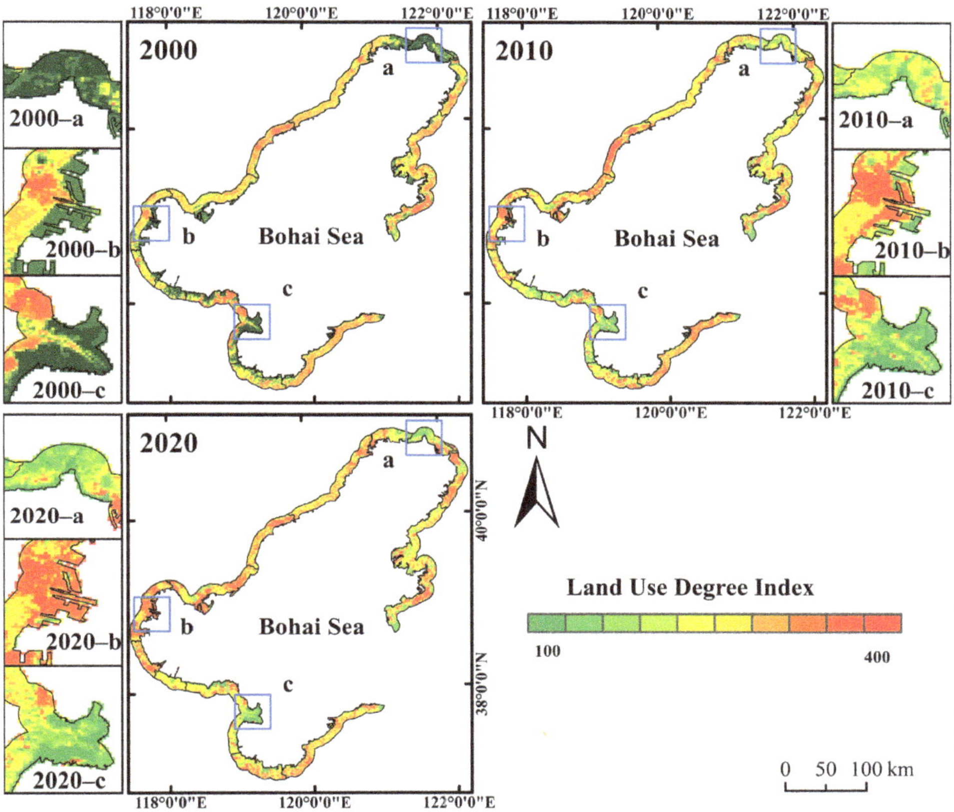

Figure 9. Spatial distribution of land use degree index in Circum-Bohai coastal zone from 2000 to 2020: (**a**) Liaohe River Delta; (**b**) Tianjin Port; (**c**) Yellow River Delta.

Table 5. Factor detection results.

Year		X1	X2	X3	X4	X5	X6	X7	X8	X9	X10	X11	X12
2000	q	0.12	0.13	0.08	0.18	0.08	0.19	0.06	0.06	0.11	0.04	0.01	0.04
	p	0	0	0	0	0	0	0	0	0	0	0	0
2010	q	0.09	0.09	0.10	0.07	0.11	0.03	0.12	0.05	0.01	0.04	0.09	0.09
	p	0	0	0	0	0	0	0	0	0	0	0	0
2020	q	0.06	0.06	0.10	0.07	0.09	0.05	0.05	0.05	0.01	0.05	0.06	0.06
	p	0	0	0	0	0	0	0	0	0	0	0	0

Unlike factor detection analysis, which examines the impact of single influencing factors on LUD in the study area, interactive detection investigates the effects of different influencing factors' interactions. As shown in Figure 10, the interaction between different influencing factors is more effective in explaining the spatial differences of LUD than a single factor, indicating that multiple factors affect the spatial heterogeneity of LUD in the study area. From 2000 to 2020, the interaction between GDP and the added value of the secondary industry, GDP and urbanization rate, GDP, and annual average precipitation effectively explained the spatial differentiation of land use in the region. The spatial differentiation characteristics of LULC in the Circum-Bohai region are mainly affected by the interaction of GDP, secondary industry added value, urbanization rate, and annual average precipitation. Although the average annual precipitation has a poor interpretation effect as a single factor, its impact on the spatial difference of LULC has been effectively demonstrated under the combined effects of GDP.

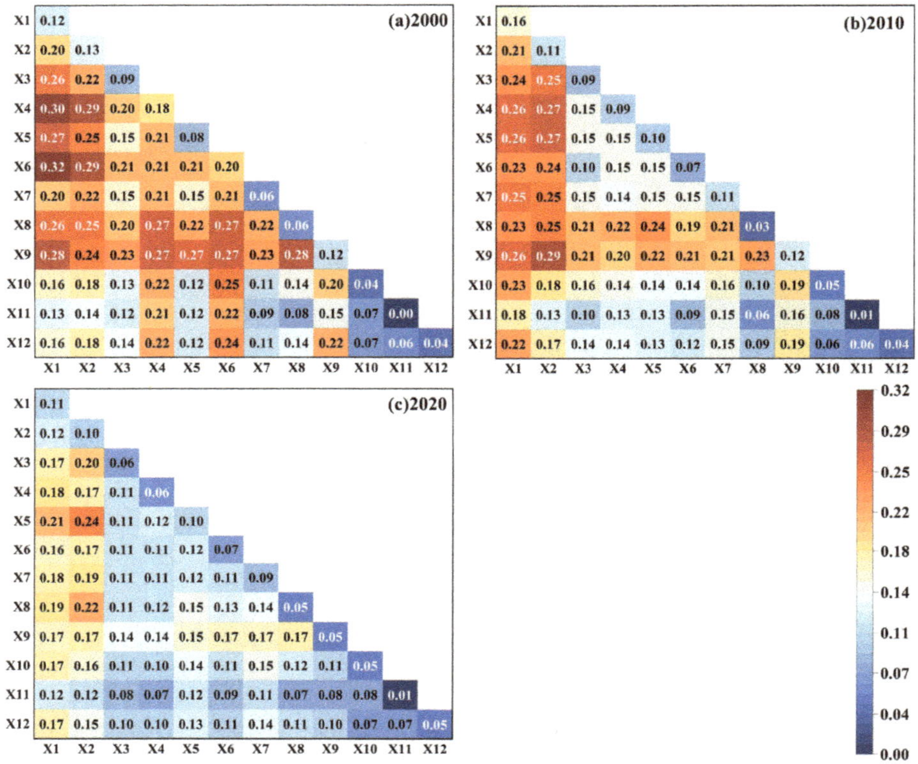

Figure 10. Interactive detection results.

5. Discussion

This study aimed to validate the effectiveness of several long-term and large-scale coastline extraction methods, such as NDWI, TCT, MNDWI, and LSWI. The results showed that the LSWI method is more effective in handling large differences between the TCT and MNDWI methods in certain areas. For instance, the TCT and MNDWI methods exhibited significant differences in artificial land, including ports, reclamation, and aquaculture, as well as the Yellow River Delta. By combining the gray values of the two exponential methods and using the OTSU method to reclassify marine land, these differences were effectively corrected, minimizing the likelihood of errors associated with using a single method. By fusing the regions with minor differences between these two methods, the classification achieved a clearer and more precise separation effect. As for morphological changes in the coastline position, the Dalian section, Bohai Bay area, and Laizhou Bay area in the Liaodong Bay region exhibited the most significant changes. This can be attributed to the deepening of China's development and utilization of marine resources, which led to the implementation of multiple coastal economic zones at the Circum-Bohai region in the early 2000s. The rate of coastline change was relatively slow from 2000 to 2005, significantly increased from 2005 to 2015, and decreased from 2015 to 2020. These changes can be attributed to China's successive policies aimed at constraining development and strengthening the protection of the Bohai Sea [55].

In the analysis of LUCC over the past two decades in the Circum-Bohai coastal zone, changes in the coastline from 2000 to 2020 were combined with a land-side buffer along the Circum-Bohai coastline. The results of LUCC were obtained through the employment of the RF classification algorithm, providing a more accurate and intuitive understanding of the region's LUCC. The coastal LUCC in the three-bay area was found to be significant, consistent with the observed changes in the coastline. This finding highlights the potential impact of coastline changes on the ecological benefits of the Circum-Bohai region and the adjacent coast's land use. Over time, there has been a gradual deepening of land use, in line with urban expansion and economic development. This trend is likely due to the increasing demand for land utilization and integration, leading to artificial breeding areas encroaching on wetland environments, and construction land encroaching on cultivated land and forest land [55].

In general, the coastal conditions of the Circum-Bohai region underwent significant changes between 2000 and 2020 due to human activities. Our findings revealed that urban expansion, artificial farming, and reclamation had not only impacted coastline changes but also influenced LUCC in the region. In other words, there is a correlation between LUCC and coastline change. Over the past two decades, changes in the location and morphology of the coastline and LUCC in the coastal zone of the Circum-Bohai region displayed pronounced spatial and temporal variations [28]. These variations were closely linked to the speed and level of regional development as well as local policies on marine development.

In regard to the LULC classification results, the OA and kappa obtained through the RF algorithm exceed the threshold of 85% recommended by the United States Geological Survey [23], with both metrics surpassing 86%. However, some errors were observed during the manual interpretation of certain areas. These errors may be attributed to the influence of image resolution and quality, as the spectral characteristics of water bodies and artificial aquaculture in some regions are similar, leading to the misclassification of these two categories. Manual visual interpretation and a comparison of various accuracy results indicated that the highest producer and user accuracies were achieved for forest land and construction land classification, suggesting that these two categories have more distinct spectral characteristics and are easier to classify.

During the computation process using GEE, computational limitations were encountered due to the large spatial span of the Circum-Bohai region and the significant number of required images for the experiment. Parallel experiments were necessary to relieve the computational pressure associated with issues such as calculation timeout, memory limit exceeding, and output errors. Future research should consider the use of higher-resolution remote sensing images, such as Sentinel-2 image data with a 10 m resolution, for LULC information extraction. Moreover, richer factors influencing LUCC should be employed for a more careful and comprehensive analysis of the driving factors impacting the coastal zone's LUCC situation.

6. Conclusions

This study aimed to explore the spatial and temporal evolution of coastline change and coastal zone LULC in the Circum-Bohai region of China from 2000 to 2020 and its driving forces. To achieve this, we first utilized the Landsat TM/OLI series data integrated in the GEE platform and employed the LSWI and the OTSU algorithm to extract the long time-series coastline of the Circum-Bohai. Furthermore, we analyzed and discussed the driving factors behind the LUCC in the study area using the geographic detector. The main findings can be summarized as follows:

(1) The length of the Circum-Bohai coastline underwent fluctuating changes from 2000 to 2020, displaying a phenomenon of initial increase followed by a decrease. The temporal changes of the coastline exhibited a slow–fast–slow trend, corresponding to periods of 2000 to 2005, 2005 to 2015, and 2015 to 2020. Spatially, there was a trend of land-to-sea expansion, with the change areas predominantly located in the three

major bay areas of Laizhou Bay, Bohai Bay, and Liaodong Bay. This change can be attributed to human activities as the main driving force.

(2) From 2000 to 2020, the types of LULC area are mainly farmland and construction land, followed by artificial wetlands and natural wetlands. The proportion of farmland and artificial wetland has remained unchanged, while the area of construction land has increased, and the natural wetlands and waterbodies have degraded. In terms of land type conversion, farmland and artificial wetland are primarily converted into construction land, while natural wetland is mainly converted into artificial wetland. From the perspective of land development, the utilization degree of Bohai Bay, Liaodong Bay, and Laizhou Bay is obviously higher than that of other regions, and it has gradually deepened over time.

(3) From 2000 to 2020, the added value of the secondary industry, urbanization rate, and population density were the single driving factors that significantly impacted the LUD in the study area. The interaction between GDP and the added value of the secondary industry has a significant impact on LULC change in the Circum-Bohai coastal zone.

The findings of this study would provide valuable data to support the management of land resources in the Circum-Bohai region and would assist relevant governments in formulating corresponding measures to achieve land and sea integration. Moreover, they would promote the scientific development of land and sea resources and improve the local eco-environment in the Circum-Bohai coastline area.

Author Contributions: Conceptualization, J.C. and W.J.; methodology, J.C.; software, W.J.; validation, J.C., M.Z. and P.W.; formal analysis, J.C.; investigation, M.Z.; resources, Y.L.; data curation, P.W.; writing—original draft preparation, W.J.; writing—review and editing, J.C.; visualization, W.J.; supervision, Y.L.; project administration, J.C.; funding acquisition, Y.L. All authors have read and agreed to the published version of the manuscript.

Funding: This research was jointly supported by the MOE Layout Foundation of Humanities and Social Sciences, grant number 17YJAZH013; the National Natural Science Foundation of China, grant number 42201077 and 42177453; and the Natural Science Foundation of Shandong Province, grant number ZR2021QD074.

Informed Consent Statement: Not applicable.

Data Availability Statement: Not applicable.

Conflicts of Interest: The authors declare no conflict of interest.

References

1. Uddin, K.; Khanal, N.; Chaudhary, S.; Maharjan, S.; Thapa, R.B. Coastal morphological changes: Assessing long-term ecological transformations across the northern Bay of Bengal. *Environ. Chall.* **2020**, *1*, 100001. [CrossRef]
2. Liu, Y.; Li, B.; Wang, Y.; Chen, L.; Li, X.; Hou, X. Evaluation of ecological connectivity in the coastal zone of Laizhou Bay-Yellow River Delta based on ecosystem service value. *Acta Ecol. Sin.* **2019**, *39*, 7514–7524. [CrossRef]
3. Cui, J.; Zhu, M.; Mi, H.; Liu, Y. Evaluation of Eco-Environmental Quality and Analysis on Spatio-Temporal Variation in Jinan, China. *Pol. J. Environ. Stud.* **2022**, *31*, 1061–1072. [CrossRef] [PubMed]
4. Yang, X.; Meng, F.; Fu, P.; Zhang, J.; Liu, Y. Instability of remote sensing ecological index and its optimisation for time frequency and scale. *Ecol. Inform.* **2022**, *72*, 101870. [CrossRef]
5. Kang, Y.; Meng, Q.; Liu, M.; Zou, Y.; Wang, X. Crop classification based on red edge features analysis of GF-6 WFV data. *Sensors* **2021**, *21*, 4328. [CrossRef]
6. Wu, Y.; Liu, Z. Research progress on methods of automatic coastline extraction based on remote sensing images. *Natl. Remote Sens. Bull.* **2019**, *23*, 582–602. [CrossRef]
7. Gens, R. Remote sensing of coastlines: Detection, extraction and monitoring. *Int. J. Remote Sens.* **2010**, *31*, 1819–1836. [CrossRef]
8. McFeeters, S.K. The use of the Normalized Difference Water Index (NDWI) in the delineation of open water features. *Int. J. Remote Sens.* **1996**, *17*, 1425–1432. [CrossRef]
9. Kauth, R.J.; Thomas, G. The tasselled cap–A graphic description of the spectral-temporal development of agricultural crops as seen by Landsat. In Proceedings of the LARS Symposia, West Lafayette, IN, USA, 29 June–1 July 1976; p. 159.
10. Feyisa, G.L.; Meilby, H.; Fensholt, R.; Proud, S.R. Automated Water Extraction Index: A new technique for surface water mapping using Landsat imagery. *Remote Sens. Environ.* **2014**, *140*, 23–35. [CrossRef]

11. Hossain, M.S.; Yasir, M.; Wang, P.; Ullah, S.; Jahan, M.; Hui, S.; Zhao, Z. Automatic shoreline extraction and change detection: A study on the southeast coast of Bangladesh. *Mar. Geol.* **2021**, *441*, 106628. [CrossRef]
12. Bamdadinejad, M.; Ketabdari, M.J.; Chavooshi, S.M.H. Shoreline extraction using image processing of satellite imageries. *J. Indian Soc. Remote Sens.* **2021**, *49*, 2365–2375. [CrossRef]
13. Xu, H. Modification of normalised difference water index (NDWI) to enhance open water features in remotely sensed imagery. *Int. J. Remote Sens.* **2006**, *27*, 3025–3033. [CrossRef]
14. Zollini, S.; Alicandro, M.; Cuevas-González, M.; Baiocchi, V.; Dominici, D.; Buscema, P.M. Shoreline extraction based on an active connection matrix (ACM) image enhancement strategy. *J. Mar. Sci. Eng.* **2019**, *8*, 9. [CrossRef]
15. Niculescu, S.; Bengoufa, S.; Mihoubi, M.; Belkessa, R.; Abbad, K. Rocky shoreline extraction using a deep learning model and object-based image analysis. *Remote Sens. Spat. Inf. Sci.* **2021**, *43*, 23–29. [CrossRef]
16. Fisher, A.; Flood, N.; Danaher, T. Comparing Landsat water index methods for automated water classification in eastern Australia. *Remote Sens. Environ.* **2016**, *175*, 167–182. [CrossRef]
17. Otsu, N. A threshold selection method from gray-level histograms. *IEEE Trans. Syst. Man Cybern.* **1979**, *9*, 62–66. [CrossRef]
18. Wang, E.; Hu, S.; Han, H.; Liu, C. Research on ice concentration in Heilong River based on the UAV low-altitude remote sensing and OTSU algorithm. *J. Hydraul. Eng.* **2022**, *53*, 68–77. [CrossRef]
19. Qu, L.A.; Chen, Z.; Li, M.; Zhi, J.; Wang, H. Accuracy improvements to pixel-based and object-based lulc classification with auxiliary datasets from Google Earth engine. *Remote Sens.* **2021**, *13*, 453. [CrossRef]
20. Wang, M.; Zhang, H.; Zhang, Y.; Lin, M.; Gong, P. Evolution analysis of land use pattern in the Changjiang River Delta coastal zone in recent 39 years. *Haiyang Xuebao* **2020**, *42*, 142–154. [CrossRef]
21. Hou, W.; Hou, X.; Sun, M.; Song, B. Land use/land cover change along low-middle latitude coastal areas of Eurasia and their driving forces from 2000 to 2010. *World Reg. Stud.* **2021**, *30*, 813–825. [CrossRef]
22. Breiman, L. Random forests. *Mach. Learn.* **2001**, *45*, 5–32. [CrossRef]
23. Liu, C.; Li, W.; Zhu, G.; Zhou, H.; Yan, H.; Xue, P. Land use/land cover changes and their driving factors in the Northeastern Tibetan Plateau based on Geographical Detectors and Google Earth Engine: A case study in Gannan Prefecture. *Remote Sens.* **2020**, *12*, 3139. [CrossRef]
24. Ma, Y.; Jiang, Q.; Meng, Z.; Li, Y.; Wang, D.; Liu, H. Classification of Land Use in Farming Area Based on Random Forest Algorithm. *Trans. Chin. Soc. Agric. Mach.* **2016**, *47*, 297–303. [CrossRef]
25. Zhang, H.; Qin, N.; Wang, J.; Li, M.; Yang, A.x.; Lu, Q. Spatiotemporal Pattern and Driving Factors of Land Use in Guangxi Coastal Zone. *Res. Soil Water Conserv.* **2022**, *29*, 367–374. [CrossRef]
26. Han, Y.; Zhu, J.; Wei, D.; Wang, F. Spatial-Temporal Effect of Sea–Land Gradient on Land Use Change in Coastal Zone: A Case Study of Dalian City. *Land* **2022**, *11*, 1302. [CrossRef]
27. Cui, J.; Zhu, M.; Liang, Y.; Qin, G.; Li, J.; Liu, Y. Land use/land cover change and their driving factors in the Yellow River Basin of Shandong Province based on google earth Engine from 2000 to 2020. *ISPRS Int. J. Geo-Inf.* **2022**, *11*, 163. [CrossRef]
28. Tang, S.; Song, L.; Wan, S.; Wang, Y.; Jiang, Y.; Liao, J. Long-Time-Series Evolution and Ecological Effects of Coastline Length in Coastal Zone: A Case Study of the Circum-Bohai Coastal Zone, China. *Land* **2022**, *11*, 1291. [CrossRef]
29. Nie, T.; Dong, G.; Jiang, X.; Lei, Y. Spatio-temporal changes and driving forces of vegetation coverage on the loess plateau of Northern Shaanxi. *Remote Sens.* **2021**, *13*, 613. [CrossRef]
30. Zhang, Z.; Hu, B.; Qiu, H.; Deng, Y. Spatio-temporal Differentiation and Driving Mechanism of Ecological Environment Vulnerability in Southwest Guangxi Karst-Beibu Gulf Coastal Zone. *J. Geo-Inf. Sci.* **2021**, *23*, 456–466. [CrossRef]
31. Lou, P.; Fu, B.; Lin, X.; Bi, L.; Ma, R.; Tang, T. Influence of Land Use Change on Ecosystem Service Value Based on GEE in the Beijing-Tianjin-Hebei Region from 1998 to 2018. *Environ. Sci.* **2019**, *40*, 5473–5483. [CrossRef]
32. Dong, J.; Xiao, X.; Menarguez, M.A.; Zhang, G.; Qin, Y.; Thau, D.; Biradar, C.; Moore, B., III. Mapping paddy rice planting area in northeastern Asia with Landsat 8 images, phenology-based algorithm and Google Earth Engine. *Remote Sens. Environ.* **2016**, *185*, 142–154. [CrossRef] [PubMed]
33. Xia, H.; Zhao, J.; Qin, Y.; Yang, J.; Cui, Y.; Song, H.; Ma, L.; Jin, N.; Meng, Q. Changes in water surface area during 1989–2017 in the Huai River Basin using Landsat data and Google earth engine. *Remote Sens.* **2019**, *11*, 1824. [CrossRef]
34. Yang, X.; Meng, F.; Fu, P.; Zhang, Y.; Liu, Y. Spatiotemporal change and driving factors of the Eco-Environment quality in the Yangtze River Basin from 2001 to 2019. *Ecol. Indic.* **2021**, *131*, 108214. [CrossRef]
35. Yang, X.; Meng, F.; Fu, P.; Wang, Y.; Liu, Y. Time-frequency optimization of RSEI: A case study of Yangtze River Basin. *Ecol. Indic.* **2022**, *141*, 109080. [CrossRef]
36. Dai, S.; Yi, X.; Luo, H.; Li, H.; Li, M.; Zheng, Q.; Hu, Y. Mapping Land Use in Hainan Island Based on Google Earth Engine and Landsat Time Series Data. *Chin. J. Trop. Crops* **2021**, *42*, 3351–3357. [CrossRef]
37. Yu, L.; Sun, L.; Zhang, D.; Liu, S.; Xie, Z.; Ping, X. Extraction of land-cover and wetland area in Bohai Rim region based on Google Earth Engine. *Chin. J. Appl. Ecol.* **2020**, *31*, 4091–4098. [CrossRef]
38. Jiang, T.; Wu, G.; Niu, P.; Cui, Z.; Bian, X.; Xie, Y.; Shi, H.; Xu, X.; Qu, K. Short-term changes in algal blooms and phytoplankton community after the passage of Super Typhoon Lekima in a temperate and inner sea (Bohai Sea) in China. *Ecotoxicol. Environ. Saf.* **2022**, *232*, 113223. [CrossRef] [PubMed]
39. Wei, F.; Han, G.; Han, M.; Zhang, J.; Li, Y.; Zhao, J. Temporal-spatial Dynamic Evolution and Mechanism of Shoreline and the Sea Reclamation in the Bohai Rim During 1980–2017. *Sci. Geogr. Sin.* **2019**, *39*, 997–1007. [CrossRef]

40. Sun, B.; Zuo, S.; Xie, H.; Li, H.; Yang, Z. Analysis of impact effects and changes of the coastline in the Bohai Bay during the past 40 years. *J. East China Norm. Univ.* **2017**, *4*, 139–148. [CrossRef]
41. Zhang, Y.; Liu, Y.; Zhang, X.; Huang, H.; Qin, K.; Bai, Z.; Zhou, X. Correlation Analysis between Land-Use/Cover Change and Coastal Subsidence in the Yellow River Delta, China: Reviewing the Past and Prospecting the Future. *Remote Sens.* **2021**, *13*, 4563. [CrossRef]
42. Liu, Y.; Hou, X.; Li, X.; Song, B.; Wang, C. Assessing and predicting changes in ecosystem service values based on land use/cover change in the Bohai Rim coastal zone. *Ecol. Indic.* **2020**, *111*, 106004. [CrossRef]
43. Zhang, D.-D.; Zhang, L. Land cover change in the central region of the lower Yangtze River based on Landsat imagery and the Google Earth Engine: A case study in Nanjing, China. *Sensors* **2020**, *20*, 2091. [CrossRef] [PubMed]
44. Chen, C.; Chen, H.; Liang, J.; Huang, W.; Xu, W.; Li, B.; Wang, J. Extraction of water body information from remote sensing imagery while considering greenness and wetness based on Tasseled Cap transformation. *Remote Sens.* **2022**, *14*, 3001. [CrossRef]
45. Song, Y.; Wang, Y.; Du, B.; Wang, J.; Dong, X. Detection of Overlapped Bullet Holes Based on Improved Otsu's Thresholding Method. *Acta Armamentarii* **2022**, *43*, 924. [CrossRef]
46. Wan, J.; Lin, S.; Mei, T.; Lin, Z.; Guo, T. Image Enhancement Algorithm of Electrowetting Display Based on Image Segmentation and Dynamic Histogram Equalization. *Acta Photonica Sin.* **2022**, *51*, 0210005. [CrossRef]
47. Wu, R.; Hu, X.; Mei, H.; He, J.; Yang, J. Spatial Susceptibility Assessment of Landslides Based on Random Forest: A Case Study from Hubei Section in the Three Gorges Reservoir Area. *Earth Sci. J. China Univ. Geosci.* **2021**, *46*, 321–330. [CrossRef]
48. Zhu, Y.; Zeng, Y.; Zhang, M. Extract of land use/cover information based on HJ satellites data and object-oriented classification. *Trans. Chin. Soc. Agric. Eng. (Trans. CSAE)* **2017**, *33*, 258–265. [CrossRef]
49. Wang, N.; Li, Q.; Du, X.; Zhang, Y.; Zhao, L.; Wang, H. Identification of main crops based on the univariate feature selection in Subei. *Natl. Remote Sens. Bull.* **2017**, *21*, 519–530. [CrossRef]
50. Pan, L.; Xia, H.; Yang, J.; Niu, W.; Wang, R.; Song, H.; Guo, Y.; Qin, Y. Mapping cropping intensity in Huaihe basin using phenology algorithm, all Sentinel-2 and Landsat images in Google Earth Engine. *Int. J. Appl. Earth Obs. Geoinf.* **2021**, *102*, 102376. [CrossRef]
51. Zhuang, D.; Liu, J. Study on the model of regional differentiation of land use degree in China. *J. Nat. Resour.* **1997**, *12*, 105–111.
52. Liu, H.; Zheng, M.; Liu, J.; Zheng, X. Sustainable land use in the trans-provincial marginal areas in China. *Resour. Conserv. Recycl.* **2020**, *157*, 104783. [CrossRef]
53. Wang, J.; Xu, C. Geodetector: Principle and prospective. *Acta Geogr. Sin.* **2017**, *72*, 116–134. [CrossRef]
54. Xu, J.; Liu, S.; Zhao, S.; Wu, X.; Hou, X.; An, Y.; Shen, Z. Spatiotemporal dynamics of water yield service and its response to urbanisation in the Beiyun river Basin, Beijing. *Sustainability* **2019**, *11*, 4361. [CrossRef]
55. Hou, X.; Zhang, H.; Li, D.; Hou, W.; Song, Y. Development trend, environmental and ecological impacts, and policy recommendations for Bohai Sea reclamation. *Acta Ecol. Sin.* **2018**, *38*, 3311–3319. [CrossRef]

remote sensing

MDPI

Article

Combining LSTM and PLUS Models to Predict Future Urban Land Use and Land Cover Change: A Case in Dongying City, China

Xin Zhao [1], Ping Wang [2,*], Songhe Gao [3], Muhammad Yasir [4] and Qamar Ul Islam [5]

1 College of Resources, Shandong University of Science and Technology, Tai'an 271000, China
2 College of Geodesy and Geomatics, Shandong University of Science and Technology, Qingdao 266590, China
3 Beijing Yuhang Intelligent Technology Co., Ltd., Beijing 100000, China
4 College of Oceanography and Space Informatics, China University of Petroleum (East China), Qingdao 266580, China
5 Department of Electrical and Computer Engineering, College of Engineering, Dhofar University, Salalah, 211, Oman
* Correspondence: skd990058@sdust.edu.cn

Abstract: Land use is a process that turns a piece of land's natural ecosystem into an artificial one. The mix of plant and man-made covers on the Earth's surface is known as land cover. Land use is the primary external force behind change in land cover, and land cover has an impact on how land use is carried out, resulting in a synergistic interaction between the two at the Earth's surface. In China's Shandong Peninsula city cluster, Dongying is a significant coastal port city. It serves as the administrative hub for the Yellow River Delta and is situated in Shandong Province, China's northeast. The changes in its urban land use and land cover in the future are crucial to understanding. This research suggests a prediction approach that combines a patch-generation land use simulation (PLUS) model and long-term short-term memory (LSTM) deep learning algorithm to increase the accuracy of predictions of future land use and land cover. The effectiveness of the new method is demonstrated by the fact that the average inaccuracy of simulating any sort of land use in 2020 is around 5.34%. From 2020 to 2030, 361.41 km^2 of construction land is converted to cropland, and 424.11 km^2 of cropland is converted to water. The conversion areas between water and unused land and cropland are 211.47 km^2 and 148.42 km^2, respectively. The area of construction land and cropland will decrease by 8.38% and 3.64%, respectively, while the area of unused land, water, and grassland will increase by 5.53%, 2.44%, and 0.78%, respectively.

Keywords: LUCC; PLUS; LSTM; deep learning; multi-scenario simulation

1. Introduction

Land Use and Land Cover Change (LUCC) is an important connection between human activities and the natural environment. The rules of land use transformation indicate the interaction between the Earth's environment system on which human beings rely for survival and the increasingly developed production system [1,2]. Modeling research is a core and important component of land change research and is the main way and means to address land usage alteration and other related mechanisms. The model research can replicate and analyze more precisely the geographical and temporal changes in land utilization and land cover change, the pattern and its evolution process, and anticipate the future change trend of key areas [3,4]. There are several models for change simulation that are employed, including dynamic, statistical, and machine learning (ML) models, to suit the high performance needs of land use modeling [5]. In order to objectively track any changes in land use, remote sensing (RS) can capture and evaluate multi-temporal satellite imagery. Data can be stored, processed, and analyzed using Geographic Information

System (GIS), and some land use analysis and simulation models can be programmed directly into GIS software. In the majority of investigations, RS and GIS can be employed alone or together [6–9]. The main models used for early land-use change prediction are the Markov chain model (MC) [10], conversion of land use and its effects (CLUE) [11], cellular automata (CA) model [12], slope, land use, exclusion, urban extension, transportation and hill shade model (SLEUTH) [13], logistic regression (LR) modeling [14], CA–Markov model [15], land transformation model (LTM) [16], land change modeler (LCM) [17], future land use simulation model (FLUS) [18]. Moreover, artificial neural networks and dynamic models are coupled to anticipate land use change [19] and weight of evidence (WOE) is utilized to quantitatively evaluate the driving force of land change [20,21]. To simulate and forecast future land use change, support vector machines are employed to build probability maps [22], among others [23–34].

The patch-generating land use simulation (PLUS) model, which combines a CA framework constructed using multiple types of irregular patch seeds with a land enlargement analytical technique, was initially envisioned in 2021 [35]. This model is more accurate in its simulations and is more in line with the landscape pattern index of the actual terrain when contrasted with prior CA frameworks [35]. To date, many scholars have made various applications based on the PLUS model. Lu et al. [36] used the PLUS model to mimic the geospatial structure of the lower Yellow River at different stages of future development. Lin et al. [37] used the Fuxian Lake basin as an example to incorporate the constrained territory of LUCC and investigate the optimum model for replicating LUCC in this basin utilizing CA–Markov, FLUS, and models. Bao et al. [38] predicted quantitative and spatial variation by coupling MC and analysis of different scenarios. The PLUS-UGB model was developed by Yang Wenjie et al. [39] based on their analysis of the spatial distribution aspects of alterations in land use in Guangzhou. Gao et al. [40] employed the PLUS model to recreate the land utilization data for Nanjing in 2025 under four scenarios and assess changes in various categories of land and land transfers under various scenarios. The numerical composition and spatial organization of production–living–ecological land were refined by Li et al. [41] using a genetic algorithm (GA). To evaluate the models' accuracy, Wang et al. [42] examined PLUS, FLUS, and MC predictions, and the PLUS model was found to be the most efficient for the 2030 terrestrial utilization study.

There are more studies related to land use simulation based on PLUS; however, the methods are mostly based on model coupling and multi-scenario design, without involving deep learning methods. This work enhances the PLUS model by predicting land utilization desire using the long short-term memory (LSTM) technique, taking into account that shifting land use is a complicated process and the trend of future land use is not stable. A type of neural network used to process sequence data is an LSTM network in deep learning models [43]. When faced with extended sequences of data, Recurrent Neural Network (RNN) can only acquire information from the next episode and lacks cognitive performance for the older period, which results in information loss [44]. LSTM is applied to the anticipated demand for land use in this research area and has made some research achievements in land use-related aspects. Xu et al. [45] used MLP and LSTM artificial neural networks as simulation models for comparison. Due to its capacity to capture more time information, LSTM outperforms MLP in modeling the dynamics of urban growth. Boulila et al. [46] realized urban expansion prediction by using the CNN-LSTM network. Wang et al. [47] developed a two-way short-term memory network-based long-time series categorization data gathering model for land use and cover. Mohanrajan et al. [48] developed a bidirectional long- and short-term memory model based on a novel visual transformer for predicting land use/land cover shifts for forested and non-forested areas in the Java Mountains, India.

The PLUS model for land use simulation requires obtaining the land development probability in the land use expansion analysis strategy and setting various parameters such as predicting land use demand in the CA model based on multiple types of random patch seeds to obtain simulation results. The quantity of land use demand has an important impact on the results of simulation forecasting; therefore, choosing the appropriate land use demand forecasting algorithm is a key step. Using conventional land use demand forecasting algorithms, such as Markov chain, linear regression method, etc., the simulation results are not accurate. Therefore, this paper introduces LSTM, a very popular model in the field of deep learning, to predict the quantity of land use demand based on time series data, which improves the simulation accuracy of the PLUS model.

Dongying is named one of the first international wetland cities, an essential element of the development plan of the Yellow River Delta Efficient Ecological Economic Zone, and the only city where the two national strategies of Yellow and Blue overlap and integrate. Understanding the regional variations in land use is vital so that sustainable development measures can be sought on this basis and ensure the coordinated growth of the economy and the environment. There may be several possibilities for the development of the future pattern of land utilization in Dongying, depending on the regional development strategy.

Taking Dongying City as the research area, this paper formulates a multi-scenario simulation scheme for land use based on the PLUS model and LSTM network to study land use evolution, which provides a variety of decision-making ideas for land resources management and helps to realize the optimal allocation of land resources. The objectives of this study include: (1) LSTM algorithm based on PyTorch deep learning framework to achieve land use demand forecasting, and (2) simulation of future land use in Dongying city under two different scenarios of natural development and ecological development priority.

2. Study Area

The capital of the Yellow River Delta in China is Dongying, a prefecture-level city in Shandong Province. The study area includes two counties (Lijin and Guangrao) and three districts (Hekou, Kenli, and Dongying) under the jurisdiction of Dongying. It lies between 36°55′ to 38°10′N and 118°07′ to 119°10′E in northern Shandong Province. It is adjacent to the Bohai Sea in the east and north, Binzhou in the west, and Zibo and Weifang in the south. The maximum vertical distance from north to south is 123 km, the maximum horizontal distance from east to west is 74 km, and the total land area is 8243 km^2 (see Figure 1).

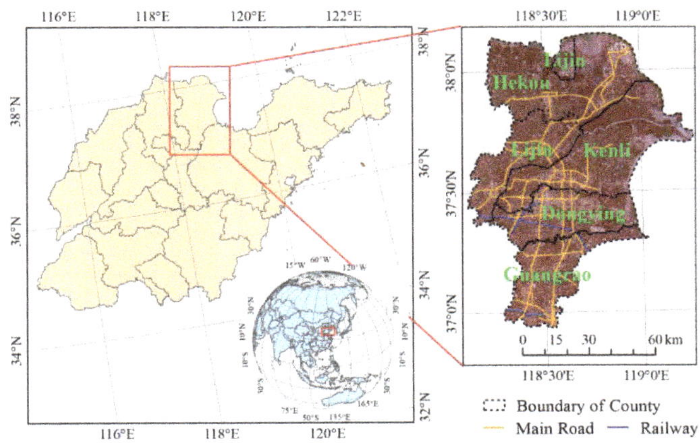

Figure 1. Location of the study area.

3. Material and Methods

3.1. Data Acquisition

China Land Cover Dataset (CLCD) is the first yearly dataset of China's land cover created by Wuhan University using the Google Earth Engine (GEE) infrastructure. The accuracy rate of the CLCD dataset was 79.31%, which was greater than the mean accuracy of the MCD12Q1 and ESACCI LC FROM the GLC and GlobalLand30 datasets [49].

Five sub-regions in the study area based on ArcGIS were randomly chosen for validation, with roughly 6000 sampling points in each area. The overall correct rate could essentially reach about 76%, and sampling results were compared with outcomes of the second national land survey to conclude that CLCD could essentially meet the accuracy requirements of the study area. The projection coordinate system was defined by ArcGIS as WGS_1984_UTM_Zone_50N, and the coordinate system for projection was Universal Transverse Mercartor Grid System (UTM) projection with 6-degree division, and the data of each year were obtained by using vector data masks. On this basis, the land use data of the study area were reclassified into cropland, grassland, water, unused land, and construction land according to the land use types required for this study. Cropland island that is directly or indirectly used for agricultural production. Grassland is the land where herbs and shrubs grow mainly and is suitable for livestock production. Water is rivers, wetlands, etc. Unused land refers to land other than cropland and construction land, mainly including barren land, saline land, swampy land, sandy land, bare land, bare rock, etc. Construction land refers to the land for building and structures, which is land for urban and rural housing and public facilities, industrial and mining land, land for energy, transportation, water conservancy, communication and other infrastructure, tourism land, military land, etc.

The kind of land use changed significantly between 2000 and 2020. (see Table 1 and Figure 2). Cropland in the research region has been gradually shrinking from 5295.27 km^2 to 4423.94 km^2 between 2000 and 2020. The amount of grassland and unused land has decreased at a rate equal to that of cropland, from 1.1% to 0.1% for grassland and from 12.1% to 2.5% for unused land, respectively, a decline of 77.39 km^2 and 712.51 km^2. Contrarily, the area of water and construction land is growing year after year, with the water growing from 282.37 km^2 to 941.94 km^2 or 3.8% to 12.7% of the total, and the construction land growing the highest, from 876.37 km^2 to 1878.03 km^2, or 11.8% to 12.7% of the total. The percentage grew between 11.8% and 25.2%. A total of 1073.93 km^2 and 735.18 km^2 of land were be converted into construction land and water, respectively, from 2000 to 2020. Of these, 678.65 km^2 and 350.76 km^2 of cropland and unused land, respectively, were be converted into construction land, making up 95.9% of the total area converted into construction land. The area of cropland and unused land converted to water is 347.81 km^2 and 304.91 km^2, respectively, accounting for 88.8% of the total area converted to water.

Table 1. Transfer matrix from 2000 to 2020 (km^2).

Land Type	Cropland	Grassland	Water	Unused Land	Construction Land
Cropland	4290.78	4.17	304.91	16.76	678.65
Grassland	39.51	4.47	15.57	4.92	22.16
Water	46.93	0.05	206.77	6.23	22.36
Unused land	45.11	0.53	347.81	153.98	350.76
Construction Land	1.59	0.01	66.89	3.79	804.10

The geographic coordinate system GCS Krasovsky 1940 (Krasovsky 1940 ellipsoid) and the projection Krasovsky 1940 Albers were used to create the vector data for Dongying, which were collected from the Resource and Environment Data Center of the Chinese Academy of Sciences (https://www.resdc.cn/, accessed on 1 September 2022). By specifying the projection with ArcGIS tools, the vector data were defined as the projection coordinate system WGS 1984 UTM_ Zone 50N.

Alteration in LUCC is influenced by several things. In general, the three main forces driving LUCC change are natural, socioeconomic, and accessible considerations. As indicated in Table 2, the PLUS model may investigate the various causes of land use change as well as the contribution of each influencing element to the growth and alteration of each category. In consideration of natural factors, 11 driving factors are selected: slope, aspect, soil erosion, digital elevation model (DEM), average annual temperature, annual accumulated temperature, annual precipitation, moisture index, aridity, net primary production (NPP), and soil organic carbon. These factors are based on various functions of land resources and are the most fundamental factors for driving force consideration. Many geographic regions have varied patterns of land use, and different elements of nature give diverse impetuses for land use change. This study region is a plain location with flat terrain, and the land is better suited for cultivating crops. This impacts how land is developed and used. The terrain and soil characteristics have a significant impact on the distribution of industrial land used for agriculture, forestry, animal husbandry, and fisheries since they can, to some extent, represent how easily different land resources can be used in the area. The degree of soil erosion and irrigation conditions both reflect the effects of the slope. Unreasonable water and soil resource use causes secondary salinization of the land, which raises groundwater levels and accelerates the destruction of the local biological environment. There is no virtuous cycle since soil erosion, frequent droughts, and flood disasters cause grassland, forest land, and water areas to alter often. Land use may also be influenced by DEM and other statistical indicators of terrain.

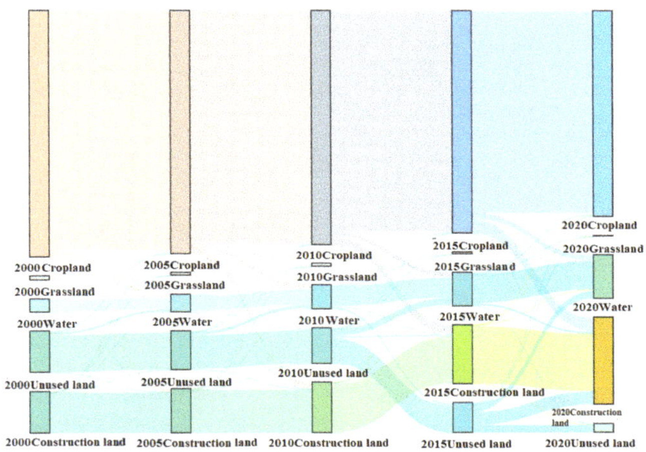

Figure 2. Dongying city 2000–2020 circulation map of various categories.

Gross domestic product (GDP) and population density are the major selection criteria when taking into account social and economic considerations. The interaction between people and society has an impact on how intensively land is used. Humans use land as a means of transportation for a variety of activities carried out in the social context. Social variables will have an impact on human behavior, and behavior, in turn, will have an impact on land use. At the end of 2000, the total population of Dongying was 1,721,300, of which 1,095,500 were involved in agriculture and 711,800 were involved in non-agricultural activities. In 2000–2020, the total population increased from 1,720,000 to 2,190,000, an increase of 27.3%, while the urban population increased from 710,000 to 1,560,000, an increase of 119.7%, a rapid increase, and the city began to enter a phase of rapid urbanization. The study area's urban population started to increase quickly, and the demand for better living conditions and environments encouraged the conversion of agricultural land to construction land, which resulted in a steady loss of agricultural land area. The need for food will rise in tandem with the increase in population density, which will cause unused land, grassland, and other land use types to be converted back into agricultural land. Socioeconomic variables show how human actions have affected the production economy. The rapid economic expansion caused by human output and social activity will further the growth of human activities.

Table 2. Drive factor data and sources.

Data Type	Data Name	Sources
Natural factors	Slope	Original acquisition
	Aspect	Original acquisition
	Soil erosion	http://www.resdc.cn
	DEM	https://earthdata.nasa.gov/
	Average annual temperature	http://www.resdc.cn
	Annual accumulated temperature	http://www.resdc.cn
	Annual precipitation	http://www.resdc.cn
	Moisture index	http://www.resdc.cn
	Aridity	http://www.resdc.cn
	NPP	https://lpdaac.usgs.gov/
	Soil organic carbon	https://data.isric.org/
Social and economic factors	population density	http://www.resdc.cn
	GDP	http://www.resdc.cn
Accessibility factors	Adjacent to highways	Euclidean distance
	Adjacent to railways	Euclidean distance
	Adjacent to water	Euclidean distance
	Adjacent to government	Euclidean distance
	Adjacent to roads I (Multilane road)	Euclidean distance
	Adjacent to roads II (Two-lane road)	Euclidean distance
	Adjacent to roads III (Mixed driving two-lane road)	Euclidean distance
	Adjacent to roads IV (Mixed two-lane or single-lane road)	Euclidean distance

Eight accessibility drivers—including proximity to roads I through IV, proximity to highways, trains, water, and government—were culled to assess the extent of their impact on land use change. The drive factor data must be combined with projection coordinate system UTM, WGS-84 geographic coordinate system, and 30 m resolution resampling. DEM estimated the slope and slope direction. Euclidean distances were calculated for impact factors such as major roads, railways, highways, water, and government distances using the spatial processing tools in ArcGIS. All driving factors were uniformly row-counted and column-counted and made the same as the land extension map.

3.2. Methodology

3.2.1. Research Framework

Figure 3 shows the workflow of this study. The flowchart is composed of three primary modules: first, the LEAS module. Utilizing a random forest to mine the effect of each driver on the growth of geographical units and produce a probability map of the development of each land use in space. This module explores the relationship between each land type and multiple drivers to uncover the conversion rules for each land type; therefore, the results of this module will indicate the development potential of each land type. Second, the CARS module. In order to simulate the evolution of multiple land use types, the CARS module incorporates a stochastic patch generation and a descending threshold mechanism to perform future site simulation based on the traditional CA model. This mechanism generates "seeds" for the development probability of each land use type when the neighborhood effect of the individual land use type equals zero. With the development probability constraint, PLUS will automatically generate the simulated patches. During the simulation, the land use demand will influence the local land use competition pattern through adaptive inertia mechanism, driving the amount of land use to meet the future demand. Using the 2010 data as the initial data, set important parameters such as spatial policy restricted area, land use requirement, transformation regulations, and neighborhood weights for the likelihood of development, and simulate the 2020 land use data. Third, model accuracy evaluation and correction module. Check the simulation results' correctness in 2020 and make any necessary PLUS model modifications.

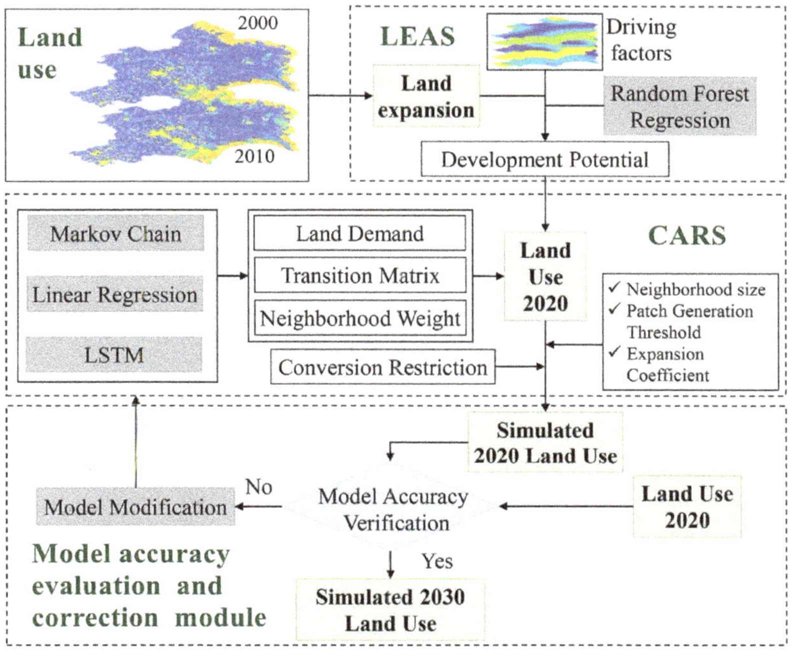

Figure 3. Land use simulation process.

In the CARS module, there are many methods for obtaining land use needs, including the conventional Markov chain and linear regression methods. Experiments have shown that using Markov chains and linear regression to predict the number of land uses in 2020 leads to poor accuracy with the PLUS prediction results compared to the actual land use in 2020. The LSTM deep learning network model, which excels in handling issues that are closely related to time series, is introduced to increase the precision of the PLUS simulation. The land use prediction in 2030 is simulated using the LSTM deep learning network model, which increases the precision of PLUS simulation.

3.2.2. Long Short-Term Memory

Recurrent neural network (RNN) in deep learning models is a class of neural networks dedicated to processing sequential data. RNN can be considered multiple replications of the same neural network, where each neural network module passes the message to the next one. Its input does not only take into account the current data input, but also the input of the previous moments, which together act as the outcome of the present instant, and the outcome of the present instant is not only output but also available for the next moment, which continues to participate in the next moment's output so that the network has a certain memory function. To solve the problem of "long-term dependence", LSTM cells were proposed to improve the memory capacity of standard recurrent cells by introducing "gates" in the cells [50]. Based on the Python environment, PyTorch is used as a deep learning framework to implement the LSTM algorithm. The model uses 5 layers of input and 5 layers of output, the length of the hidden layer output feature vector is 64, and finally, the accuracy is improved using the method of averaging three sets of LSTM results, and the structure diagram is shown in Figure 4. The most important component of the LSTM is the state cell $s_i^{(t)}$. Here, forget gates are used to manage the self-loop weights (or corresponding time constants) $f_i^{(t)}$, and the sigmoid cell sets the weights to values between 0 and 1.

$$f_i^{(t)} = \sigma\left(b_i^f + \sum_j U_{i,j}^f x_j^{(t)} + \sum_j W_{i,j}^f h_j^{(t-1)}\right) \tag{1}$$

where $x^{(t)}$ is the present input vector, h^t is the present hidden layer vector, and h^t provides the results from each LSTM cell. b^f, U^f, W^f are the bias, input weights, and loop weights of the forgetting gate, respectively. As a result, the internal state of the LSTM cells is updated using a conditional self-loop weight $f_i^{(t)}$ according to Equation (2).

$$s_i^{(t)} = f_i^{(t)} s_i^{(t-1)} + g_i^{(t)} \sigma(b_i + \sum_j U_{i,j} x_j^{(t)} + \sum_j W_{i,j} h_j^{(t-1)}) \tag{2}$$

where b, U, W are the bias in the LSTM cell, the input weights, and the cyclic weights of the forgetting gate, respectively. The external input gate cell $g_i^{(t)}$ is updated similarly to the forgetting gate, but with its parameters.

$$g_i^{(t)} = \sigma(b_i^g + \sum_j U_{i,j}^g x_j^{(t)} + \sum_j W_{i,j}^g h_j^{(t-1)}) \tag{3}$$

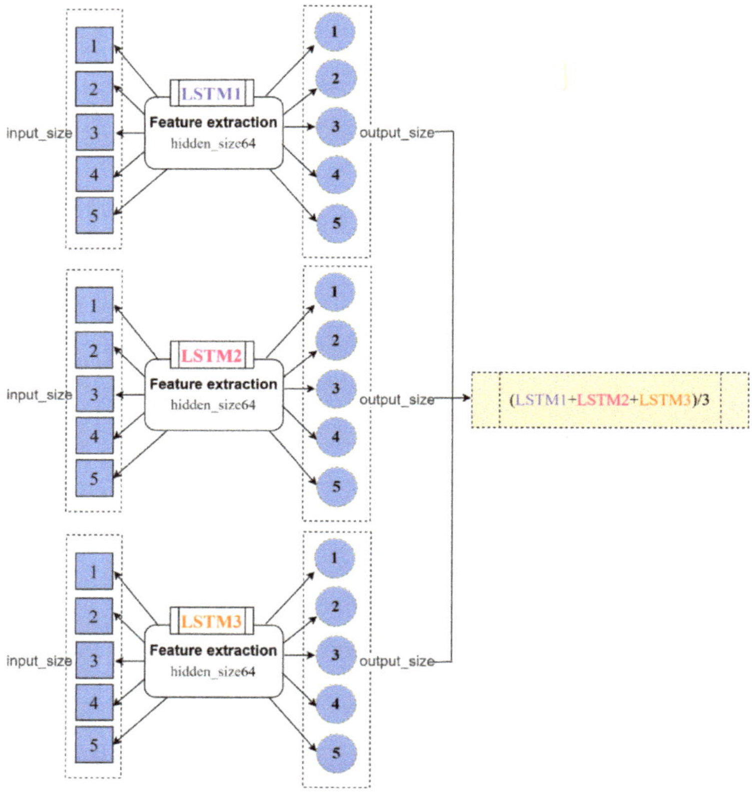

Figure 4. Structure of LSTM prediction method.

The output of the LSTM cell $h_i^{(t)}$ can also be closed by an output gate $q_i^{(t)}$ (using a sigmoid cell as a gate).

$$h_i^{(t)} = \tan h(s_i^{(t)})q_i^{(t)} \tag{4}$$

$$q_i^{(t)} = \sigma(b_i^o + \sum_j U_{i,j}^o x_j^{(t)} + \sum_j W_{i,j}^o h_j^{(t-1)}) \tag{5}$$

where b^o, U^o, W^o are the bias, input weights, and loop weights of the forgetting gates, respectively. Three gates $s_i^{(t)}$ can be selected as additional inputs to the first cell.

The loss function is employed to quantify the degree to which the model's predicted value deviates from the actual value. Generally speaking, a model performs better the loss function. Loss functions for various models are typically different. mean absolute error loss (also called L_1 loss) and mean squared error loss (also called L_2 loss) are the two loss functions frequently employed in regression issues [51].

3.2.3. Patch-Generating Land Use Simulation Model

A raster-based patch-generating land use simulation model was created to assist in the identification of probable drivers and their various contributions to change, allowing for a better understanding of the relationships underlying LUCC changes. The model contains a brand-new data mining approach for locating land use change laws. To better replicate and model actual landscape patterns, this model combines a CA model with a patch-generating simulation technique [35].

1. *Land expansion analysis strategy*

The random forest classification (RFC) technique is used by LEAS to investigate the connection between changing land use types and various factors. The RFC algorithm is a decision-tree-based integrated classifier that can handle multidimensional input data and output the growth probability $P_{i,j}^d$ of type k at cell i.

$$P_{i,j}^d(x) = \frac{\sum_{n=1}^{M} I(h_n(x) = d)}{M} \tag{6}$$

Other forms of utilization are transformed to type k if the value of d is 1, and other transformations are indicated if it is 0; M is the total number of decision trees, and x is a composite vector of several drivers. $h_n(x)$ is the prediction type of the nth decision tree of vector x.

The LEAS module extracts the ultimate potential growth of each land use in space by analyzing the effects of each driver on the growth of geographic units based on random forests. While ignoring the source type of land use, LEAS can examine the expanding zones of each changing land use to determine the transitional rules for all land use types. By avoiding the analysis of transition types, which increases exponentially with the number of land use types, the analysis process of land use change is effectively simplified.

2. *CA model based on multi-type random patch seeding*

The cellular automata model based on multi-type random patch seeds (CARS) is a model that includes the plaque formation mechanism of random seeds based on multiple land uses. The CA model is a scenario-driven land-use simulation model that integrates "top-down (e.g., global land use demand)" and "bottom-up (e.g., local land-use competition mechanisms)". In the simulation process, the land use demand will affect the local land use competition mode through the adaptive inertia mechanism, driving the amount of land use to meet future demand.

To simulate the evolution of multiple land use types, the CARS module combines random seed generation and a descending threshold to perform future land use simulation operations based on the traditional CA model. When the neighborhood effect of a single land-use type equals 0, the mechanism generates "seeds" to the development probability of each land-use type. Subject to the development probability, PLUS will automatically generate simulated plaques.

3.2.4. Accuracy Verification

In this study, the superior value FoM is employed to describe the LUCC model's accuracy. The ratio of the intersection of observed and predicted changes to the sum of observed and predicted changes is known as figure of merit (FoM) [52]. The value figure might be between 0% and 100%.

$$FoM = B/(A + B + C + D) \quad (7)$$

where A represents the incorrect region due to observed changes that are predicted to be persistent, B represents the correct region due to observed changes that are predicted to be variable, C represents the incorrect region due to observed changes that are predicted to be incorrectly acquired categories, and D represents the incorrect region due to observed tenacity that is predicted to be variable.

LUCC models with larger variability tend to have higher prediction accuracy and higher FoM. To account for model error, both observed and predicted variation must be considered. The importance of describing map differences based on quantitative inconsistencies and positional inconsistencies. For evaluating the accuracy of simulated variation, the FoM index is superior to the widely used kappa coefficient [53]. It was discovered through validation that FoM index values frequently fall between 0 and 0.59 and are typically lower than 0.3 [18].

4. Results

4.1. Land Use Multi-Scenario Simulation

Scenario simulation analysis is a kind of scenario description of future land use development based on historical data. By presetting different development goals, it simulates the future land use pattern in the study area under different scenarios. For different characteristics of the study area, different scenario simulation analyses can be used to achieve the optimal allocation of land resources and can provide a variety of decision-making thinking on land resource management. Influenced by regional development strategies, there may be multiple possibilities for the evolution of future land use patterns in the Yellow River Delta. Since the Shandong Yellow River Delta National Nature Reserve is divided into two areas, north and south, the southern area is located at the current mouth of the Yellow River and belongs to the Kenli district, while the northern area is located at the mouth of the former Yellow River channel after its diversion in 1976 and belongs to Hekou district. Therefore, ecological development priority scenarios were simulated for the Hekou district and Kenli district, and natural development scenarios were simulated for the Dongying district, Guangrao County and Lijin County.

4.2. PLUS Simulation of Natural Development Scenario

According to the expansion patterns and change characteristics of various land use types in the study area throughout history and following the land use structure change pattern, the natural development scenario takes into account the relationship between natural, socioeconomic, and accessibility factors and land use changes. For Dongying district, Guangrao, and Lijin counties, the number of land demand rasters in 2030 was predicted using LSTM networks based on historical images of land use in 2000 and 2010, and natural development scenarios were simulated for Dongying district, Guangrao, and Lijin county, respectively. Assuming a natural development scenario, Table 3 displays the real land use data for 2020 and the predicted data for 2030.

Table 3. Land use area under natural development scenarios in 2020 and 2030 (km^2).

Land Type	2020 Dongying District	2030 Dongying District	2020 Guangrao County	2030 Guangrao County	2020 Lijin County	2030 Lijin County
Cropland	539.78	481.52	782.60	720.11	897.45	850.63
Grassland	0.78	5.60	0.11	2.64	0.01	0.19
Water	148.24	126.05	77.10	46.31	53.82	51.25
Unused land	34.41	84.15	13.50	47.81	15.00	52.70
Construction Land	423.66	449.55	297.14	353.58	231.51	243.02

In Dongying district, cropland will shrink from 539.78 km^2 to 481.52 km^2 between 2020 and 2030, accounting for 41.98% of the total area instead of 47.06%. From 0.78 km^2 to 5.6 km^2, or 0.07% of the total area, the grassland area rose. From 12.93% of the total area to 10.99%, or 148.24 km^2 to 126.05 km^2, the water area fell. Unused land increased in size from 34.41 km^2 to 84.15 km^2, making up 7.34% of the overall area. Building land expanded from 423.66 km^2 to 449.55 km^2, accounting for 39.2% of the whole area, up from 36.94%. According to a comprehensive analysis, cropland has the biggest change among various land use types, with a decrease of 5.08%. The second is unused land, up 4.34%. Thirdly, construction land increased by 2.28% and water area decreased by 1.94%. The grassland is relatively stable and only increased by 0.42%.

From 2020 to 2030, the cropland area of Guangrao county will decrease from 782.60 km^2 to 720.11 km^2, and the percentage will drop from 66.86% to 61.52% in the overall area. The water area fell from 53.82 km^2 to 51.25 km^2, representing 4.28% of the total area from 4.49%; the grassland area rose from 0.01 km^2 to 0.09 km^2, representing 0.01% of the total area; the undeveloped land rose in size from 15 km^2 to 52.7 km^2, or 4.4% of the total area, up from 1.25%; the construction land expanded from 231.51 km^2 to 243.02 km^2, making up 20.29% of the whole area as opposed to 19.33%. Further analysis revealed that cropland has the biggest change among various land use types, with a decrease of 5.34%. The second is construction land, up 4.82%. Thirdly, the unprofitable land increased by 2.93% and the water area decreased by 2.63%. The grassland is relatively stable and only increases by 0.22%.

The cropland area in Lijin county will drop from 897.45 km^2 to 850.63 km^2 between 2020 and 2030, while its share of the overall area will fall from 74.93% to 71.02%. The area of grassland increased from 0.01 km^2 to 0.09 km^2, making up 0.01% of the overall area; 4.28% of the total area, down from 4.49%, was made up of water, which declined from 53.82 km^2 to 51.25 km^2; From 1.25% to 4.4% of the total area, the amount of undeveloped land expanded from 15 km^2 to 52.7 km^2; construction land expanded from 231.51 km^2 to 243.02 km^2, making up 20.29% of the whole area as opposed to 19.33%. According to a comprehensive analysis, cropland has the biggest change among various land use types, with a decrease of 3.91%. The second is unused land, up 3.15%. Thirdly, construction land increased by 0.96% and water area decreased by 0.21%. The grassland is relatively stable and only increases by 0.01%.

4.3. PLUS Simulation of Ecological Development Priority Scenario

The concept of ecological priority and green development is the basis of the priority ecological development program. While optimizing the pattern of territorial space development and protection, ecological civilization is integrated into the overall process and support level of environmental protection in the Yellow River Delta. The number of land demand rasters for Hekou district and Kenli district in 2030 is estimated using an LSTM network based on historical land use pictures from 2000 to 2010. Table 4 displays the land use outcomes of 2030 ecological development priority scenarios. The ecological source area and high habitat quality area are taken as the ecology preservation restricted area.

Table 4. Land use area under ecological development priority scenarios in 2020 and 2030 (km^2).

Land Type	2020 Hekou District	2030 Hekou District	2020 Kenli District	2030 Kenli District
Cropland	998.44	959.85	1204.68	1140.27
Grassland	0.53	6.66	7.8	51.65
Water	410.83	577.86	251.27	321.63
Unused land	76.79	200.55	45.97	211.74
Construction Land	456.38	198.05	469.46	253.9

To extract the ecological source site, the core area needs to be found. The morphological spatial grid will segment the foreground objects of the input image. The statistical number of core area surface after segmentation is 68,113 rasters, and the core area of the original data is severely fragmented. The statistical area of the core area is sorted from large to small, and the surface with an area larger than 0.7 km^2 is selected and combined and extracted, that is, about 86.56% of the core area is extracted as ecological source land, which greatly reduces the fragmentation degree. This ecological source area is mainly concentrated in the Yellow River Delta Reserve and the Yellow River Estuary Ecotourism Area.

Running the habitat quality module, the degradation degree image was generated for the first time, and the maximum value of degradation was 0.0818, at which time the obtained habitat quality map had an extremely uneven distribution of index values, causing aggregation. Therefore, it is necessary to adjust the habitat parameter to half of the maximum degradation value, i.e., 0.0404, and run the program again to obtain the habitat quality map. The habitat quality values were separated at equal intervals, and the habitat quality was classified into five habitat quality classes: low, lower, medium, higher, and high. Areas with high habitat quality and ecological source sites were taken as ecological development protection sites, i.e., restricted areas under the ecological priority scenario.

Under the priority scenario of ecological development in the Hekou district from 2020 to 2030, the area of cropland decreases from 998.44 km^2 to 959.85 km^2, representing 51.39% to 49.4%. In terms of the percentage of the overall area, grassland grows from 0.53 km^2 to 6.66 km^2, or from 0.03% to 0.34%. Water covered an area that expanded from 410.83 km^2 to 577.86 km^2, or 21.14% to 29.74% of the total area. Unused land increased in size from 76.79 km^2 to 200.55 km^2, or 3.95% to 10.32% of the total area. Construction land's area shrunk from 456.38 km^2 to 198.05 km^2 or 23.49% to 10.2% of the total area. Comprehensive analysis, the biggest change in various land use types is construction land, down 13.29%; followed by water up 8.6%, unused land up 6.37%, cropland down 1.99%, and grassland up 0.31%.

In the Kenli district, cropland will shrink from 1204.68 km^2 to 1140.27 km^2 between 2020 and 2030, and its share of the whole area will drop from 60.87% to 57.61%. In contrast, grassland will grow from 7.8 km^2 to 51.65 km^2, accounting for 2.61% of the total area, up from 0.39%. Water coverage grew from 251.27 km^2 to 321.63 km^2, or 16.25%, from 12.7% of the total area. The area of undeveloped land increased from 45.97 km^2 to 211.74 km^2, or 10.7% of the total area, up from 2.32%. Construction land reduced in size from 469.46 km^2 to 253.9 km^2, accounting for 12.83% of the total area as opposed to 23.72%. According to a thorough investigation, cropland decreased by 3.26%, cropland decreased by 10.89%, cropland increased by 3.26%, grassland increased by 2.22%, unused land increased by 8.38%, water increased by 3.55%, and building land decreased by 10.89%.

4.4. Land Use Conversion from 2020 to 2030

The interaction between different types of land use is intricate and competitive in the land system. As water areas are increased, other land uses must be scaled back, and when biological lands are protected, cropland and construction sites must bear the burden of all development. Therefore, to address the needs of protecting cropland, the environment, and urban co-development, the relevant departments must develop pertinent policies, beginning with sustainable development, and the protection of the ecological environment must be put in the top priority position to avoid the issue of the ecological environment degrading while the financial and cultural development is going on. To secure the harmonious development of social and economic development and ecological conservation, it is also vital to avoid destroying the current ecological base via the disorderly extension of construction land and to put the protection of cropland into practice.

Figure 5 shows the transition between land-use categories from 2020 to 2030. Figure 6 visualizes the actual and simulated land use distribution in 2020 and 2030. The area of land that will be transferred to water between 2020 and 2030 will be the most, at 613.3 km^2, followed by cropland, unused land, and construction land, at 514.35 km^2, 489.49 km^2, and 391.3 km^2, respectively. The area of land that will be transferred to grassland land will be the least, at 59.47 km^2. The conversion between cropland and water, with an area of 424.11 km^2, followed by the conversion between construction land and cropland, with an area of 361.41 km^2, makes up the majority of the transfer of land use. The remaining land types are primarily concentrated in the conversion between unused land and cropland, with an area of 211.47 km^2 each. The remaining land types are mostly concentrated in the area that results from the transition of cropland, unused land, and water, which is 42.37 km^2, 211.47 km^2, and 148.42 km^2, respectively.

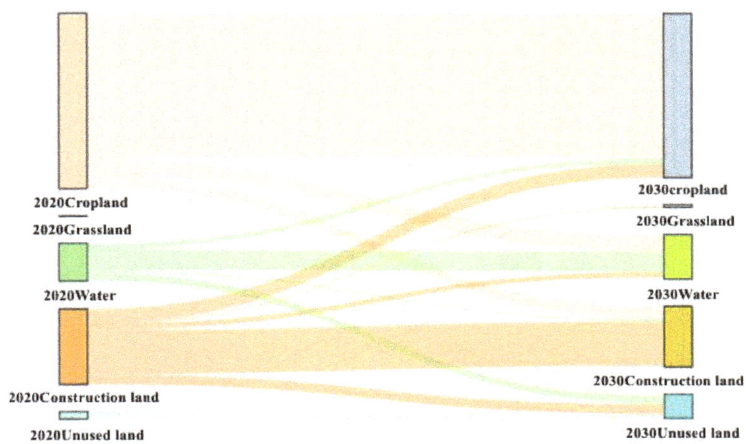

Figure 5. Dongying city 2020–2030 circulation map of various categories.

According to the analysis of the full study region in Figure 6, the cropland area will decrease from 4421.57 km^2 to 4153.08 km^2 from 2020 to 2030, representing 55.83% of the total area from 59.47%. The grassland area increased by 0.9%, from 9.23 km^2 to 66.74 km^2, or 0.12% of the total area. From 12.66% to 15.1% of the total area, the water area rose from 940.97 km^2 to 1123.24 km^2. From 2.5% of the total area to 8.03% of the total area, the unused land area expanded from 185.54 km^2 to 597.05 km^2. The area of construction land decreased from 1877.46 km^2 to 1498.38 km^2, and the proportion in the total area decreased from 25.25% to 20.14%. According to a comprehensive analysis, construction land decreased by 8.38%, unused land increased by 5.53%, cropland decreased by 3.64%, water area increased by 2.44%, and grassland increased by 0.78%.

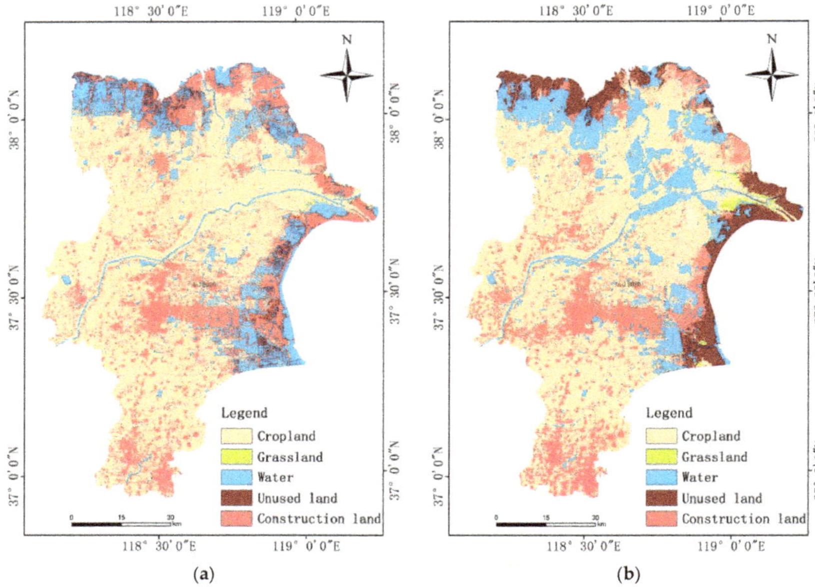

Figure 6. Multi-scenario simulation analysis of land use: (**a**) actual data in 2020; (**b**) simulated data in 2030.

4.5. Contribution of Driving Factors

The LEAS module in the PLUS model employs the random forest model method to determine the likelihood that each type of land use would develop, as well as the role that each driver played in the conversion of that type of land use into another type of land use (see Figure 7). The sampling rate is 0.1, and the number of decision trees is set to 50.

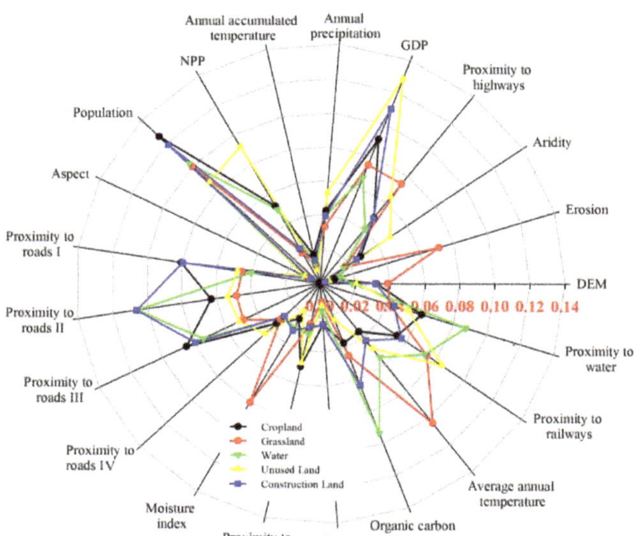

Figure 7. Contribution of driving factors to expansion of various land use.

The driver of the largest contribution to cropland area expansion is population density, followed by GDP and distance to tertiary roads. Roads III is a type of road with a low road surface level, whose main function is to connect local counties, towns and villages, remote suburban areas, or functional area locations. This contribution indicates that the cropland in the study area is closely related to human distribution, which is mostly distributed around rural areas and positively correlated with the distance to tertiary roads while being far from areas with higher GDP. The largest contribution of grassland expansion was the average annual temperature, followed by population density and moisture index. This indicates that the distribution of grasslands is highly correlated with temperature, with suitable temperature and wetness being hot areas for grassland distribution and negatively correlated with population density. The greatest contribution of water area expansion is the distance to secondary roads and population density. Water areas are also closely related to human distribution and are mostly located in less densely populated areas with easy access. The largest expansion of unused land is GDP. Changing land use will change GDP. GDP fluctuates from high to low as an element or asset, showing that a change in land use causes a market value to drop to zero. Population density, GDP, and proximity to secondary roads are the three factors that affect the expansion of construction land the most. According to actual urban development, the majority of construction land is dispersed in regions with high densities of people, businesses, and vehicles. Future construction land will also expand in areas with high population concentration and well-developed transportation arteries.

5. Discussion

5.1. Uniqueness of the Study Area

The Yellow River Delta is rich in natural resources and intact ecosystems, carrying a major mission of national environmental and energy security, and harboring huge potential for economic growth. However, the accelerated urbanization and increasingly frequent human activities in recent years have led to the intensification of wetland landscape fragmentation, biodiversity reduction, and environmental pollution, which, together with the reduction in incoming sediment, soil salinization, and wetland ecosystem reverse succession, have made the originally fragile and sensitive ecosystem of the Yellow River Delta face serious challenges and pose a significant risk to the region's ability to sustain its development. It seriously jeopardizes the region's ability to develop economically and socially sustainably. In this context, the possible future land use scenarios of the Yellow River Delta are simulated and predicted based on the PLUS model using the LSTM network, which is important for reconciling the conflicts between conservation and development and formulating scientific regional development plans.

5.2. Improved PLUS Model Based on LSTM

Based on the Python environment, the LSTM algorithm is implemented using Python as a deep learning framework. The model uses 5 layers of input, and 5 layers of output, and the length of the hidden layer output feature vector is 64. Finally, the method of averaging three sets of LSTM results is used to improve accuracy. Set the loss function to which the predicted value and the true value are different, the better the loss function, the better the performance of the model. Different models generally use different loss functions. Mean squared error loss is one of the most commonly used loss functions in machine learning and deep learning regression tasks and is also known as L_2 loss. Mean absolute error is another commonly used loss function, also known as L_1 loss.

FoM values are large for areas with large land use changes. A small land use change leads to a small FoM value. The land use change in Dongying is relatively small; therefore, the overall FoM value is small. Analyze the impact of three different methods of Markov chain, linear regression, and LSTM network in predicting land use demand results on the accuracy of PLUS simulation results in 2020 under the same parameter settings (see Table 5). The results have an overall high accuracy rate, with an average error of about 5.34%, indicating that the accuracy of the simulation has achieved the desired effect. Cropland has the highest simulation accuracy; grassland and unused land have lower accuracy rates; water and construction land have intermediate accuracy rates. The simulation accuracy increases as the FoM value increases. The maximum simulation accuracy is achieved by the enhanced PLUS model based on the LSTM L_1 Loss network, and the FoM value is 0.1335. The enhancement of the PLUS model based on the LSTM network is effective, according to experiments. For erratic time series with a greater number of fixed components, LSTM networks typically perform better. Since the L_1 Loss function does not depend on the output deviating from the true value, it is more suited to preserving the model's stability. It computes the absolute value of the difference between the estimated value and the true value. To improve precision, three groups of LSTM data are averaged.

Table 5. Comparison of land use PLUS simulation accuracy in 2020.

Prediction Method of Land Use Demand	Markov Chain	Linear Regression	LSTM L_1 Loss	LSTM L_2 Loss
2020FoM	0.0852	0.0928	0.1335	0.1293

From Table 6, it can be seen that the overall accuracy of the simulation of the PLUS improved model based on the LSTM network is high, and the average error is about 5.34%, indicating that the accuracy of the simulation has achieved the desired effect. Among them, the accuracy of cropland simulation is the highest. Grassland and unused land have lower accuracy and the accuracy rate of water and construction land is in the middle.

Table 6. Comparison between simulated area and actual area in 2020.

Number of Land Use Rasters	Cropland	Grassland	Water	Unused Land	Construction Land
Actual number in 2020	4915486	10257	1046601	206312	2086697
MC-based predicts number of simulations	4900257	111142	943662	679400	1645803
LR-based predicts number of simulations	5258616	49011	796388	713674	1345696
LSTM L_1 loss-based predicts number of simulations	4940619	11282	1094461	221467	1997524
Error (%)	0.51%	9.99%	4.57%	7.34%	−4.27%

5.3. Multi-Scenario Analysis

A type of scenario description of future land use development based on historical data is scenario simulation analysis. It stimulates the future land use pattern in the research area under various scenarios by presetting various development targets. Several scenario simulation analyses can be utilized to achieve optimal land resource allocation and can offer a variety of decision-making thoughts on land resource management depending on the characteristics of the studied area. The regional development strategy may have a significant impact on how the Yellow River Delta's future land use pattern develops. Due to the north and south divisions of the Shandong Yellow River Delta National Natural Reserve, the southern area is located at the current mouth of the Yellow River and belongs to the Kenli district, while the northern area is located at the mouth of the old Yellow River channel after its diversion in 1976 and belongs to the Hekou district. Therefore, ecological development priority scenarios were simulated for Hekou and Kenli districts, and natural development scenarios were simulated for the Dongying districts, Guangrao county, and Lijin county.

6. Conclusions

This study uses a combination of the PLUS model and the LSTM network to predict the land use of different administrative districts in Dongying city in 2030 under different development scenarios. The evolution of land use in Dongying city from 2020 to 2030 is analyzed. Comparing and quantifying the three methods of MC, LR, and LSTM for predicting the quantity of land use demand under the same parameters, the accuracy of PLUS simulation verifies that the precision of the PLUS-enhanced model prediction based on the LSTM network achieves the desired effect.

A simulation of land usage under multiple scenarios (including natural development and ecological development priorities) and a simulation of changes to the environment in 2030 are presented. The cropland area is extensively converted to the construction land area in the Dongying district, Guangrao county, and Lijin county under the natural development scenario, resulting in a significant increase in the construction land area and a decrease in the cropland area of about 167 km^2. However, in Hekou and Kenli districts, the area of cropland and construction land has greatly decreased under the priority scenario of ecological development, while the water area has significantly risen by about 103 km^2. Cropland will be changed to water in an area of 424.11 km^2 and construction land will be converted to cropland in an area of 361.41 km^2 between 2020 and 2030. It can be seen that the sub-regional scenario simulation scheme proposed in this paper can effectively predict future ecological protection and high-quality development in Dongying City.

Dongying City takes a high level of ecological environment protection to promote high-quality economic development as its future goal. How to model and optimize the spatial allocation of land resources in Dongying City based on more accurate multi-scenario models is the focus of future research. Improving the efficiency of land resource allocation and output, promoting the quality of urbanization, and coordinating regional development in China can provide new impetus for the next round of economic growth.

Author Contributions: Conceptualization, X.Z., P.W., S.G. and M.Y.; Methodology, X.Z.; software, S.G.; validation, M.Y. and P.W.; formal analysis, X.Z.; investigation, S.G.; resources, M.Y.; data curation, S.G.; writing—original draft preparation, X.Z. and Q.U.I.; writing—review and editing, P.W. and X.Z.; visualization, M.Y.; supervision, S.G. All authors have read and agreed to the published version of the manuscript.

Funding: This paper is supported by College of Resources, Shandong University of Science and Technology.

Data Availability Statement: All data used during the study are available at https://zenodo.org/record/4417810.

Conflicts of Interest: The authors declare no conflict of interest.

References

1. Brown, D.G.; Walker, R.; Manson, S.; Seto, K. Modeling land use and land cover change. In *Land Change Science: Observing, Monitoring and Understanding Trajectories of Change on the Earth's Surface*; Kluwer Academic Publisher: London, UK, 2004; pp. 395–409. [CrossRef]
2. Ilyasd, M.; Yasir, M.; Hossain, M.S. Urban Area Extraction and Land Use Land Cover Monitoring of Charsadda District, Pakistan. *Earth Sci. Malays. (ESMY)* **2022**, *6*, 96–102. [CrossRef]
3. Baker, W.L. A review of models of landscape change. *Landsc. Ecol.* **1989**, *2*, 111–133. [CrossRef]
4. Lambin, E.F. Modelling and monitoring land-cover change processes in tropical regions. *Prog. Phys. Geogr.* **1997**, *21*, 375–393. [CrossRef]
5. Aburas, M.M.; Ahamad, M.S.S.; Omar, N.Q. Spatio-temporal simulation and prediction of land-use change using conventional and machine learning models: A review. *Environ. Monit. Assess.* **2019**, *191*, 205. [CrossRef]
6. Alghais, N.; Pullar, D. Modelling future impacts of urban development in Kuwait with the use of ABM and GIS. *Trans. GIS* **2018**, *22*, 20–42. [CrossRef]

7. Islam, K.; Jashimuddin, M.; Nath, B.; Nath, T.K. Land use classification and change detection by using multi-temporal remotely sensed imagery: The case of Chunati wildlife sanctuary, Bangladesh. *Egypt. J. Remote Sens. Space Sci.* **2018**, *21*, 37–47. [CrossRef]
8. Ji, W.; Ma, J.; Twibell, R.W.; Underhill, K. Characterizing urban sprawl using multi-stage remote sensing images and landscape metrics. *Comput. Environ. Urban Syst.* **2006**, *30*, 861–879. [CrossRef]
9. Mundia, C.N.; Aniya, M. Analysis of land use/cover changes and urban expansion of Nairobi city using remote sensing and GIS. *Int. J. Remote Sens.* **2005**, *26*, 2831–2849. [CrossRef]
10. Huang, W.; Liu, H.; Luan, Q.; Jiang, Q.; Liu, J.; Liu, H. Detection and prediction of land use change in Beijing based on remote sensing and GIS. *Int. Arch. Photogramm. Remote Sens. Spat. Inf. Sci* **2008**, *37*, 75–82.
11. Liu, M.; Hu, Y.; Chang, Y.; He, X.; Zhang, W. Land use and land cover change analysis and prediction in the upper reaches of the Minjiang River, China. *Environ. Manag.* **2009**, *43*, 899–907. [CrossRef] [PubMed]
12. Araya, Y.H.; Cabral, P. Analysis and modeling of urban land cover change in Setúbal and Sesimbra, Portugal. *Remote Sens.* **2010**, *2*, 1549–1563. [CrossRef]
13. KantaKumar, N.L.; Sawant, N.G.; Kumar, S. Forecasting urban growth based on GIS, RS and SLEUTH model in Pune metropolitan area. *Int. J. Geomat. Geosci.* **2011**, *2*, 568–579.
14. Park, S.; Jeon, S.; Choi, C. Mapping urban growth probability in South Korea: Comparison of frequency ratio, analytic hierarchy process, and logistic regression models and use of the environmental conservation value assessment. *Landsc. Ecol. Eng.* **2012**, *8*, 17–31. [CrossRef]
15. Al-Sharif, A.A.; Pradhan, B. Monitoring and predicting land use change in Tripoli Metropolitan City using an integrated Markov chain and cellular automata models in GIS. *Arab. J. Geosci.* **2014**, *7*, 4291–4301. [CrossRef]
16. Tayyebi, A.; Pekin, B.K.; Pijanowski, B.C.; Plourde, J.D.; Doucette, J.S.; Braun, D. Hierarchical modeling of urban growth across the conterminous USA: Developing meso-scale quantity drivers for the Land Transformation Model. *J. Land Use Sci.* **2013**, *8*, 422–442. [CrossRef]
17. Mishra, V.N.; Rai, P.K.; Mohan, K. Prediction of land use changes based on land change modeler (LCM) using remote sensing: A case study of Muzaffarpur (Bihar), India. *J. Geogr. Inst. Jovan Cvijic SASA* **2014**, *64*, 111–127. [CrossRef]
18. Liu, X.; Liang, X.; Li, X.; Xu, X.; Ou, J.; Chen, Y.; Li, S.; Wang, S.; Pei, F. A future land use simulation model (FLUS) for simulating multiple land use scenarios by coupling human and natural effects. *Landsc. Urban Plan.* **2017**, *168*, 94–116. [CrossRef]
19. Qiang, Y.; Lam, N.S. Modeling land use and land cover changes in a vulnerable coastal region using artificial neural networks and cellular automata. *Environ. Monit. Assess.* **2015**, *187*, 57. [CrossRef]
20. Abdullahi, S.; Pradhan, B. Sustainable brownfields land use change modeling using GIS-Based weights-of-evidence approach. *Appl. Spat. Anal. Policy* **2016**, *9*, 21–38. [CrossRef]
21. Abdullahi, S.; Pradhan, B.; Mojaddadi, H. City compactness: Assessing the influence of the growth of residential land use. *J. Urban Technol.* **2018**, *25*, 21–46. [CrossRef]
22. Mustafa, A.; Rienow, A.; Saadi, I.; Cools, M.; Teller, J. Comparing support vector machines with logistic regression for calibrating cellular automata land use change models. *Eur. J. Remote Sens.* **2018**, *51*, 391–401. [CrossRef]
23. Zhang, Y.; Luo, J.; Zhang, Y.; Huang, Y.; Cai, X.; Yang, J.; Mao, D.; Li, J.; Tuo, X.; Zhang, Y. Resolution Enhancement for Large-Scale Real Beam Mapping Based on Adaptive Low-Rank Approximation. *IEEE Trans. Geosci. Remote Sens.* **2022**, *60*, 1–21. [CrossRef]
24. Wang, P.; Yu, P.; Lu, J.; Zhang, Y. The mediation effect of land surface temperature in the relationship between land use-cover change and energy consumption under seasonal variations. *J. Clean. Prod.* **2022**, *340*, 130804. [CrossRef]
25. Zhang, S.; Bai, X.; Zhao, C.; Tan, Q.; Luo, G.; Wang, J.; Li, Q.; Wu, L.; Chen, F.; Li, C. Global CO_2 consumption by silicate rock chemical weathering: Its past and future. *Earth's Future* **2021**, *9*, e2020EF001938. [CrossRef]
26. Xu, L.; Liu, X.; Tong, D.; Liu, Z.; Yin, L.; Zheng, W. Forecasting urban land use change based on cellular automata and the PLUS model. *Land* **2022**, *11*, 652. [CrossRef]
27. Liu, X.; Kong, M.; Tong, D.; Zeng, X.; Lai, Y. Property rights and adjustment for sustainable development during post-productivist transitions in China. *Land Use Policy* **2022**, *122*, 106379. [CrossRef]
28. Tong, D.; Chu, J.; Han, Q.; Liu, X. How land finance drives urban expansion under fiscal pressure: Evidence from Chinese cities. *Land* **2022**, *11*, 253. [CrossRef]
29. Liu, X.; Li, Z.; Fu, X.; Yin, Z.; Liu, M.; Yin, L.; Zheng, W. Monitoring House Vacancy Dynamics in The Pearl River Delta Region: A Method Based on NPP-VIIRS Night-Time Light Remote Sensing Images. *Land* **2023**, *12*, 831. [CrossRef]
30. Liu, Z.; Xu, J.; Liu, M.; Yin, Z.; Liu, X.; Yin, L.; Zheng, W. Remote sensing and geostatistics in urban water-resource monitoring: A review. *Mar. Freshw. Res.* **2023**. [CrossRef]
31. Sun, R.; Fu, L.; Cheng, Q.; Chiang, K.-W.; Chen, W. Resilient Pseudorange Error Prediction and Correction for GNSS Positioning in Urban Areas. *IEEE Internet Things J.* **2023**. [CrossRef]
32. Liu, Y.; Zhang, K.; Li, Z.; Liu, Z.; Wang, J.; Huang, P. A hybrid runoff generation modelling framework based on spatial combination of three runoff generation schemes for semi-humid and semi-arid watersheds. *J. Hydrol.* **2020**, *590*, 125440. [CrossRef]
33. Xu, D.; Zhu, D.; Deng, Y.; Sun, Q.; Ma, J.; Liu, F. Evaluation and empirical study of Happy River on the basis of AHP: A case study of Shaoxing City (Zhejiang, China). *Mar. Freshw. Res.* **2023**. [CrossRef]
34. Zhao, X.; Wang, P.; Yasir, M.; Liu, Z. Decision support system based on spatial and temporal pattern evolution of ecological environmental quality in the Yellow River Delta from 2000 to 2020. *Soft Comput.* **2022**, *26*, 11033–11044. [CrossRef]

35. Liang, X.; Guan, Q.; Clarke, K.C.; Liu, S.; Wang, B.; Yao, Y. Understanding the drivers of sustainable land expansion using a patch-generating land use simulation (PLUS) model: A case study in Wuhan, China. *Comput. Environ. Urban Syst.* **2021**, *85*, 101569. [CrossRef]
36. Lu, C.; Qi, X.; Zheng, Z.; Jia, K. PLUS-model based multi-scenario land space simulation of the Lower Yellow River Region and its ecological effects. *Sustainability* **2022**, *14*, 6942. [CrossRef]
37. Lin, Z.; Peng, S. Comparison of multimodel simulations of land use and land cover change considering integrated constraints-A case study of the Fuxian Lake basin. *Ecol. Indic.* **2022**, *142*, 109254. [CrossRef]
38. Bao, S.; Yang, F. Spatio-temporal dynamic of the land use/cover change and scenario simulation in the southeast coastal shelterbelt system construction project region of China. *Sustainability* **2022**, *14*, 8952. [CrossRef]
39. Yang, W.; Du, J.; He, H.; Yang, W.; Chen, L. Urban growth boundary delimitation in Guangzhou city based on PLUS-UGB multi-scenarios simulation. *Geospat. Inf* **2022**, *20*, 38–42.
40. Gao, L.; Tao, F.; Liu, R.; Wang, Z.; Leng, H.; Zhou, T. Multi-scenario simulation and ecological risk analysis of land use based on the PLUS model: A case study of Nanjing. *Sustain. Cities Soc.* **2022**, *85*, 104055. [CrossRef]
41. Li, X.; Fu, J.; Jiang, D.; Lin, G.; Cao, C. Land use optimization in Ningbo City with a coupled GA and PLUS model. *J. Clean. Prod.* **2022**, *375*, 134004. [CrossRef]
42. Wang, J.; Zhang, J.; Xiong, N.; Liang, B.; Wang, Z.; Cressey, E.L. Spatial and temporal variation, simulation and prediction of land use in ecological conservation area of Western Beijing. *Remote Sens.* **2022**, *14*, 1452. [CrossRef]
43. Rumelhart, D.E.; Hinton, G.E.; Williams, R.J. Learning representations by back-propagating errors. *Nature* **1986**, *323*, 533–536. [CrossRef]
44. Yu, Y.; Si, X.; Hu, C.; Zhang, J. A review of recurrent neural networks: LSTM cells and network architectures. *Neural Comput.* **2019**, *31*, 1235–1270. [CrossRef]
45. Xu, T.; Zhou, D.; Li, Y. Integrating ANNs and Cellular Automata–Markov Chain to Simulate Urban Expansion with Annual Land Use Data. *Land* **2022**, *11*, 1074. [CrossRef]
46. Boulila, W.; Ghandorh, H.; Khan, M.A.; Ahmed, F.; Ahmad, J. A novel CNN-LSTM-based approach to predict urban expansion. *Ecol. Inform.* **2021**, *64*, 101325. [CrossRef]
47. Wang, H.; Zhao, X.; Zhang, X.; Wu, D.; Du, X. Long time series land cover classification in China from 1982 to 2015 based on Bi-LSTM deep learning. *Remote Sens.* **2019**, *11*, 1639. [CrossRef]
48. Mohanrajan, S.N.; Loganathan, A. Novel vision transformer–based bi-LSTM model for LU/LC prediction—Javadi Hills, India. *Appl. Sci.* **2022**, *12*, 6387. [CrossRef]
49. Yang, J.; Huang, X. The 30 m annual land cover dataset and its dynamics in China from 1990 to 2019. *Earth Syst. Sci. Data* **2021**, *13*, 3907–3925. [CrossRef]
50. Hochreiter, S.; Schmidhuber, J. Long short-term memory. *Neural Comput.* **1997**, *9*, 1735–1780. [CrossRef]
51. Tsokos, C.P.; Welch, R. Bayes discrimination with mean square error loss. *Pattern Recognit.* **1978**, *10*, 113–123. [CrossRef]
52. Perica, S.; Foufoula-Georgiou, E. Model for multiscale disaggregation of spatial rainfall based on coupling meteorological and scaling descriptions. *J. Geophys. Res. Atmos.* **1996**, *101*, 26347–26361. [CrossRef]
53. Pontius, R.G.; Boersma, W.; Castella, J.-C.; Clarke, K.; De Nijs, T.; Dietzel, C.; Duan, Z.; Fotsing, E.; Goldstein, N.; Kok, K. Comparing the input, output, and validation maps for several models of land change. *Ann. Reg. Sci.* **2008**, *42*, 11–37. [CrossRef]

remote sensing

MDPI

Article

A Reference-Free Method for the Thematic Accuracy Estimation of Global Land Cover Products Based on the Triple Collocation Approach

Pengfei Chen [1,2], Huabing Huang [1,2], Wenzhong Shi [3,*] and Rui Chen [1,2]

1 School of Geospatial Engineering and Science, Sun Yat-sen University, and Southern Marine Science and Engineering Guangdong Laboratory (Zhuhai), Zhuhai 519082, China
2 Key Laboratory of Comprehensive Observation of Polar Environment (Sun Yat-sen University), Ministry of Education, Zhuhai 519082, China
3 Smart Cities Research Institute, Department of Land Surveying and Geo-Informatics, The Hong Kong Polytechnic University, Hong Kong 999077, China
* Correspondence: john.wz.shi@polyu.edu.hk

Abstract: Global land cover (GLC) data are an indispensable resource for understanding the relationship between human activities and the natural environment. Estimating their classification accuracy is significant for studying environmental change and sustainable development. With the rapid emergence of various GLC products, the lack of high-quality reference data poses a severe risk to traditional accuracy estimation methods, in which reference data are always required. Thus, meeting the needs of large-scale, fast evaluation for GLC products becomes challenging. The triple collocation approach (TCCA) is originally applied to assess classification accuracy in earthquake damage mapping when ground truth is unavailable. TCCA can provide unbiased accuracy estimation of three classification systems when their errors are conditionally independent. In this study, we extend the idea of TCCA and test its performance in the accuracy estimation of GLC data without ground reference data. Firstly, to generate two additional classification systems besides the original GLC data, a k-order neighbourhood is defined for each assessment unit (i.e., geographic tiles), and a local classification strategy is implemented to train two classifiers based on local samples and features from remote sensing images. Secondly, to reduce the uncertainty from complex classification schemes, the multi-class problem in GLC is transformed into multiple binary-class problems when estimating the accuracy of each land class. Building upon over 15 million sample points with remote sensing features retrieved from Google Earth Engine, we demonstrate the performance of our method on WorldCover 2020, and the experiment shows that screening reliable sample points during training local classifiers can significantly improve the overall estimation with a relative error of less than 4% at the continent level. This study proves the feasibility of estimating GLC accuracy using the existing land information and remote sensing data, reducing the demand for costly reference data in GLC assessment and enriching the assessment approaches for large-scale land cover data.

Keywords: land cover; reference-free method; accuracy estimation; local classification strategy; triple collocation approach (TCCA)

1. Introduction

Since the Global Land Cover Characterisation dataset was first released in the early 1990s, tremendous efforts have been made to produce global land cover (GLC) products towards better accuracy and resolution [1,2]. Benefiting from remote sensing and artificial intelligence techniques, GLC production efficiency has significantly improved in the last decades. More than 10 GLC products with up to 10 m resolution are available now [3–7]. These GLC products provide unprecedented details of the Earth's surface and promote a batch of studies on environmental, biological and social sciences on a global scale [8–11].

The application scope of a GLC product is significantly affected by its uncertainty, which many factors can influence, including the definition of the classification scheme (e.g., the hierarchy and description of land classes) [12], mapping scale (e.g., the minimum mapping unit) [13], input data (e.g., satellite image) and classification methods [14]. Thematic accuracy assessment is a fundamental procedure for producers and users to learn the uncertainty in GLC data and thus make rational decisions [12,15]. Currently, most GLC datasets are assessed by comparison with ground reference data. The confusion matrix, together with some related descriptive metrics such as the overall accuracy (OA), user accuracy (UA) and producer accuracy (PA), is used to characterise the accuracy [5,16,17]. Reference data are commonly collected based on a stratified random sampling design, in which the true class for each sample point is determined by either field survey or visual interpretation using high-resolution satellite images [15,18]. However, given the huge amount of sample points needed for GLC assessment, obtaining sufficient high-quality reference data on a global scale is always laborious and time-consuming [19–21].

Considerable efforts have been made to reduce the cost of reference data collection, such as utilising open data sources such as OpenStreetMap or establishing web-based platforms to collect crowdsourcing data from global volunteers for the validation of specific GLC products [22–25]. A recent example can be seen in the validation of GlobeLand30, in which an online system GLCVal was designed and used by approximately 30 countries and international organisations for collaborative assessment [26]. Crowdsourcing also indicates a promising direction for GLC assessment with the increasing popularity of geotagged social media and related mobile apps [27,28]. However, significant concerns are raised about the reliability of crowdsourcing data because of the varying expertise of volunteers, and the uneven distribution of volunteers would possibly lead to very few sample points in some regions [29,30].

Considering the limitation of sample-based assessment, some scientists seek to estimate accuracy without standard reference data. Given the availability of vast global and regional land cover products, various cross-comparison strategies were proposed to investigate the spatial, semantic or areal consistency between different GLC datasets [31–33]. The conversion of GLC products with heterogeneous classification systems is the first step to enabling cross-comparison [34]. Considerable efforts have been made to harmonise different land cover products, for example, by converting them into a standard classification system or developing fuzzy approaches based on the semantic similarity in the definition of land classes [32,33]. However, these methods require many subjective rules and present several difficulties in dealing with complex land classes, which makes the conversion of various GLC products still a challenging practice [35–37].

Instead of rigorous accuracy estimation, some studies have attempted to identify potentially erroneous land objects without reference data, in which models based on outlier detection techniques were widely adopted. A typical example can be seen in Radoux and Defourny (2010) [38], where probabilistic outliers in terms of spectral features were detected based on an iterative trimming method. This method was also extended to extract reliable samples for GLC mapping [39]. To overcome the possible failure of statistical assumptions in detecting outlying land objects, Chen et al. developed a series of reference-free methods by adopting proximity-based outlier detection techniques and proposed several quantitative measures of the reliability of land cover vector data [40] and raster data [21]. Although these methods are useful as an auxiliary tool to assist the manual inspection and the improvement of land cover products, they cannot estimate the confusion matrix and fail to provide explicit information of accuracy.

Another batch of model-based studies took the ground truth as an unobserved (latent) variable. Inference models, such as the latent class model (LCM) and triple collocation approach (TCCA (The acronym TCCA is used following the original study of Pierdicca et al. [41])), were adapted to estimate the thematic accuracy based on additional labels for each land case [13,41–44]. Specially, TCCA was established to estimate the confusion matrixes of three classification systems without reference data. The estimation was proven unbiased when the three systems' errors were conditionally independent [41]. These pioneering works verified the feasibility of estimating land cover accuracy without ground reference data, and TCCA seems promising because it can provide rich accuracy information (e.g., the confusion matrix) rather than a single accuracy metric. However, a critical issue emerges when applying TCCA to the assessment of GLC. Given that a GLC product could be deemed as a single snapshot of the surface at a specific point in time, no other independent versions of GLC follow the same classification scheme and specification that can directly serve as the additional two systems to build up the foundation of TCCA. Furthermore, given the massive data and unbalanced distribution of GLC, the performance of TCCA in GLC accuracy estimation remains unclear.

This study aims to investigate strategies for using TCCA when only a single classification dataset is available and, thus, proposes a feasible workflow for estimating GLC accuracy without reference data. To reduce the influence of complex classification schemes, the multi-class problem in GLC is transformed into multiple binary-class problems when estimating the accuracy of each land class. Taking geographic tiles as the basic assessed unit, two local classifiers are independently trained for each land class of each assessed tile based on samples randomly selected from its neighbouring tiles. To ensure the basic assumption on conditional independence of TCCA, a Gaussian density function is applied to weight the number of samples from neighbouring tiles of different orders. These local classifiers are then applied to predict the land labels of the assessed unit, and the resultant local classifications and the original one are taken as the input of TCCA to estimate the classification accuracy. We also explore the effectiveness of the outlier detection technique in extracting reliable local samples, which might reduce errors in local classifications that are inherited from the original data and thus avoid the potential violation of conditional independence. Building upon over 15 million sample points retrieved from Google Earth Engine (GEE) and model parameters tuned from a regional land cover dataset, LandCover-Net Africa (LCN-AFR), we test the performance of our approach on high-resolution GLC datasets, European Space Agency (ESA) WorldCover 2020. Our approach is expected to provide a low-cost solution for a rapid investigation of the spatial accuracy of GLC and enrich the methodological framework of GLC mapping.

2. Methods

2.1. Mathematical Foundation of TCCA

The triple collocation technique was initially developed for modelling the errors in ocean wind speed retrievals [45] and subsequently adopted in the assessment of many geophysical variables, such as soil moisture [46], sea surface salinity [47] and precipitation [48]. Concerning discrete variables in classification problems, Pierdicca et al. [41] conceived the TCCA model to estimate the confusion matrixes of three classification systems without ground truth data.

Let X, Y and Z denote the three systems, and Θ be the unobserved ground truth. The sample points associated with each X, Y, Z system and Θ are denoted as x, y, z and θ, respectively. The basic assumption of TCCA is that the errors of X, Y, Z are conditionally independent (e.g., $P(x|y,\theta) = P(x,\theta)$), and thus the following equation is obtained:

$$P(x,y,\theta) = P(x|y,\theta)P(y,\theta) = \frac{P(x,\theta)P(y,\theta)}{P(\theta)} \tag{1}$$

By marginalising over parameter θ, we could obtain the joint probability $P(x,y)$, which is equivalent to the confusion matrix of X, Y systems. The element of the confusion matrix XY can be expressed as follows:

$$p_{i,j}^{XY} = \sum_{k=1}^{N} \frac{p_{i,k}^{X\Theta} p_{i,k}^{Y\Theta}}{p_k^{\Theta}} \tag{2}$$

where $p_{i,k}^{X\Theta}$ denotes the joint probability for a sample point of a true class k being labelled as class i in system X, and p_k^{Θ} is the probability for a sample point belonging to class k in ground truth Θ.

According to Equation (2), the target confusion matrixes $X\Theta$, $Y\Theta$ and $Z\Theta$ in accuracy estimation should satisfy the following equations after some matrix operations:

$$\begin{aligned} X\Theta \cdot P \cdot X\Theta^T &= XZ \cdot YZ^{-1} \cdot XY^T \\ Y\Theta \cdot P \cdot Y\Theta^T &= YZ \cdot XZ^{-1} \cdot XY \\ Z\Theta \cdot P \cdot Z\Theta^T &= YZ^T \cdot XY^{-1} \cdot XZ \end{aligned} \tag{3}$$

where P is a diagonal matrix composed of $\frac{1}{p_k^{\Theta}}$ $(k = 1, \ldots, N)$. Additionally, considering the prevalence of each class in Θ, one can obtain the following constraints:

$$p_k^{\Theta} = \sum_{i=1}^{N} p_{i,j}^{X\Theta} = \sum_{i=1}^{N} p_{i,j}^{Y\Theta} = \sum_{i=1}^{N} p_{i,j}^{Z\Theta} \tag{4}$$

Considering the joint probability $P(x, y, z, \theta)$ similar to the deduction in Equations (1) and (2), one can obtain additional constraints:

$$p_{i,j,k}^{X,Y,Z} = \sum_{m=1}^{N} \frac{p_{i,m}^{X,\Theta} p_{j,m}^{Y,\Theta} p_{k,m}^{Z,\Theta}}{\left(p_m^{\Theta}\right)^2} \tag{5}$$

Finally, together with Equations (3)–(5), the target confusion matrixes $X\Theta$, $Y\Theta$ and $Z\Theta$ can be obtained with some algebraic operations, and more computational details can be found in the original paper [41].

2.2. Solution for TCCA Applied to GLC Assessment

To distinguish our method from the original TCCA, we termed our method GLC-TCCA in the following contents. The workflow of GLC-TCCA is shown in Figure 1. The core of GLC-TCCA is to generate additional two classification system based on local classifiers trained by samples from neighbouring tiles of the assessed one. Indeed, training local classifiers is not unusual in current land cover production [39,49,50]. Given the heterogeneity of the surface, a local classifier is expected to achieve better classification accuracy than a global one by adaptively learning the regional characteristics of the land. Sample for training local classifiers are obtained by either manual interpretation or derived from the existing land cover products [51]. While manual interpretation generally provides reliable samples but involves a huge workload [52], deriving samples from existing datasets has been demonstrated to be much more efficient and allows the resultant classifications to follow the same classification scheme, making it increasingly popular in large-scale land cover mapping [51,53]. The processes will be introduced step by step in the following subsections.

2.2.1. Data Partition

The GLC dataset is first partitioned using specific geographical tiles (e.g., a 3×3 degree grid). This step aims to determine the assessed unit in the estimation and reduce the data size processed at once.

2.2.2. Neighbourhood Construction

Training sample points are randomly selected from the training pool of each tile t, which is here defined as its neighbourhood set $\Phi_k(t) = \{\phi_1, \ldots, \phi_k\}$, where ϕ_i is the i-order neighbourhood of t_i as shown in Figure 1, and k is the maximum order considered here. Unlike the common definition of the neighbourhood as the adjacent 3×3 tiles [51,54], the adoption of a high-order neighbourhood would introduce a certain amount of sample points from distant tiles into the training phase. According to Tobler's First Law of Geography [55], points that are distant from t would be less related to the ones from t. Given the basic assumption of

TCCA, that is, the errors of the three systems should be conditionally independent to generate unbiased estimation [41], using a high-order neighbourhood set is expected to increase further the independence between $\Phi_k(t)$ and t, while keeping the effectiveness of local classifiers as local features are still captured. However, as high-order neighbourhoods would dominate $\Phi_k(t)$ with an increase in k, this study adopts a Gaussian density function to determine the number of sample points m_i from the i-order neighbourhood, which can be written as:

$$m_i = M \times \frac{\mathrm{Norm}(u, \sigma^2)}{\sum_{i=1}^{k} \mathrm{Norm}(u, \sigma^2)} \tag{6}$$

where the $\mathrm{Norm}(u, \sigma^2)$ refers to the Gaussian density function with the expectation of u and standard deviation of σ. M is the total number of sample points for training. For simplicity, σ is set to 1, and k is set to 9 in this study, while the optimal value of u is tuned by the simulation experiment discussed in Section 3.2.

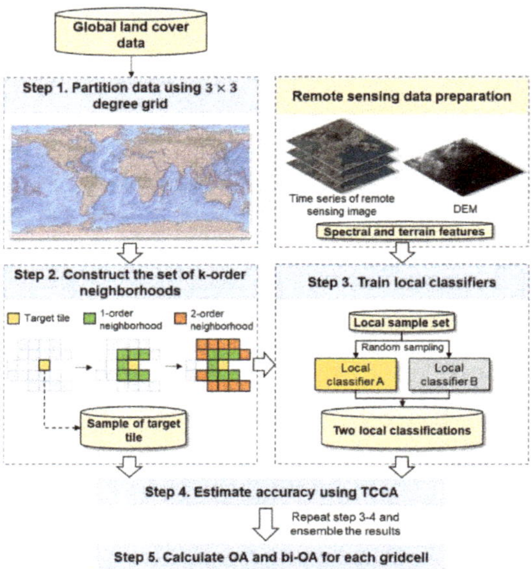

Figure 1. Workflow of the proposed GLC-TCCA.

2.2.3. Training Two Local Classifiers

By taking the sample points' class label as the dependent variable, two different classifiers were trained on a balanced dataset using features from remote sensing data. To reduce the correlation between the two local classification systems, the classifiers were trained using exclusive points. It should be notable that since the two classifiers have different mechanisms and are trained on different bases, their errors tend to be uncorrelated [44,56].

The local classifiers were subsequently applied to t_i so that two additional systems could be obtained for further analysis using TCCA. Given the widely existing spatial heterogeneity in large-scale land cover data [57,58], the classification errors in different tiles are also likely to present different distributions. In this regard, the correlation between the original classification and the one predicted by local classifier could be largely reduced using the sample from the defined neighbourhood in this study.

Finally, to increase the robustness of our method, the training and analysis phases were repeated multiple times with different random samples, and the results were combined to obtain the final estimation of the thematic accuracy.

2.2.4. Estimating the Accuracy for a Single Class

TCCA is theoretically capable of handling multi-class problems with the complexity of approximately $O(n^2)$, which would significantly increase with a larger number of classes [41]. Therefore, in this study, considering various classification schemes in GLC projects, we transform the multi-class problem into multiple binary-class problems by iteratively taking one class as positive and the rest as negative. This operation has at least two advantages. Firstly, the model uncertainty from local classifiers and TCCA would be vastly reduced as the number of classes decreases [59]. Secondly, the sample space, which determines the possible outcome of an experiment, will significantly change as some classes are merged [60]. As a result, the correctness of the reformed dichotomous data would be theoretically higher than the raw one, which might help to reduce the potential error correlation between the local classification systems and the original one to improve the effectiveness of TCCA.

2.2.5. Estimating the Overall Accuracy

One issue raised by the transformation mentioned above is the ensembling of accuracy metrics: one can only obtain the accuracy estimation for a single class for each binary-class case, while the accuracy information (e.g., the OA) of the entire data is not directly measured. The present study takes OA as an example to demonstrate the calculation of the accuracy of the entire data using the OA estimations from the binary-class case of each class.

Let CM_t denote the confusion matrix of the sample from tile t, which can be expressed as follows:

$$CM_t = \begin{bmatrix} n_{11} & n_{12} & \cdots & n_{1N} \\ n_{21} & n_{22} & \cdots & n_{2N} \\ \cdots & \cdots & \cdots & \cdots \\ n_{N1} & \cdots & \cdots & n_{NN} \end{bmatrix} \quad (7)$$

where N is the number of classes of the target GLC. The OA for class m after transforming to the binary-class problem could be expressed as follows:

$$OA_m = \frac{n_{mm} + \sum\limits_{i \neq m, j \neq m}^{N} n_{ij}}{\sum\limits_{i=1, j=1}^{N} n_{ij}} \quad (8)$$

We term the OA in Equation (8) as bi-OA in the following content to distinguish it from the generic OA metric. Therefore, let OA_{all} denote the OA of CM_t; the sum of bi-OAs could be written as follows:

$$\sum_{m=1}^{N} OA_m = \frac{N \sum\limits_{i=1}^{N} n_{ii} + (N-2) \sum\limits_{i=1, j=1, i \neq j}^{N} n_{ij}}{\sum\limits_{i=1, j=1}^{N} n_{ij}} = 2 * OA_{all} + N - 2 \quad (9)$$

Thus, one can follow Equation (9) to calculate OA_{all} based on the bi-OA estimations.

2.3. Testing Conditional Independence

Conditional independence is a crucial assumption in GLC-TCCA. To find out the optimal couple of classifiers that satisfied TCCA's assumption on conditional independence, we need to first test the conditional independence among the local classification systems and the prediction. In this study, the degree of conditional independence was quantified based on the following equation:

$$\rho_1 = \frac{S_1^* - S_1^X S_1^Y}{\sqrt{S_1^X(1-S_1^X)S_1^Y(1-S_1^Y)}} \tag{10}$$

for the target class and

$$\rho_0 = \frac{S_0^* - S_0^X S_0^Y}{\sqrt{S_0^X(1-S_0^X)S_0^Y(1-S_0^Y)}} \tag{11}$$

for the rest data in the dichotomous problem. The superscripts X and Y denote the two classification systems to be compared. $S_1^* = P(X=1, Y=1|\Theta=1)$ and $S_0^* = P(X=0, Y=0|\Theta=0)$. When ρ_1 and ρ_0 are substantially close to 0, the classification system X and Y could be considered conditionally independent [56,61,62]. For simplicity, we defined a single indicator CI to quantify the correlation independence between two classification systems:

$$CI = \sqrt{|\rho_1\rho_0|} \tag{12}$$

3. Experiments

Three experiments were sequentially conducted in this article. The first experiment (Section 3.1) tested the sensitivity of TCCA when data were significantly imbalanced. The results would be beneficial in determining the minimum proportion of sample size that could be handled by TCCA in a reliable manner. The second experiment (Section 3.2) was designed to test the influences of different couples of local classifiers and to determine the optimal parameter in the Gaussian density function. The third experiment (Section 3.3) demonstrated the practices of GLC-TCCA on real-life GLC assessment.

3.1. Sensitivity of TCCA on Extremely Imbalanced Data

TCCA has been proven to be unbiased in addressing moderately unbalanced data (e.g., when the prevalence is 0.8 and 0.2 for two classes) [41]. However, given that data can be extremely unbalanced in GLC (e.g., built-up accounts for only 0.7% in WorldCover 2020), to what extent TCCA can deal with such unbalance should be further explored.

In this section, we reconsidered the scenario simulated in the original paper of TCCA, in which three classifiers X, Y and Z, with a false alarm rate and misdetection rates of (8%,12%), (10%, 30%) and (20%, 40%), respectively, were applied to unbalanced data with 4000 sample points (3200 for class 1 and 800 for class 2) [41]. We examined the effectiveness of TCCA by varying the prevalence of class 2 from 0.2 to 0.005. Particularly, as the uncertainty of TCCA would be large with a small sample size (e.g., less than 1000), we reset the total number of simulated samples to 400,000 and fixed the sampling rate to 0.5. Thus, each test had at least 1000 points for the minor class. Furthermore, we monitored the estimation of commonly used metrics, including OA, PA and UA, in terms of their mean absolute percentage error (MAPE) in this experiment. Mathematically, MAPE could be written as:

$$MAPE = \frac{|Y - \hat{Y}|}{Y} \tag{13}$$

where $\hat{Y}$ is the estimate, and Y is the actual value.

The results of the MAPE measurements for different accuracy metrics are shown in Figure 2. TCCA achieved especially low MAPE for OA in all the cases, whereas the estimations of PA and UA were less accurate and indicated large deviations when the minor class was assigned a prevalence lower than 0.01. When the size of the minor class is reduced to a comparable magnitude to the model uncertainty of TCCA, even a slight error in estimating the element of the confusion matrix will introduce significant deviations for PA and UA of the minor class. The estimation for OA is more stable than the one for other metrics because OA in an unbalanced scenario is primarily determined by the true-positive number of the major class, which would not change significantly because of the relatively small error in the elements of the confusion matrix. Thus, to maintain the reliability of the results, we skipped the estimation of classes with a prevalence lower than 0.01 and focused on the OA estimation in the following experiments. It is also noteworthy that, according to the study of Pierdicca et al. [41], TCCA becomes less stable when the sample size is small. Therefore, we further excluded the estimation of classes with a sample size lower than 100 to control the reliability in the following experiments.

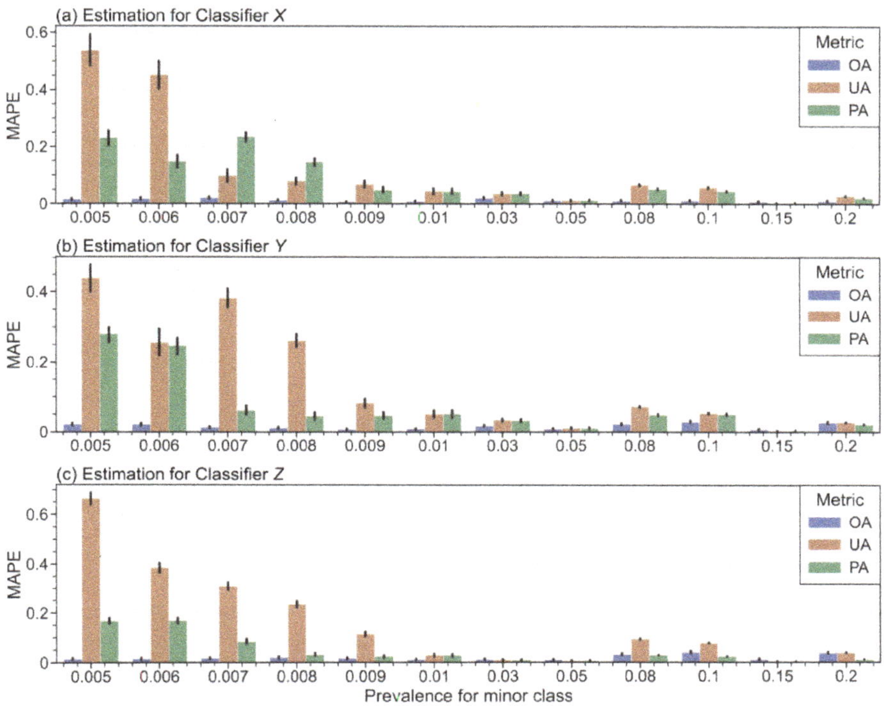

Figure 2. Values of MAPE with different prevalence.

3.2. *Tuning GLC-TCCA on a Real-Life Dataset with Known Ground Truth*

3.2.1. Experimental Setup

In this section, we used LCN-AFR to test the performance of our proposed approach and tune parameters for the further case study. LCN-AFR is an annual training dataset at a 10 m resolution for land cover classification, which consists of 1980 chips of 256 × 256 pixels across 89 tiles (3 × 3 degree) in Africa [63]. The whole dataset was produced under a seven-class scheme and manually validated by three independent experts. More details on LCN-AFR can be found in Alemohammad and Booth (2020) [63]. The present study filtered the most reliable pixels with a consensus score of 100, which can be used as ground truth for validation. Furthermore, to save the computational cost, we randomly sampled 5% from the filtered result and finally obtained a collection of over 3.8 million pixels for further experiments.

The workflow of this experiment is shown in Figure 3. Firstly, we trained a preliminary classifier based on 5% random pixels and the Sentinel-2 time series associated with each chip in LCN-AFR. A total of 240 spectral features were extracted from the Sentinel-2 images, following the instructions in the original paper on LCN-AFR [63]. Then, the trained preliminary classifier was applied to the whole collection to generate predictions for each pixel. Finally, we applied GLC-TCCA to the prediction in each tile, following the strategies introduced in Section 2.2. Additional concerns were given to explore the influence of two initial conditions for TCCA in this experiment, including the choice of classifiers and the parameter u in Gaussian density function used for sample collection. It is worth noting that tree-based models and neural network (NN) are widely used in GLC mapping [6,9], and we selected two representative methods, namely, the random forest model (RF) (Scenario A) and the Multi-layer Perceptron classifier (MLP) (Scenario B), as the preliminary classifiers. This allowed us to test our methods under different scenarios.

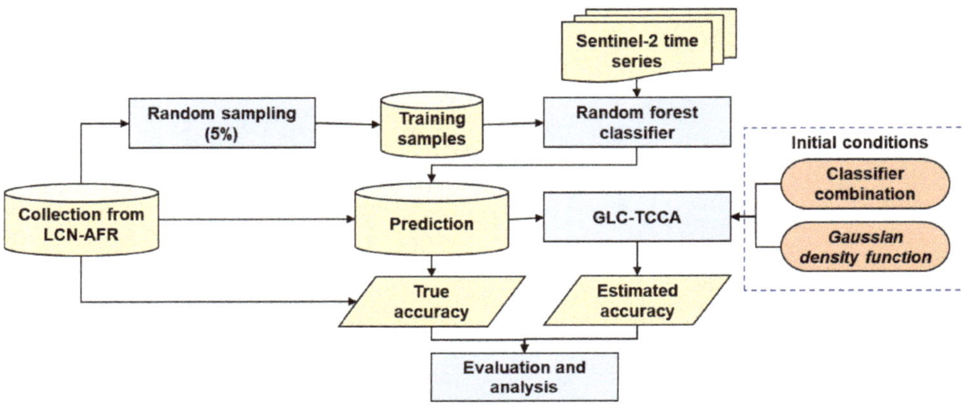

Figure 3. Processes for tuning GLC-TCCA on LCN-AFR dataset.

3.2.2. Choice of Classifiers

We chose four basic classifiers as candidates for the two local classifiers in this study, including decision tree (DT), random forest (RF), support vector machine (SVM), and Gaussian Naive Bayes (GNB).

We ranged the parameter u from 1 to 6. We first calculated the CI values between the systems produced by each classifier and the prediction. Notably, considering the computation cost and effectiveness of training, a balanced training set of 10,000 points (i.e., 5000 for the target class and 5000 for the rest) was selected during the training of the local classifier for each land class.

As shown in Table 1, DT and GNB show a lower average value of CI (<0.1) in all cases, whilst the values associated with RF and SVM are substantially larger than 0, which indicates that RF and SVM might not satisfy the assumption of conditional independence in TCCA. To further test the applicability of GNB and DT, we calculated the CI between them. Additionally, to demonstrate the effectiveness of the exclusion of assessed tile when constructing the neighbourhood set, we created a control group in which the assessed tile was included in the neighbourhood set and recorded the corresponding CI values. As shown in Figure 4, the exclusion of the assessed tile from the neighbourhood set could significantly reduce the CI values, especially when the parameter u is small. When u became larger, the Gaussian density function would assign less weight to the assessed tile in the control group. Therefore, the difference in CI between the control and experimental group would diminish. Moreover, it can be seen in Figure 4(a-3,b-3) that the increment of CI in control group is relatively smaller than those in other scenarios, which indicates that the errors in the outcomes of the selected local classifiers tend to be uncorrelated even though the training sample might inherit some errors from the assess tiles.

Table 1. Statistics of CI between the system produced by different local classifiers and values of u. Std refers to standard deviation.

Preliminary Classifier	Sampling Parameter	Local Classifier							
		DT		RF		SVM		GNB	
		Mean	Std	Mean	Std	Mean	Std	Mean	Std
RF	$u = 1$	0.084	0.042	0.411	0.117	0.387	0.114	0.082	0.057
	$u = 2$	0.081	0.037	0.408	0.112	0.375	0.112	0.073	0.050
	$u = 3$	0.068	0.036	0.391	0.108	0.361	0.108	0.062	0.047
	$u = 4$	0.061	0.035	0.349	0.103	0.326	0.116	0.053	0.040
	$u = 5$	0.056	0.033	0.327	0.123	0.298	0.115	0.050	0.032
	$u = 6$	0.054	0.031	0.311	0.125	0.284	0.123	0.045	0.030
MLP	$u = 1$	0.097	0.047	0.459	0.118	0.530	0.116	0.010	0.065
	$u = 2$	0.088	0.044	0.466	0.113	0.523	0.121	0.010	0.062
	$u = 3$	0.073	0.041	0.441	0.105	0.493	0.124	0.092	0.052
	$u = 4$	0.076	0.032	0.404	0.122	0.451	0.119	0.082	0.050
	$u = 5$	0.073	0.035	0.362	0.147	0.406	0.151	0.070	0.046
	$u = 6$	0.064	0.031	0.344	0.155	0.386	0.146	0.076	0.052

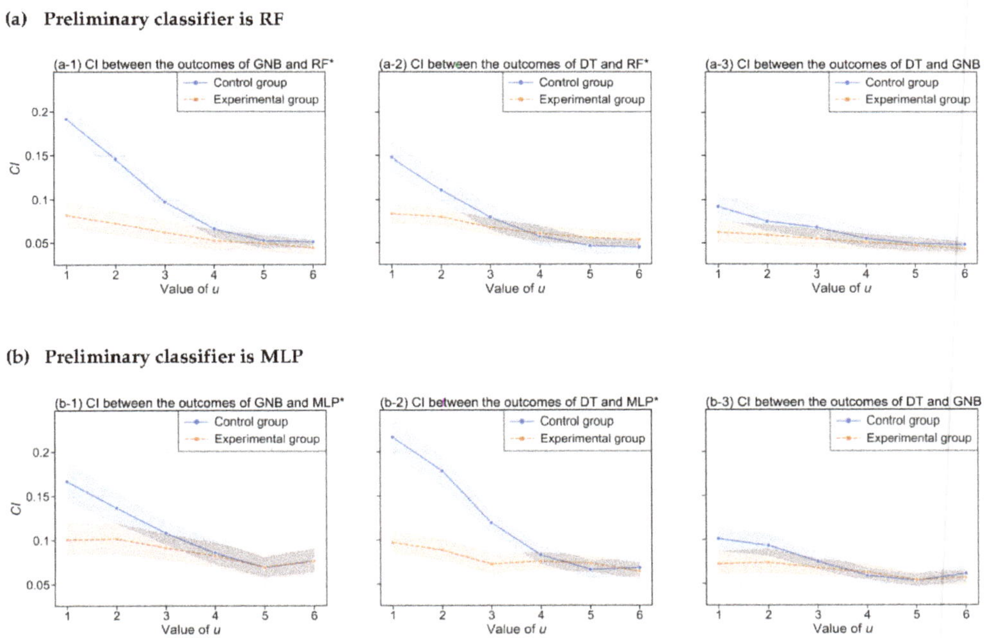

Figure 4. Statistics of CI between the preliminary classifier (i.e., RF or MLP) and local classifiers (i.e., DT and GNB) in different scenarios. * denotes the preliminary classifier.

According to the superior performance in the simulation, DT and GNB were selected as the two local classifiers for the following experiments:

3.2.3. Selection of the Parameter u in the Gaussian Density Function

To choose the optimal value of u in the Gaussian density function, we investigated the influence of different values of u on estimation accuracy. The resultant estimations were evaluated in terms of the mean absolute error (MAE) and signed deviation (SD), which are expressed as follows:

$$\begin{aligned} \text{MAE}(\hat{\theta}) &= \tfrac{1}{n}\sum_{i=1}^{n}|\theta_i - \hat{\theta}_i| \\ \text{SD}(\hat{\theta}) &= \theta_i - \hat{\theta}_i \end{aligned} \tag{14}$$

where $\hat{\theta}_i$ refers to the estimate of the true accuracy metric θ_i (e.g., OA) of grid i. Compared with MAE, which measures the average magnitude of errors in the estimates, the mean of SD (MSD) captures the average differences between truth and estimate. An MSD value smaller than zero may imply a general overestimation.

As shown in Figure 5(a-1,b-1), overestimations can be observed in the control group in terms of their MSD substantially lower than 0 when u is small (e.g., $u = 1$ or 2). The reason could be that local classifiers tend to inherit classification errors from the assessed tile in the control group. As a result, the local classifiers would produce similar predictions that contain correlated errors to the data being evaluated. According to Pierdicca et al. (2017) [41], such a correlation would lead to overestimating the accuracy of systems with similar outcomes. This inference is also supported by Table 1 and Figure 4 as the highest values of CI generally appear at $u = 1$. In contrast, the overestimation was largely reduced with the exclusion of the assessed tile as the MSD is close to 0 in the experimental group. These findings, again, justify the construction of the neighbourhood set in this study.

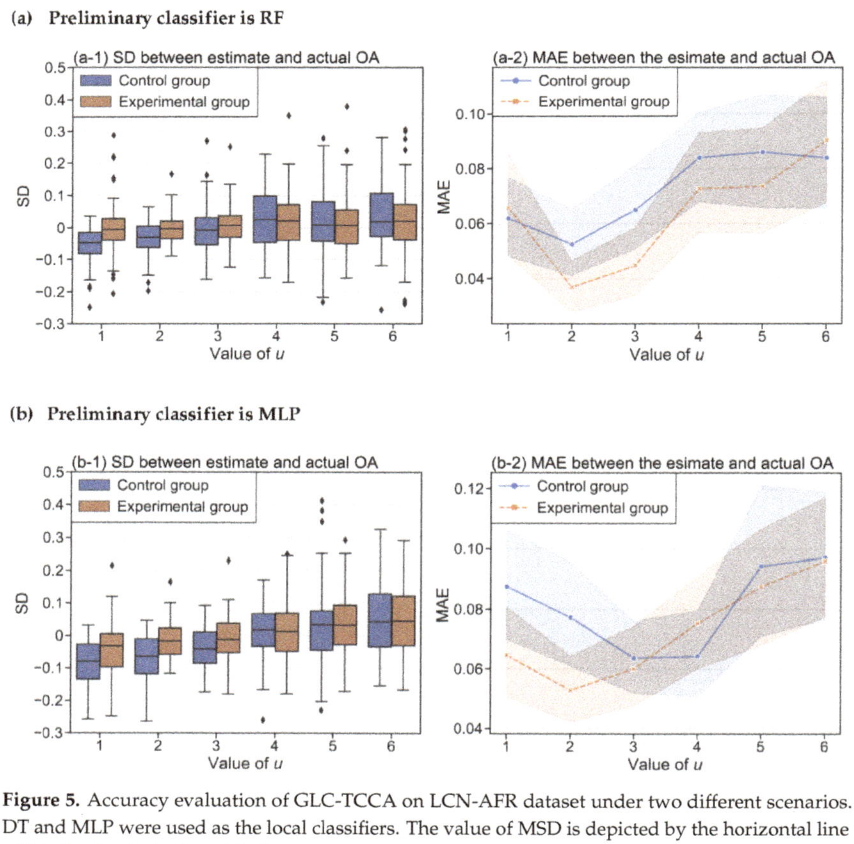

Figure 5. Accuracy evaluation of GLC-TCCA on LCN-AFR dataset under two different scenarios. DT and MLP were used as the local classifiers. The value of MSD is depicted by the horizontal line within the box in (**a-1**) and (**b-1**).

As shown in Figure 5(a-2,b-2), the MAE of the estimation declines at first and then increases with the increment of u. This can also be explained by the fact that the local classifiers tend to inherit errors that might be correlated to those in the assessed tile when the training sample concentrates in a low-order neighbourhood. However, when u becomes too large, the effectiveness of local classifiers is lowered. As TCCA generally trusts the systems that present similar outcomes more [41], TCCA would judge the target system (i.e., the outcome of the preliminary classifier) to be less accurate when it is significantly different to the other two local systems. In sum, regardless of the preliminary classifier used, the smallest MAE always appear at $u = 2$. Therefore, given that LCN-AFR has very similar data properties (e.g., resolution and classification scheme) to WorldCover 2020, we finally adopted $u = 2$ in the Gaussian density function and used DT and GNB as the local classifiers in the following experiments on WorldCover 2020.

3.3. GLC-TCCA Applied to WorldCover 2020

In this section, we demonstrated the performance of GLC-TCCA on WorldCover 2020, a 10 m resolution GLC dataset produced by ESA. WorldCover 2020 is claimed to have an overall accuracy of 74.4%. WorldCover 2020 is also available on GEE, which significantly offers easy access and data processing. Details about the accuracy of WorldCover 2020 can be found in the European Space Agency (2021) [64].

3.3.1. Data Preparation

To generate a sufficient number of global sample points for applying GLC-TCCA, we randomly picked points based on a density of approximately one point per 9 km^2, considering memory and computation limitations in GEE. To train the local classifiers, a total of 98 remote sensing features were extracted from Sentinel-1 (S1), Sentinel-2 (S2) and AW3D30 Version 3.2. Details of the feature extraction could be found in the Supplementary file. It should be noted that different feature combinations are expected to influence the results of GLC-TCCA, however, exploring the best combination is out of the scope of this study.

Given that S1 and S2 do not cover the full surface of the Earth, some sample points would be invalid because of the lack of image coverage. Thus, they were discarded in this study. For example, Greenland and the northernmost regions of Canada were excluded from this study due to the lack of S1 images in VV-VH mode. Finally, we obtained over 15 million points covering 2207 3 × 3 degree tiles (Figure 6) for the following experiments.

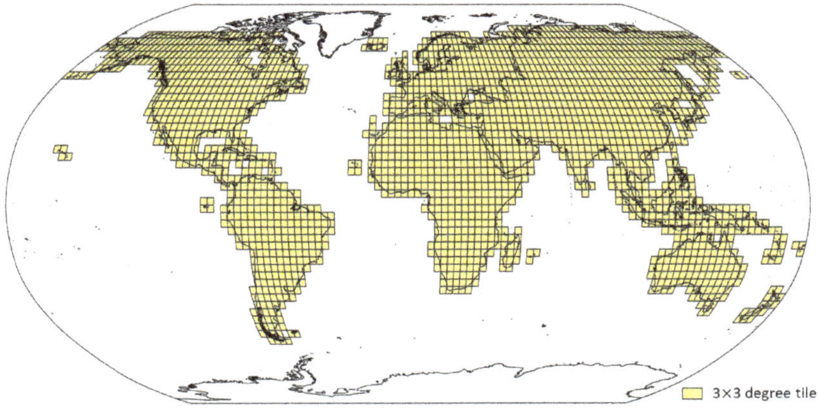

Figure 6. Distribution of 3 × 3 degree tiles.

3.3.2. Estimation of WorldCover 2020 at the Continent Level

Given the lack of rigorous reference data (i.e., grid-specific accuracy information) for WorldCover 2020, we utilised the accuracy information for each continent and the global accuracy map reported in the product report of WorldCover 2020 to verify the performance of our method.

The accuracy map estimated by GLC-TCCA and the official one are compared in Figure 7. Given that numeric data are unavailable for the official accuracy map, we cannot make a direct quantitative comparison. However, through a quick visual comparison, we found that the estimation of GLC-TCCA captured most of the spatial patterns of the accuracy map. For example, significant high-accuracy areas (connected by a blue dashed line) and low-accuracy areas (connected by a red dashed line) were successfully identified by GLC-TCCA. The estimated OA values for each continent are summarised in Table 2. In most cases, the estimations of GLC-TCCA are close to the reported OA provided by ESA, with a mean absolute error (MAE) of 3.40, and a mean absolute percentage error (MAPE) of 4.71%. These findings prove the effectiveness of the proposed GLC-TCCA in providing a generally reliable estimation and capturing the most spatial variation of the accuracy of WorldCover 2020.

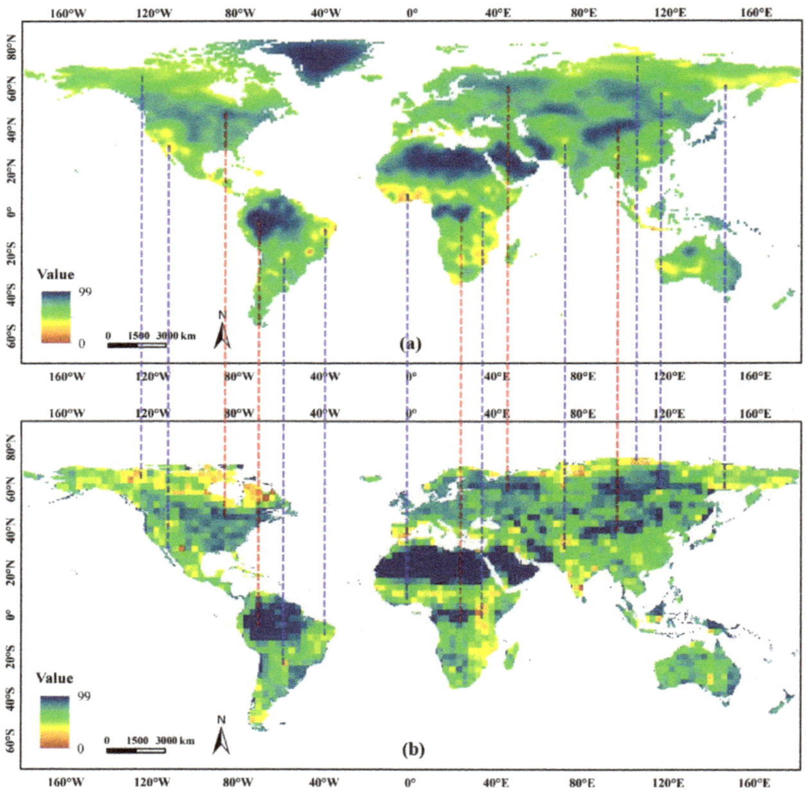

Figure 7. Visual comparison between the official accuracy map of WorldCover 2020: (**a**) and the estimated one of GLC-TCCA (**b**). (**a**) is retrieved from the product validation report of WorldCover 2020. The red dashed lines connect areas with high accuracy in both (**a**) and (**b**), whilst the blue dashed lines connect those low-accuracy areas.

Table 2. Comparison of the reported and estimated OA at the continent level.

Continent	Reported OA (Official)	Estimated OA	Absolute Error	Absolute Percentage Error
South America	76.10	80.04	3.94	5.18%
Europe	76.80	77.14	0.34	0.44%
Asia	80.70	77.59	5.11	3.85%
Africa	73.60	76 41	2.81	3.82%
North America	72.20	70.96	1.24	1.71%
Oceania	67.50	76.47	8.97	13.29%
Average			3.40	4.71%

There are at least two reasons that explain the mismatches between Figures 7a and 6b: (1) Model errors exist in the results of GLC-TCCA, and (2) Given that two accuracy layers are produced with different spatial resolutions (i.e., 100 m for (a) and 3 × 3 degree tile for (b)), some local extremum in (a) tend to be smoothed in (b), which is believed to cause a certain level of mismatch, such as the one in eastern South America.

It is noteworthy that certain grids at high latitudes have smaller areas due to the distortion caused by projection and occupation by the sea, resulting in their reduced weights in the calculation of overall accuracy. The northeast regions of North America, for instance, visually exhibit a greater difference in accuracy; however, this does not significantly diminish the overall accuracy of North America, as shown in Table 2.

The largest absolute error was observed in the estimation of Oceania. The reason for the relatively lower effectiveness of GLC-TCCA in Oceania are as follows: (1) A large proportion of grids in Oceania is close to the sea, and their neighbourhoods are relatively broken, containing fewer neighbouring tiles than the ones in other continents. As shown in Table 3, the average number of neighbours of Oceania is only 146, which is significantly lower than the global average of 223. In that sense, sample points might concentrate in low-order neighbourhoods, and an overestimation could be expected, as discussed in Section 3.2. The same issue also occurs in South America, where the number of neighbours is only 172. (2) As Oceania presents the lowest true OA of 67.5% among all the continents, sample points from the neighbourhood set are more likely to be wrongly labelled. Thus, classifications produced by local classifiers tend to inherit more correlated errors from the original data, which is expected to result in an overestimation, according to Pierdicca et al. [41].

Table 3. Number of neighbours in the neighbourhood set Φ_9 at the continent level.

Continent	Average Number of Neighbouring Tiles
South America	172
Europe	245
Asia	255
Africa	237
North America	196
Oceania	146
Global	223

3.3.3. Improving GLC-TCCA with the Screening of Reliable Sample

The final section verifies the feasibility and effectiveness of the proposed GLC-TCCA. However, as mentioned above, some estimates are less reliable because of the high proportion of erroneous sample points. To that end, we added an outlier detection phase based on isolation forest (IForest) for the filtration of reliable sample points before training local classifiers. Compared with other outlier detection techniques, IForest has been proven robust and efficient in handling large data volumes and high-dimensional problems given its ensemble framework and nearly linear complexity [65].

For GLC data, erroneous pixels may be limited to common outliers that present abnormal dispersions from others and possibly form clusters when the number of errors reaches a certain level. In this regard, we discarded the top 10% sample points in the anomaly score estimated by IForest. We chose 10% as the threshold because OA for GLC products is commonly around 80, and we conservatively assumed that 50% of the errors might form clusters. Thus, the remaining sample can still capture most land characteristics that allow training effective local classifiers.

As shown in Figure 8, the overestimation of OA in Oceania and South America was significantly reduced, whereas the estimated accuracy of other continents remained consistent with the previous estimates that were produced without IForest. After using IForest, the MAE was reduced from 3.40 to 2.54, and MAPE was reduced from 4.71% to 3.45%.

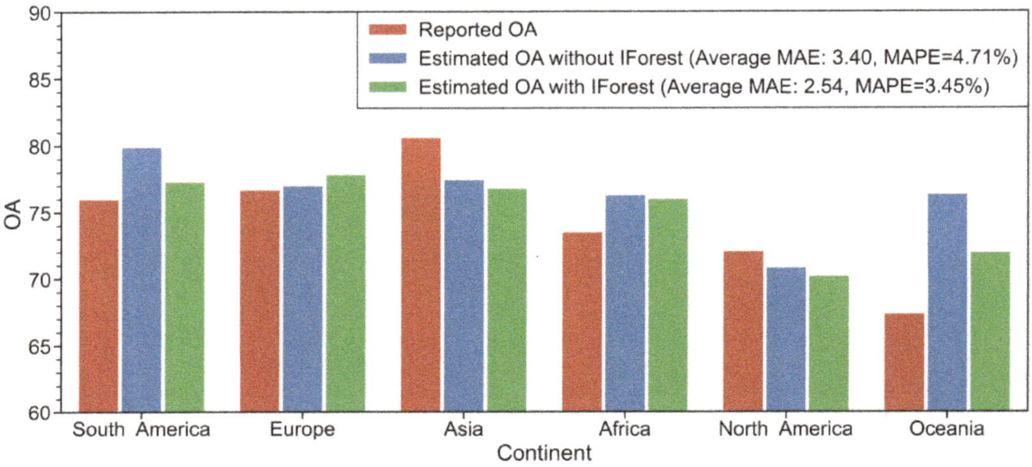

Figure 8. Comparison of the estimates with and without using IForest for sample screening.

3.3.4. Class-Specific Accuracy Analysis Based on the Estimates of Improved GLC-TCCA

The most significant deviation seems to occur in the *Grassland* of Oceania. According to the product report of WorldCover 2020, the true bi-OA of *Grassland* in Oceania was computed to be 72%, which is the lowest among all the cases we considered in Figure 9. Therefore, approximately 28% of the samples are erroneous when training the local classifiers to evaluate the accuracy of *Grassland*. Although IForest was applied to reduce the influence of the errors from the original data, the influence cannot be fully avoided when the quantity of erroneous pixels is too large. Furthermore, *Grassland* accounts for 54.64% of the original data in Oceania. This low accuracy of *Grassland* and its high proportion in Oceania would increase the error correlation between the three input systems for GLC-TCCA and thus result in less reliable estimates.

We further calculated the bi-OA for each class based on Equation (8) and compared the result with its truth derived from the product report of WorldCover 2020. As shown in Figure 9, the low value of MAPE for each continent proves the general correctness of the bi-OA estimates. However, deviations can be observed in some minor classes, such as *Moss and Lichen* in Asia (0.71% of the original data in Asia) and *Herbaceous wetland* in Europe (0.71% of the original data in Europe). As discussed in Section 3.1, lower prevalence will result in a larger deviation when solving the binary-classification problem using TCCA. However, these deviations would have less effect on the estimation of the entire dataset because the absolute sample sizes of these classes are small.

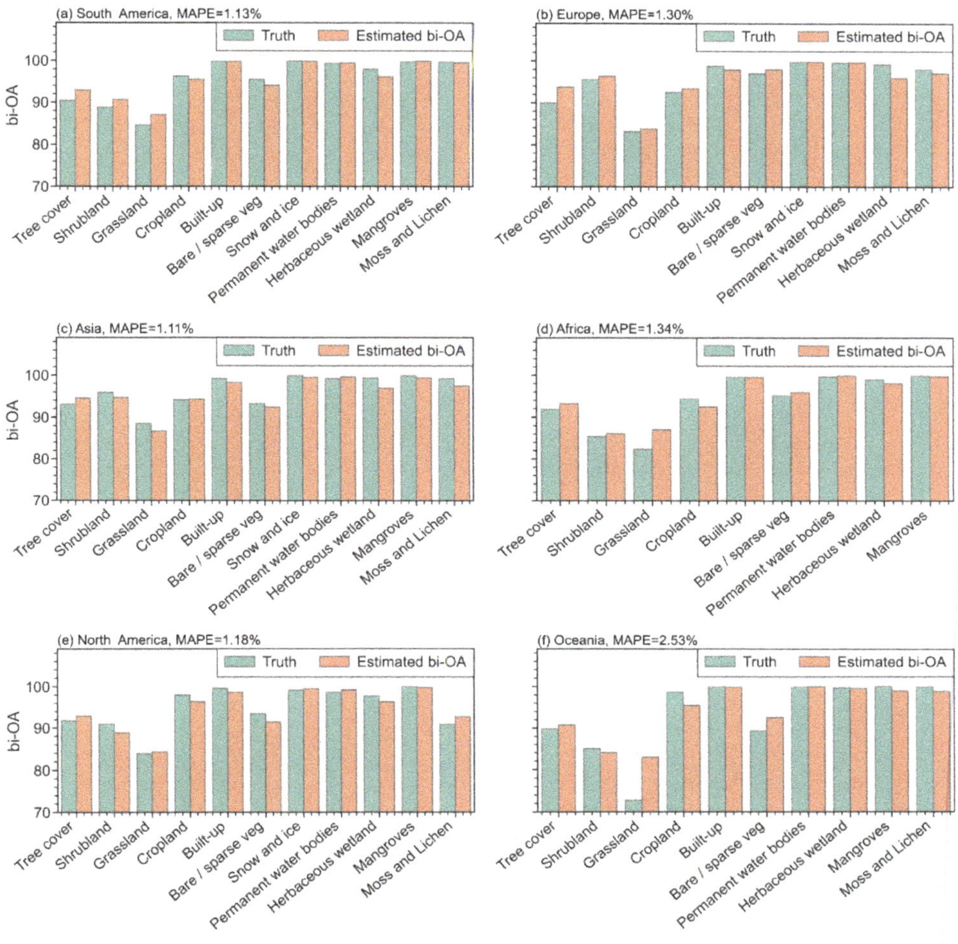

Figure 9. Comparison of the reported and estimated class-specific accuracy at the continent level.

4. Discussion

The essence of GLC-TCCA is to estimate the confusion matrixes of multiple classifications based on their agreements. The overall correctness of the data should be carefully checked before applying GLC-TCCA in practice. The estimation could be invalid if the accuracy of the original data is too low. In that situation, both local classifiers are likely to produce erroneous predictions, which will lead to a high proportion of correlated errors between local classifications and thus violate the assumption of their conditional independence given the ground truth. However, given that the majority of existing GLC data are reported to have a similar OA to WorldCover 2020, which is over 70%, we believe GLC-TCCA could also apply to most GLC products, given its performance in WorldCover 2020.

It is important to note that the accuracy of each classification does not affect the results of GLC-TCCA, as long as the assumption of conditional independence is satisfied [41]. In this study, the construction of neighbourhoods plays a crucial role in consolidating this assumption by enabling the utilization of spatial heterogeneity to reduce the potential dependency between the assessed data and local classifications. This step also ensures the effectiveness of local classifiers by using local samples and avoiding potential correlation between local classifications caused by very low accuracy as discussed above. In addition, we weakened the impact of classification errors on local classifiers by screening reliable samples and conducting multiple times of training, which would further reduce the potential dependency between the original data and local classifications. The effectiveness of these operations was demonstrated by the significantly reduced CI values shown in Figure 4 and the improved estimates shown in Figure 8.

The computational cost could be a critical concern when using GLC-TCCA in practice. Retrieving massive global sample points with their spectral and terrain features is the first step for conducting GLC-TCCA. Fortunately, with the help of modern cloud computing platforms, such as GEE, the remote sensing data processing cost can be largely reduced. Furthermore, although GLC-TCCA involves many computational procedures, such as training local classifiers and outlier detectors, its ensemble framework makes most of these procedures parallelisable. Thus, the time cost can be greatly reduced by distributing the estimation task.

5. Conclusions

While various GLC products are rapidly produced and updated with advanced classification techniques, their assessment seems to "fall behind" in providing timely accuracy reports to support decision-making. Despite the high reliability, traditional sample-based assessment approaches are often restrained by the lack of ground truth and criticised for the huge workload and low efficiency. This study develops a new reference-free method termed GLC-TCCA, particularly for estimating the thematic accuracy of GLC products. Compared with traditional assessment methods that rely heavily on ground reference data, GLC-TCCA makes full use of the original classification information and does not rely on any external data except remote sensing images. According to the experiment on WorldCover 2020, GLC-TCCA can provide accurate estimates for individual land classes and the whole dataset at the continent level with a relative error of approximately 4%.

The reference-free characteristics of GLC-TCCA make it a useful tool for GLC producers and users. From the producers' perspective, GLC-TCCA allows a quick survey of spatial accuracy of large-scale land cover datasets without conducting situ measurement. Furthermore, the spatial accuracy layer produced by GLC-TCCA is valuable for producers to target poorly classified regions and thus take more efficient interventions (e.g., training sophisticated local classification models or scrutinising poorly classified regions) to improve the data accuracy before delivery to the users. From users' perspective, the results of GLC-TCCA could serve as a third-party inspection to validate the reported accuracy of land cover data and allow users' self-services to obtain classification accuracy, which is only empirically estimated and reported by the producers.

Improvements can be made in future works. Firstly, as many grid cells are located on the edge of a continent or island, their neighbours' actual number and spatial coverage could be significantly smaller than those located in the land's interior. To address that issue, adaptive neighbourhoods could be developed to train better local classifiers. Secondly, only spectral and terrain features are used in this study. Given the massive land classification information in existing GLC data and crowdsourcing data, how auxiliary information could be used to enhance GLC-TCCA needs to be further explored. Thirdly, the effect of the accuracy of the original data on the reliability of estimations needs to be further quantified by experiments and simulations. Lastly, stable and efficient software needs to be developed in future work to enhance the practical utilisation of GLC-TCCA.

Supplementary Materials: The following supporting information can be downloaded at: https://www.mdpi.com/article/10.3390/rs15092255/s1.

Author Contributions: Conceptualization, W.S.; data curation, P.C. and R.C.; formal analysis, P.C. and H.H.; funding acquisition, P.C.; investigation, P.C.; methodology, P.C.; resources, P.C. and H.H.; software, P.C. and R.C.; supervision, W.S.; validation, P.C. and H.H.; visualization, P.C.; writing—

original draft, P.C.; writing—review and editing, H.H. and W.S. All authors have read and agreed to the published version of the manuscript.

Funding: This research was funded by the National Natural Science Foundation of China, grant number 42101351, and Guangdong Basic and Applied Basic Research Foundation, grant number 2023A1515011174.

Conflicts of Interest: The authors declare no conflict of interest. The funders had no role in the design of the study; in the collection, analyses, or interpretation of data; in the writing of the manuscript; or in the decision to publish the results.

References

1. Gong, P.; Zhang, W.; Yu, L.; Li, C. New research paradigm for global land cover mapping. *J. Remote Sens.* **2016**, *20*, 1002–1016. [CrossRef]
2. Loveland, T.R.; Reed, B.C.; Brown, J.F.; Ohlen, D.O.; Zhu, Z.; Yang, L.; Merchant, J.W. Development of a global land cover characteristics database and IGBP DISCover from 1 km AVHRR data. *Int. J. Remote Sens.* **2000**, *21*, 1303–1330. [CrossRef]
3. Bossard, M.; Feranec, J.; Otahel, J. *CORINE Land Cover Technical Guide: Addendum 2000*; European Environment Agency: Copenhagen, Denmark, 2000.
4. Brown, C.F.; Brumby, S.P.; Guzder-Williams, B.; Birch, T.; Hyde, S.B.; Mazzariello, J.; Czerwinski, W.; Pasquarella, V.J.; Haertel, R.; Ilyushchenko, S.; et al. Dynamic World, Near real-time global 10 m land use land cover mapping. *Sci. Data* **2022**, *9*, 251. [CrossRef]
5. Chen, J.; Chen, J.; Liao, A.; Cao, X.; Chen, L.; Chen, X.; He, C.; Han, G.; Peng, S.; Lu, M.; et al. Global land cover mapping at 30 m resolution: A POK-based operational approach. *ISPRS J. Photogramm. Remote Sens.* **2015**, *103*, 7–27. [CrossRef]
6. Gong, P.; Wang, J.; Yu, L.; Zhao, Y.; Zhao, Y.; Liang, L.; Niu, Z.; Huang, X.; Fu, H.; Liu, S.; et al. Finer resolution observation and monitoring of global land cover: First mapping results with Landsat TM and ETM+ data. *Int. J. Remote Sens.* **2013**, *34*, 2607–2654. [CrossRef]
7. Hansen, M.C.; Potapov, P.V.; Pickens, A.H.; Tyukavina, A.; Hernandez-Serna, A.; Zalles, V.; Turubanova, S.; Kommareddy, I.; Stehman, S.V.; Song, X.-P.; et al. Global land use extent and dispersion within natural land cover using Landsat data. *Environ. Res. Lett.* **2022**, *17*, 034050. [CrossRef]
8. Belward, A.S.; Skøien, J.O. Who launched what, when and why; trends in global land-cover observation capacity from civilian earth observation satellites. *ISPRS J. Photogramm. Remote Sens.* **2015**, *103*, 115–128. [CrossRef]
9. Li, X.; Chen, G.; Liu, X.; Liang, X.; Wang, S.; Chen, Y.; Pei, F.; Xu, X. A New Global Land-Use and Land-Cover Change Product at a 1-km Resolution for 2010 to 2100 Based on Human–Environment Interactions. *Ann. Assoc. Am. Geogr.* **2017**, *107*, 1040–1059. [CrossRef]
10. Mantyka-Pringle, C.S.; Visconti, P.; Di Marco, M.; Martin, T.G.; Rondinini, C.; Rhodes, J.R. Climate change modifies risk of global biodiversity loss due to land-cover change. *Biol. Conserv.* **2015**, *187*, 103–111. [CrossRef]
11. Straume, K. The social construction of a land cover map and its implications for Geographical Information Systems (GIS) as a management tool. *Land Use Policy* **2014**, *39*, 44–53. [CrossRef]
12. Congalton, R.G.; Gu, J.; Yadav, K.; Thenkabail, P.; Ozdogan, M. Global Land Cover Mapping: A Review and Uncertainty Analysis. *Remote Sens.* **2014**, *6*, 12070–12093. [CrossRef]
13. Chen, P.; Shi, W.; Kou, R.; Wan, Y. A quantitative investigation of the uncertainty associated with mapping scale in the production of land-cover/land-use data. *Int. J. Remote Sens.* **2018**, *39*, 8798–8817. [CrossRef]
14. Tchuente, A.T.K.; Roujean, J.-L.; Faroux, S. ECOCLIMAP-II: An ecosystem classification and land surface parameters database of Western Africa at 1km resolution for the African Monsoon Multidisciplinary Analysis (AMMA) project. *Remote Sens. Environ.* **2010**, *114*, 961–976. [CrossRef]
15. Tsendbazar, N.; Herold, M.; Li, L.; Tarko, A.; de Bruin, S.; Masiliunas, D.; Lesiv, M.; Fritz, S.; Buchhorn, M.; Smets, B.; et al. Towards operational validation of annual global land cover maps. *Remote Sens. Environ.* **2021**, *266*, 112686. [CrossRef]
16. Nelson, M.D.; Garner, J.D.; Tavernia, B.G.; Stehman, S.V.; Riemann, R.I.; Lister, A.J.; Perry, C.H. Assessing map accuracy from a suite of site-specific, non-site specific, and spatial distribution approaches. *Remote Sens. Environ.* **2021**, *260*, 112442. [CrossRef]
17. Foody, G.M. Explaining the unsuitability of the kappa coefficient in the assessment and comparison of the accuracy of thematic maps obtained by image classification. *Remote Sens. Environ.* **2020**, *239*, 111630. [CrossRef]
18. Olofsson, P.; Foody, G.M.; Herold, M.; Stehman, S.V.; Woodcock, C.E.; Wulder, M.A. Good practices for estimating area and assessing accuracy of land change. *Remote Sens. Environ.* **2014**, *148*, 42–57. [CrossRef]
19. Zhao, Y.; Gong, P.; Yu, L.; Hu, L.; Li, X.; Li, C.; Zhang, H.; Zheng, Y.; Wang, J.; Zhao, Y.; et al. Towards a common validation sample set for global land-cover mapping. *Int. J. Remote Sens.* **2014**, *35*, 4795–4814. [CrossRef]
20. Stehman, S.V.; Foody, G.M. Key issues in rigorous accuracy assessment of land cover products. *Remote Sens. Environ.* **2019**, *231*, 111199. [CrossRef]
21. Chen, P.; Huang, H.; Shi, W. Reference-free method for investigating classification uncertainty in large-scale land cover datasets. *Int. J. Appl. Earth Obs. Geoinf.* **2022**, *107*, 102673. [CrossRef]
22. Fonte, C.C.; Martinho, N. Assessing the applicability of OpenStreetMap data to assist the validation of land use/land cover maps. *Int. J. Geogr. Inf. Sci.* **2017**, *31*, 2382–2400. [CrossRef]

23. Fritz, S.; See, L.; Perger, C.; McCallum, I.; Schill, C.; Schepaschenko, D.; Duerauer, M.; Karner, M.; Dresel, C.; Laso-Bayas, J.-C.; et al. A global dataset of crowdsourced land cover and land use reference data. *Sci. Data* **2017**, *4*, 170075. [CrossRef] [PubMed]
24. Saah, D.; Johnson, G.; Ashmall, B.; Tondapu, G.; Tenneson, K.; Patterson, M.; Poortinga, A.; Markert, K.; Quyen, N.H.; Aung, K.S.; et al. Collect Earth: An online tool for systematic reference data collection in land cover and use applications. *Environ. Model. Softw.* **2019**, *118*, 166–171. [CrossRef]
25. Stehman, S.V.; Fonte, C.C.; Foody, G.M.; See, L. Using volunteered geographic information (VGI) in design-based statistical inference for area estimation and accuracy assessment of land cover. *Remote Sens. Environ.* **2018**, *212*, 47–59. [CrossRef]
26. Chen, J.; Chen, L.; Chen, F.; Ban, Y.; Li, S.; Han, G.; Tong, X.; Liu, C.; Stamenova, V.; Stamenov, S. Collaborative validation of GlobeLand30: Methodology and practices. *Geo-Spat. Inf. Sci.* **2021**, *24*, 134–144. [CrossRef]
27. Bayas, J.C.L.; See, L.; Bartl, H.; Sturn, T.; Karner, M.; Fraisl, D.; Moorthy, I.; Busch, M.; van der Velde, M.; Fritz, S. Crowdsourcing LUCAS: Citizens Generating Reference Land Cover and Land Use Data with a Mobile App. *Land* **2020**, *9*, 446. [CrossRef]
28. Bayas, J.C.L.; See, L.; Fritz, S.; Sturn, T.; Perger, C.; Dürauer, M.; Karner, M.; Moorthy, I.; Schepaschenko, D.; Domian, D.; et al. Crowdsourcing In-Situ Data on Land Cover and Land Use Using Gamification and Mobile Technology. *Remote Sens.* **2016**, *8*, 905. [CrossRef]
29. Fonte, C.C.; Antoniou, V.; Bastin, L.; Estima, J.; Arsanjani, J.J.; Bayas, J.-C.L.; See, L.; Vatseva, R. Assessing VGI Data Quality. In *Mapping and the Citizen Sensor*; Foody, G., See, L., Fritz, S., Mooney, P., Olteanu-Raimond, A.-M., Fonte, C.C., Antoniou, V., Eds.; Ubiquity Press: London, UK, 2017; pp. 137–163.
30. Koukoletsos, T.; Haklay, M.; Ellul, C. Assessing Data Completeness of VGI through an Automated Matching Procedure for Linear Data. *Trans. GIS* **2012**, *16*, 477–498. [CrossRef]
31. Gao, Y.; Liu, L.; Zhang, X.; Chen, X.; Mi, J.; Xie, S. Consistency Analysis and Accuracy Assessment of Three Global 30-m Land-Cover Products over the European Union using the LUCAS Dataset. *Remote Sens.* **2020**, *12*, 3479. [CrossRef]
32. Hua, T.; Zhao, W.; Liu, Y.; Wang, S.; Yang, S. Spatial Consistency Assessments for Global Land-Cover Datasets: A Comparison among GLC2000, CCI LC, MCD12, GLOBCOVER and GLCNMO. *Remote Sens.* **2018**, *10*, 1846. [CrossRef]
33. Pérez-Hoyos, A.; García-Haro, F.; San-Miguel-Ayanz, J. Conventional and fuzzy comparisons of large scale land cover products: Application to CORINE, GLC2000, MODIS and GlobCover in Europe. *ISPRS J. Photogramm. Remote Sens.* **2012**, *74*, 185–201. [CrossRef]
34. Liu, L.; Zhang, X.; Gao, Y.; Chen, X.; Shuai, X.; Mi, J. Finer-Resolution Mapping of Global Land Cover: Recent Developments, Consistency Analysis, and Prospects. *J. Remote Sens.* **2021**, *2021*, 5289697. [CrossRef]
35. Foody, G.M. Global and Local Assessment of Image Classification Quality on an Overall and Per-Class Basis without Ground Reference Data. *Remote Sens.* **2022**, *14*, 5380. [CrossRef]
36. Herold, M.; Mayaux, P.; Woodcock, C.; Baccini, A.; Schmullius, C. Some challenges in global land cover mapping: An assessment of agreement and accuracy in existing 1 km datasets. *Remote Sens. Environ.* **2008**, *112*, 2538–2556. [CrossRef]
37. Yang, H.; Li, S.; Chen, J.; Zhang, X.; Xu, S. The Standardization and Harmonization of Land Cover Classification Systems towards Harmonized Datasets: A Review. *ISPRS Int. J. Geo-Inf.* **2017**, *6*, 154. [CrossRef]
38. Radoux, J.; Defourny, P. Automated Image-to-Map Discrepancy Detection using Iterative Trimming. *Photogramm. Eng. Remote Sens.* **2010**, *76*, 173–181. [CrossRef]
39. Radoux, J.; Lamarche, C.; Van Bogaert, E.; Bontemps, S.; Brockmann, C.; Defourny, P. Automated Training Sample Extraction for Global Land Cover Mapping. *Remote Sens.* **2014**, *6*, 3965–3987. [CrossRef]
40. Chen, P.; Shi, W.; Kou, R. Reference-Free Measurement of the Classification Reliability of Vector-Based Land Cover Mapping. *IEEE Geosci. Remote Sens. Lett.* **2019**, *16*, 1090–1094. [CrossRef]
41. Pierdicca, N.; Anniballe, R.; Noto, F.; Bignami, C.; Chini, M.; Martinelli, A.; Mannella, A. Triple Collocation to Assess Classification Accuracy without a Ground Truth in Case of Earthquake Damage Assessment. *IEEE Trans. Geosci. Remote Sens.* **2017**, *56*, 485–496. [CrossRef]
42. Baraldi, A.; Bruzzone, L.; Blonda, P. Quality assessment of classification and cluster maps without ground truth knowledge. *IEEE Trans. Geosci. Remote Sens.* **2005**, *43*, 857–873. [CrossRef]
43. Steele, B.M. Maximum posterior probability estimators of map accuracy. *Remote Sens. Environ.* **2005**, *99*, 254–270. [CrossRef]
44. Foody, G.M. Latent Class Modeling for Site- and Non-Site-Specific Classification Accuracy Assessment without Ground Data. *IEEE Trans. Geosci. Remote Sens.* **2011**, *50*, 2827–2838. [CrossRef]
45. Stoffelen, A. Toward the true near-surface wind speed: Error modeling and calibration using triple collocation. *J. Geophys. Res. Oceans* **1998**, *103*, 7755–7766. [CrossRef]
46. Gruber, A.; Su, C.-H.; Crow, W.T.; Zwieback, S.; Dorigo, W.A.; Wagner, W. Estimating error cross-correlations in soil moisture data sets using extended collocation analysis. *J. Geophys. Res. Atmos.* **2016**, *121*, 1208–1219. [CrossRef]
47. Hoareau, N.; Portabella, M.; Lin, W.; Ballabrera-Poy, J.; Turiel, A. Error Characterization of Sea Surface Salinity Products Using Triple Collocation Analysis. *IEEE Trans. Geosci. Remote Sens.* **2018**, *56*, 5160–5168. [CrossRef]
48. Li, C.; Tang, G.; Hong, Y. Cross-evaluation of ground-based, multi-satellite and reanalysis precipitation products: Applicability of the Triple Collocation method across Mainland China. *J. Hydrol.* **2018**, *562*, 71–83. [CrossRef]
49. Gong, P.; Li, X.; Wang, J.; Bai, Y.; Chen, B.; Hu, T.; Liu, X.; Xu, B.; Yang, J.; Zhang, W.; et al. Annual maps of global artificial impervious area (GAIA) between 1985 and 2018. *Remote Sens. Environ.* **2019**, *236*, 111510. [CrossRef]

50. Zhang, X.; Liu, L.Y.; Wu, C.S.; Chen, X.D.; Gao, Y.; Xie, S.; Zhang, B. Development of a global 30 m impervious surface map using multisource and multitemporal remote sensing datasets with the Google Earth Engine platform. *Earth Syst. Sci. Data* **2020**, *12*, 1625–1648. [CrossRef]
51. Zhang, X.; Liu, L.; Chen, X.; Gao, Y.; Xie, S.; Mi, J. GLC_FCS30: Global land-cover product with fine classification system at 30 m using time-series Landsat imagery. *Earth Syst. Sci. Data* **2021**, *13*, 2753–2776. [CrossRef]
52. Zhu, Z.; Gallant, A.L.; Woodcock, C.E.; Pengra, B.; Olofsson, P.; Loveland, T.R.; Jin, S.; Dahal, D.; Yang, L.; Auch, R.F. Optimizing selection of training and auxiliary data for operational land cover classification for the LCMAP initiative. *ISPRS J. Photogramm. Remote Sens.* **2016**, *122*, 206–221. [CrossRef]
53. Arsanjani, J.J.; Tayyebi, A.; Vaz, E. GlobeLand30 as an alternative fine-scale global land cover map: Challenges, possibilities, and implications for developing countries. *Habitat Int.* **2016**, *55*, 25–31. [CrossRef]
54. Zhang, H.K.; Roy, D.P. Using the 500 m MODIS land cover product to derive a consistent continental scale 30 m Landsat land cover classification. *Remote Sens. Environ.* **2017**, *197*, 15–34. [CrossRef]
55. Tobler, W.R. A Computer Movie Simulating Urban Growth in the Detroit Region. *Econ. Geogr.* **1970**, *46*, 234–240. [CrossRef]
56. Foody, G.M. Assessing the accuracy of land cover change with imperfect ground reference data. *Remote Sens. Environ.* **2010**, *114*, 2271–2285. [CrossRef]
57. Tateishi, R.; Uriyangqai, B.; Al-Bilbisi, H.; Ghar, M.A.; Tsend-Ayush, J.; Kobayashi, T.; Kasimu, A.; Hoan, N.T.; Shalaby, A.; Alsaaideh, B. Production of Global Land Cover Data–GLCNMO. *Int. J. Digit. Earth* **2011**, *4*, 22–49. [CrossRef]
58. Yu, W.; Li, J.; Liu, Q.; Zeng, Y.; Zhao, J.; Xu, B.; Yin, G. Global Land Cover Heterogeneity Characteristics at Moderate Resolution for Mixed Pixel Modeling and Inversion. *Remote Sens.* **2018**, *10*, 856. [CrossRef]
59. Sahare, M.; Gupta, H. A Review of Multi-Class Classification for Imbalanced Data. *Int. J. Adv. Comput. Res.* **2012**, *2*, 160.
60. Chernoff, E.J. Sample space partitions: An investigative lens. *J. Math. Behav.* **2009**, *28*, 19–29. [CrossRef]
61. Branscum, A.; Gardner, I.; Johnson, W. Estimation of diagnostic-test sensitivity and specificity through Bayesian modeling. *Prev. Veter-Med.* **2005**, *68*, 145–163. [CrossRef]
62. Georgiadis, M.P.; Johnson, W.O.; Gardner, I.A.; Singh, R. Correlation-Adjusted Estimation of Sensitivity and Specificity of Two Diagnostic Tests. *J. R. Stat. Soc. Ser. C* **2003**, *52*, 63–76. [CrossRef]
63. Alemohammad, H.; Booth, K. LandCoverNet: A Global Benchmark Land Cover Classification Training Dataset. *arXiv* **2020**, arXiv:2012.03111.
64. European Space Agency. *Product Validation Report (D12-PVR)*; WorldCover_PVR_v1.0; European Space Agency: Paris, France, 2021.
65. Liu, F.T.; Ting, K.M.; Zhou, Z.-H. Isolation Forest. In Proceedings of the 2008 Eighth IEEE International Conference on Data Mining, Pisa, Italy, 15–19 December 2008; IEEE: Pisa, Italy, 2008; pp. 413–422.

remote sensing

MDPI

Article

Efficient Deep Semantic Segmentation for Land Cover Classification Using Sentinel Imagery

Anastasios Tzepkenlis, Konstantinos Marthoglou and Nikos Grammalidis *

Centre for Research and Technology Hellas, Information Technologies Institute, 57001 Thessaloniki, Greece
* Correspondence: ngramm@iti.gr

Abstract: Nowadays, different machine learning approaches, either conventional or more advanced, use input from different remote sensing imagery for land cover classification and associated decision making. However, most approaches rely heavily on time-consuming tasks to gather accurate annotation data. Furthermore, downloading and pre-processing remote sensing imagery used to be a difficult and time-consuming task that discouraged policy makers to create and use new land cover maps. We argue that by combining recent improvements in deep learning with the use of powerful cloud computing platforms for EO data processing, specifically the Google Earth Engine, we can greatly facilitate the task of land cover classification. For this reason, we modify an efficient semantic segmentation approach (U-TAE) for a satellite image time series to use, as input, a single multiband image composite corresponding to a specific time range. Our motivation is threefold: (a) to improve land cover classification performance and at the same time reduce complexity by using, as input, satellite image composites with reduced noise created using temporal median instead of the original noisy (due to clouds, calibration errors, etc.) images, (b) to assess performance when using as input different combinations of satellite data, including Sentinel-2, Sentinel-1, spectral indices, and ALOS elevation data, and (c) to exploit channel attention instead of the temporal attention used in the original approach. We show that our proposed modification on U-TAE (mIoU: 57.25%) outperforms three other popular approaches, namely random forest (mIoU: 39.69%), U-Net (mIoU: 55.73%), and SegFormer (mIoU: 53.5%), while also using fewer training parameters. In addition, the evaluation reveals that proper selection of the input band combination is necessary for improved performance.

Keywords: convolutional neural networks; deep neural networks; land cover classification; machine learning; random forest; Sentinel-1; Sentinel-2; transformer models

1. Introduction

Modern techniques in Earth observation and remote sensing combine multi-temporal data from different satellites to extract important information that is useful in decision making [1], as well as in land cover/land use classification [2].

The EU's Copernicus Sentinel constellation offers both optical (e.g., Sentinel-2 [3]) as well as synthetic aperture radar (SAR, Sentinel-1 [4]) data. Similarly, other satellite-based products (e.g., Landsat-5, Landsat-7, and Landsat-8 [5] provided in cooperation between NASA and the U.S. Geological Survey) supply high-quality, free data capable of estimating parameters related to land use and environmental issues [6]. The free and open data policies adopted by the EU Copernicus program and by NASA and the U.S. Geological Survey can be considered as a milestone of space technology [7].

Land cover and land use (LCLU) is rapidly changing due to natural and anthropogenic factors, including natural and man-made disasters, as well as human activities. Important cues regarding serious environmental problems and risks can be provided to managing authorities by monitoring LCLU changes. For this reason, accurately assessing LCLU maps and their alterations is crucial for the effective management of natural resources and the continuous monitoring of environmental changes [8]. Towards this aim, LCLU monitoring models were formulated to evaluate changes in land cover and use patterns for diverse applications, e.g., land cover and use changes were used to monitor coastal zone areas [9] and wetland zones [10].

The rapid development in space technologies and the growing number of Earth observation satellites' sensors produce an expanding amount of data. However, producing land cover maps from these data is a complicated, labour intensive, and time-consuming process [11]. Since land cover maps are difficult to create, they usually have limited temporal resolution. For instance, EU Corine land cover products [12] are only available for years 1990, 2000, 2006, 2012, and 2018, while the recent European Space Agency (ESA) WorldCover products [13] are only available for years 2020 and 2021. Therefore, it is significant to create machine learning models that can accurately create land cover maps from satellite data to enhance the monitoring of land cover changes.

Conventional machine learning techniques are commonly employed for managing these medium resolution satellite image time series. Prior work on land cover classification utilised algorithms, such as random forest [14,15], hidden Markov models [16,17], or support vector machines [18], to classify manually designed features, such as spectral statistics and phenological metrics [19,20]. As consumer computing power is significantly increasing [21], and the cost is decreasing, deep learning methods are gaining popularity due to their ability to extract more representative features by processing large amounts of data. For instance, convolutional neural networks (CNNs) were adopted from the wider field of computer vision to establish spatial representations of satellite imagery or image time series [22–24]. In addition, network architectures based on CNNs, such as the well-known U-Net model that was initially recommended for biomedical image segmentation [25], are able to automatically produce meaningful pixel-based (semantic) image segmentation. For this reason, the U-Net model and its variants were already utilised in many remote sensing problems to produce accurate segmentation masks, including land cover and use prediction [26–30].

The transformer architecture [31], originally proposed to handle sequential data in natural language processing (NLP) tasks, captures long-range dependencies using self-attention layers, instead of traditional CNNs or recurrent neural networks (RNNs) that can encode efficiently only local dependencies, i.e., between neighbouring elements.

Due to their efficiency, transformers were used extensively in numerous computer vision tasks. A systematic review of recent advances in remote sensing based on transformers is made in [32]. In particular, the work of [33] introduces vision transformers for image recognition tasks, representing an image as a sequence of patches and processing it via a conventional transformer encoder, similar to those used in NLP tasks.

In semantic segmentation, Zheng et al. [34] proposed the segmentation transformer (SETR), which attained state-of-the-art results for standard semantic segmentation benchmarking datasets, demonstrating that using transformers on this task is a viable option. In order to improve the computational efficiency on large images and to obtain multi-scale features, architectures such as the pyramid vision transformer (PVT) [35] or Swin transformer [36] were proposed. Extending this work, the SegFormer architecture [37] managed to achieve a new state-of-the-art performance level regarding efficiency, accuracy, and robustness across three semantic segmentation datasets that are available to the public. This improved performance is mainly due to a novel positional-encoding-free and hierarchical transformer encoder and a lightweight All-MLP decoder that produces a potent, powerful representation without involving complicated and computationally intensive modules.

Garnot et al. [38] introduced a modified version of the original transformer encoder by Vaswani for classifying crops in predefined agricultural parcels, based on the Sentinel-2 image time series. The architecture, named PSE + TAE, used pixel-set encoders to extract learned statistics of spectra distribution in the spatial dimensions of the parcels and is shown to compare favourably with other transformer-, CNN-, and RNN-based architectures. However, further modifications in the temporal encoder transformer were introduced in [39], in order to avoid inessential computations and parameters, while preserving a significant level of expressiveness and flexibility. This highly optimised version of the encoder transformer, namely the lightweight temporal encoder transformer (LTAE), was integrated within a U-NET-like architecture, namely U-TAE, in [40] for semantic segmentation of the multispectral satellite time series. Although this approach can also be applied for land cover classification, the use of the satellite image time series as an input has significant disadvantages: (a) this data can be very noisy due to clouds, calibration errors, etc., and (b) memory requirements are significantly increased as the number of input images and channels increases. To address these issues, we propose to use the Google Earth Engine platform [41] to preprocess satellite data within a specific time range from multiple sources, such as Sentinel-2, Sentinel-1, and ALOS elevation, in order to create image composites with reduced noise, by using a temporal median filter. Furthermore, we modify the U-TAE approach to use channel attention for calculating weights to different channels, instead of the temporal attention used in the original approach.

The main contributions can be summarised as follows:

1. We create Sentinel-2 and Sentinel-1 composite images at 12 coastal, riparian or lakeside locations in Greece corresponding to specific time ranges within the year 2020. The resulting images, each containing 17 channels, along with associated land cover annotation from the ESA WorldCover product, will be freely provided as an open dataset for training DL models for land cover classification;
2. We use this dataset to train a modified U-TAE approach, which uses band attention instead of temporal attention;
3. We evaluate the performance obtained by selecting as input different band combinations and;
4. We perform a comparative performance evaluation of the proposed approach with two state-of-the-art deep semantic segmentation (U-NET, SegFormer) architectures and one traditional ML algorithm (random forest).

2. Materials and Methods

2.1. Overview

An overview of the proposed approach is illustrated in Figure 1. More specifically, Sentinel-2 and Sentinel-1 composite images are created and pre-processed using Google Earth Engine from 12 coastal, riparian, or lakeside locations in Greece corresponding to specific time ranges within the year 2020. The resulting images are split into 256 × 256 tiles and associated labels are obtained from the ESA WorldCover product. Training of the different ML/DL approaches is performed using input data from 11 regions, while the last region is used for testing. The following subsections will describe in more detail this methodology.

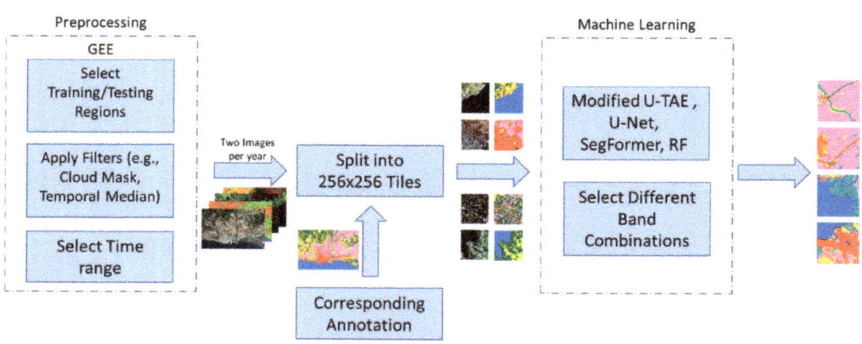

Figure 1. Flowchart of the proposed land cover classification approach.

2.2. Regions of Interest

From both a physical and anthropogenic geographical perspective, Greek landscapes represent three different, albeit highly variable and historically changing, geographical entities [42], namely (a) island, (b) coastal, and (c) inland landscapes. In order to create the training set, eleven region of interests (ROIs) were selected, namely three island cities (Chalcida in Euboea, Mytilene in Lesvos, and Agios Nikolaos in Crete), four coastal cities (Kavala, Alexandroupoli, Preveza, and Northeastern Laconia) and four lakeside or riparian inland regions (Edessa, Kastoria, Kozani, and Karditsa), as shown in Figure 2. The ROI used for testing is the greater region of Thessaloniki, which includes Thermaikos Gulf and the Axios Delta National Park. The area was chosen as the Greek pilot site for the EPIPELAGIC project [43], as it is a coastal area that contains three river deltas and has different land uses as well as various economic activities, which in many cases have significant environmental impacts.

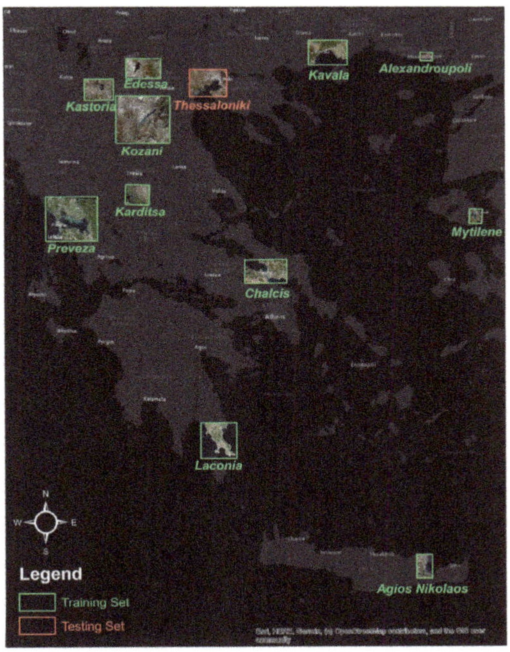

Figure 2. The selected training sites (green polygon) as well as the testing site (red polygon).

2.3. Remote Sensing Data Selection

We opt to use satellite data from Sentinel-2 (Level 2A) [3] and Sentinel-1 (GRD) [4] missions, within the year 2020 (two median six months images, January to June and July to December) as well as associated spectral indices and digital elevation model (DEM) products are used for the creation of the training and testing datasets. The corresponding annotations are obtained from the ESA WorldCover v100 project.

Specifically, we use Sentinel-2 Level-2A images with 12 multispectral bands (resolutions of 10, 20, and 60 m/pixel) representing surface reflectance. However, similarly to [38], the noisy atmospheric bands (i.e., bands for aerosols and water vapor with 60 m/pixel) are omitted since they cannot provide any useful information, as on all images, cloud masks are applied.

Additionally, the Sentinel-1 GRD (S1) data used are the backscatter intensities that can be measured from each of two polarization channels, namely VH and VV. The active microwave data from the synthetic aperture radar (SAR) has enormous potential for mapping and monitoring land cover, notably in identifying the water bodies and wetland vegetation [4]. This is due to the capabilities of day and night observation, as well as cloud penetration.

Furthermore, to improve the overall land cover classification, three indices derived from the Sentinel-2 mission were included: the normalised difference vegetation index (*NDVI*) [44], the normalised difference built-up index (*NDBI*) [45], and the normalised difference water index (*NDWI*) [46], which are calculated based on the Equations (1)–(3):

$$NDVI = \frac{NIR - RED}{NIR + RED} \tag{1}$$

$$NDBI = \frac{SWIR1 - NIR}{SWIR1 + NIR} \tag{2}$$

$$NDWI = \frac{GREEN - NIR}{GREEN + NIR}. \tag{3}$$

NDVI provides important cues regarding the existence of live green vegetation. It is often used to monitor drought, forecast agricultural production, and assist in forecasting fire zones and desert offensive maps. *NDBI* distinguishes urban areas with higher reflectance in the shortwave infrared spectral range and therefore it is used to specify human settlements, as well as other infrastructure. Finally, high values of *NDWI* are used to identify water bodies.

Additionally, the 30 m resolution digital surface model (DSM) from the ALOS World 3D—30 m (AW3D30) v3.2 product released in 2021 from the Japan Aerospace Exploration Agency (JAXA) was selected to supply auxiliary data, namely elevation and slope, that can be useful for detecting land cover. Figure 3 illustrates different bands from the input data set used for testing.

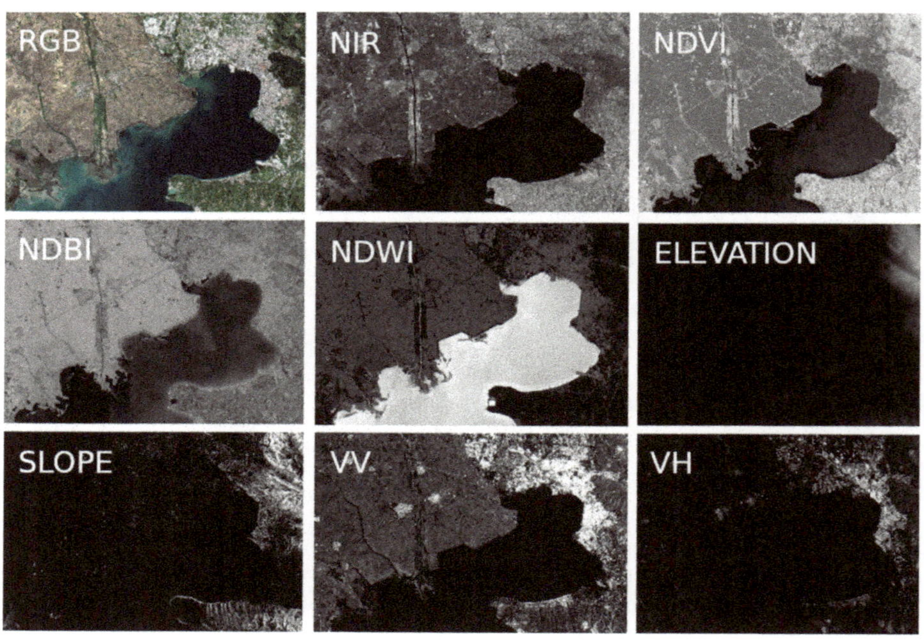

Figure 3. Visualisation of different bands from the input dataset.

The annotation data used for training and evaluation was obtained from the European Space Agency (ESA) WorldCover v100 v3.2 product [13], released in 2021. This product provides a global land cover map for the year 2020 at 10 m resolution based on Sentinel-1 and Sentinel-2 data. It consists of eleven land cover classes, namely trees, shrubland, grassland, cropland, built-up, barren/sparse vegetation, snow and ice, open water, herbaceous wetland, mangroves, and moss and lichen. Three classes (snow and ice, mangroves, and moss and lichen) are not present in the ROIs used for this study and were omitted. As shown in Figure 4, the dataset is greatly unbalanced, with more than half of the dataset belonging to two classes (trees and open water), while three classes (herbaceous wetland, built-up, and barren/sparse vegetation) are underrepresented (0.8%, 2.8%, and 3.1% of the dataset, respectively).

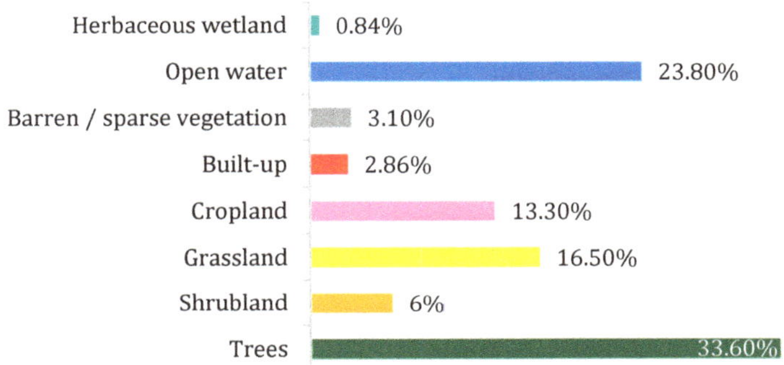

Figure 4. Bar chart showing the distribution of the eight classes in the dataset.

2.4. *Remote Sensing Preprocessing*

Google Earth Engine (GEE) is a platform for scientific analysis and visualisation of geospatial datasets. GEE also provides application programming interfaces (in JavaScript and Python) and other tools for the analysis of large datasets [41]. Pre-processing was performed remotely using Python in GEE, resulting in a single multichannel image for each ROI, which was then downloaded locally.

Specifically, regarding the Sentinel-2 Level 2A multispectral imagery, a two-step preprocessing procedure is used: first, all available images in the year 2020 with less than 5% cloud coverage are selected and any cloud and cirrus pixels are masked-out using the QA60 cloud mask. At the second stage, a temporal median filter is applied to each pixel of the selected images, resulting in a single, noise-free composite multispectral image. Similarly, regarding Sentinel-1 SAR GRD imagery, a temporal median filter is applied to each pixel of all available images in the year 2020, resulting in a single multichannel image composite. Finally, two additional channels are generated from the DSM band of the ALOS AW3D30 product, i.e., the elevation (ELEVATION) value in meters provided for each pixel, as well as the slope (SLOPE) in degrees, which was computed using a built-in GEE function. All bands and products were resampled to 10 m/pixel resolution to achieve the same resolution of the WorldCover product.

2.5. *Land Cover Classification Algorithms*

In the following, we briefly introduce our proposed modified version of U-TAE as well as two other popular semantic segmentation algorithms, namely the CNN-based U-Net and the transformer-based SegFromer. We also present the well-known random forest algorithm, which is often used for land cover classification. In more detail:

(a) U-Net with temporal attention encoder (U-TAE) [40]: This model encodes a multitemporal image sequence in the following steps: (1) a shared multi-level spatial convolutional encoder embeds each image in a simultaneous and independent way, (2) a temporal attention encoder creates a single feature map for every level by stacking the temporal dimensions of the resulting sequence. Specifically, in order to reduce the memory and computational requirements, for every pixel it produces temporal attention masks at the lowest resolution, which are then spatially interpolated at all resolutions. (3) A convolutional decoder calculates features at every resolution level and the final predicted segmentation mask is produced as the output of the highest resolution level;

The architecture uses group normalisation with 4 groups. The input of the original network was modified by changing the shape of the image time sequence from T × C × H × W to C × 1 × H × W, with T the number of temporal instances, C the number of channels, and H × W the height and width of each image. The first two dimensions were permuted to take advantage of the attention module of the U-TAE architecture and to ensure channel-wise attention.

(b) U-Net [25]: This is a U-shaped architecture with an encoder and a decoder that extends the fully convolutional networks segmentation for semantic segmentation [47]. Through a step-by-step downsampling operation, high-level features from the encoder are extracted, while the decoder gradually upsamples these features and combines the output with skip connections to return the feature map to the size of the input. The use of skip connections is important to enable feature reusability and stabilise training and convergence;

(c) SegFormer [37]: This is a hierarchical transformer architecture that extends the segmentation transformer (SETR) proposed in [34]. In the encoding stage, efficient transformer modules are used, while in the decoding stage, multilayer perceptrons (MLPs) are applied. Specifically, a transformer encoder that has a hierarchical structure outputs multiple features, each divided by ascending powers of two without positional encoding. This increases performance even if the training and testing resolutions are different. A lightweight MLP decoder aggregates information from different layers, combining both local and global attention to produce powerful representations. Advantages of the proposed algorithm include: (i) the hierarchical transformer structure, which significantly reduces the computational cost without restricting the effective receptive field, (ii) the positional-encoding free encoder, and (iii) a simple, straightforward, and very efficient decoder;

(d) The random forest (RF) algorithm [48] is an ensemble learning algorithm, i.e., combines multiple ML models to obtain the final solution to classification or regression problems. In this case, we used multiple decision trees for land cover classification. Decision tree-based classifiers were widely studied over the past two decades and were used in many practical applications, including remote sensing, due to several advantages: the concept is intuitively appealing, training is relatively simple, and classification is fast. Unlike many machine learning models that function as "black boxes", a decision tree is an explainable machine learning algorithm and its logic can be fully understood by simply visualising the decision tree. Random forest algorithms for classification use a learning method that builds a set of decision trees during training and outputs the average prediction of the individual trees. Random decision forests are preferred because they avoid the tendency of decision trees to overfit on the training set. Although some degree of explainability is lost as the number of decision trees increases, it is still possible to determine feature importance in a trained RF model.

2.6. Implementation Details and Metrics

In the case of the three DL algorithms, the training process was implemented using PyTorch and was run on a NVIDIA RTX 3080 with 12 GB of memory. To overcome the memory constraints and requirements for images of fixed size, all training and testing images are split into non-overlapping square tiles of size 256 × 256. Thus, 4258 training and 204 square tiles were created and all DL algorithms were trained on all available tiles of size 256 × 256.

The U-TAE model was run using the official code repository, where the required changes to the model architecture were made to use channel attention instead of temporal attention. The initial learning rate value is 10^{-3}, with an Adam optimiser and cross-entropy loss. Due to memory constraints, the U-TAE model was run using a batch size of 2, with the exception of the 17B combination, which used batch size of 1 (Table 1).

Table 1. Input band combination dictionary.

Code	Input Channels
3B	B2, B3, B4
4B	B2, B3, B4, B8
6B	B2, B3, B4, B8, B11, B12
10B	B2, B3, B4, B5, B6, B7, B8, B8A, B11, B12
IND	NDVI, NDBI, NDWI
S1	VV, VH
DEM	ELEVATION, SLOPE
13B	6B + IND + S1 + DEM
17B	10B + IND + S1 + DEM

The SegFormer and the U-Net models were run using the mmsegmentation codebase [49], for efficient development and experimentation. The SegFormer models had a starting learning rate of 6×10^{-5}, a polynomial learning rate schedule, and an AdamW optimizer [50]. The U-Net model used a SGD optimiser with a starting learning rate of 10^{-2}. The batch size is 8 and cross-entropy loss for both models.

Data normalisation was performed on all deep learning algorithms by calculating (X-mean)/standard deviation, where X is the corresponding value, mean and standard deviation are concerning each corresponding image. This normalisation technique was seen to improve the accuracy of the results in preliminary tests that were omitted from the paper.

During training of the three deep learning models, no augmentation techniques were used. Furthermore, the U-Net and SegFormer models were adapted to accommodate multichannel image training, as the original techniques were developed for 3-channel RGB images.

Regarding the RF algorithm, in order to reduce the memory and computational requirements, a sampling grid is defined in all training images and only pixels on this sampling grid are considered for training. Experiments showed that the optimal grid cell size was equal to 64. The parameters of the random forest classifier were selected using the GridSearch function. For example, when using 17 bands, the optimal parameters are: (a) number of trees: 41, (b) minimum samples in order to split a tree: 10, and (c) minimum number of leaf samples on a leaf node: 2.

To evaluate the results of the algorithms, two metrics were used: overall accuracy (*OA*) and mean intersection-over-union (*mIoU*). In semantic segmentation tasks, *mIoU* is widely used as an evaluation metric. It is a measure of how well the predicted segmentation map aligns with the ground truth segmentation map. Another important metric in every classification task is overall accuracy (*OA*). *OA* is the ratio of the correctly classified samples to the total number of samples, as shown in Equation (4). More specifically:

$$OA = \frac{TP + TN}{TP + TN + FP + FN} \quad (4)$$

where *TP* (*FP*) is the number of true (false) positives and *TN* (*FN*) is the number of true (false) negatives. On the other hand, *mIoU* is defined as the average of the intersection-over-union (*IoU*) values between each predicted class and ground truth class across all pixels in an image, i.e.,

$$mIoU = \frac{1}{C} \sum_{1}^{C} IoU_C \quad (5)$$

where C is the total number of classes and IoU:

$$IoU = \frac{TP}{TP + FP + FN}. \quad (6)$$

3. Results

3.1. Experimental Results

In the experiments, we use different band combinations as input. For more clarity, in Table 1, we provide a dictionary of codes describing each band combination used.

The experimental results for the models are presented in Table 2. The highest mIoU and OA values were obtained using the U-TAE and U-Net models with the 13B combination, while the SegFormer model and its variations demonstrate generally inferior performance.

Table 2. Comparisons between the three deep learning models and random forest, with different input band combinations. Highest overall is U-TAE with 13B combination (red), followed by U-NET with 13B combination (blue).

	3B		6B		13B		17B		
	OA(%)	mIoU(%)	OA(%)	mIoU(%)	OA(%)	mIoU(%)	OA(%)	mIoU(%)	Params (In Millions)
Random Forest	39	22.28	60	31.16	68	39.69	68	38.89	0.24
SegFormer B0	63.04	37.91	82.78	49.99	84.15	53.50	77.65	46.14	3.7
SegFormer B2	71.29	40.09	80.28	48.96	80.28	47.39	80.96	48.96	27.5
SegFormer B5	75.84	40.30	82.40	49.61	83.48	52.80	80.28	48.92	84.6
U-NET	74.02	43.44	80.90	51.17	83.99	55.73	79.85	49.05	29.0
U-TAE	71.08	38.88	76.67	47.54	84.89	57.25	80.26	47.27	1.1

The U-TAE model yields better results than the other two DL models, with much fewer parameters. In Figure 5a we can see that with 13 band input the U-TAE model both outperforms the other two in mIoU score, and is lighter in parameters. In Figure 5b, the chart shows the mIoU scores of Table 2, where U-TAE is clearly outperforming the other models.

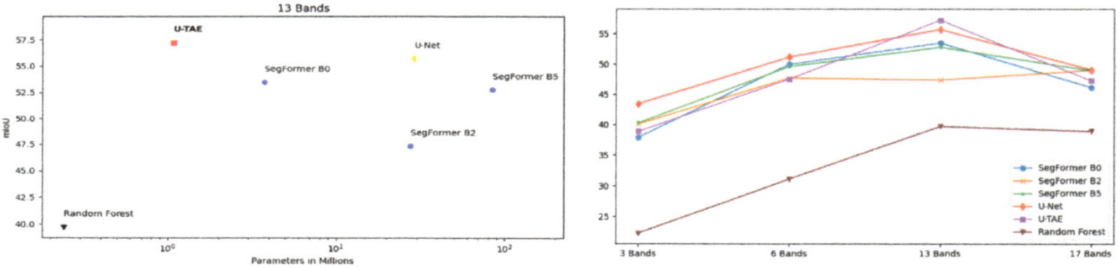

(a) (b)

Figure 5. (**a**) The mIoU score for each model in regard to its parameters. (**b**) The effect of band input on each model's mIoU score. The highest overall result is achieved by U-TAE.

It is also noteworthy that increasing the input bands from 13 to 17 actually decreases the score. That means that not all the information in the input is beneficial to the accuracy of the model, and the addition of certain bands can actually reduce it. All models show a decline in the metrics when moving from Sentinel-2 6 bands to 10 bands, suggesting that more is not always better. In the following ablation study, we will try to further investigate this phenomenon.

Aside from the metrics, it is important that the model also has good recall, precision, and overall more concise class predictions. In Figure 6a–d we see that the U-TAE model has the best recall and precision compared to the other models. Additionally, the U-TAE model predicts the classes with less representation in the dataset, i.e., barren/sparse vegetation, built-up, and herbaceous wetland, with fewer misclassifications, i.e., FPs and FNs.

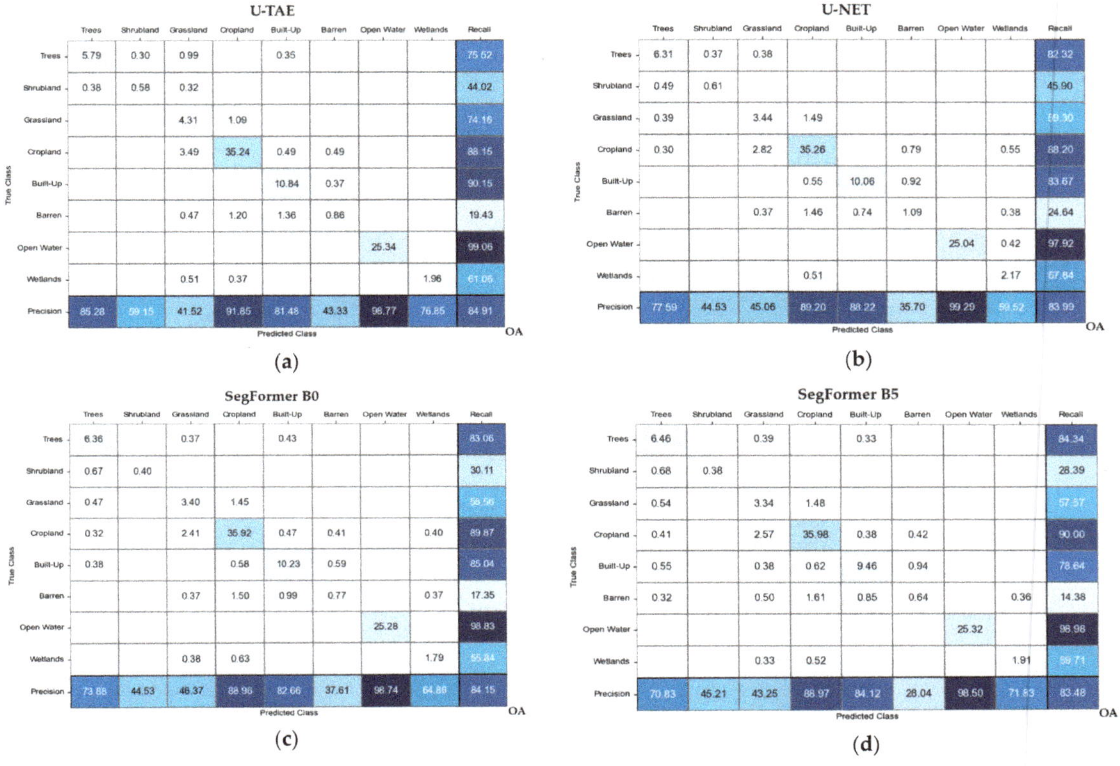

Figure 6. From left to right, top to bottom. Confusion matrices for the best models of (**a**) SegFormer B0, (**b**) SegFormer B5, (**c**) U-NET, and (**d**) U-TAE. The last row is the class precision, and the last column is the class recall. Percents smaller than 0.3% are omitted for clarity.

Figures 7 and 8 illustrate the ground truth segmentation map of the testing set (Thessaloniki), and the predicted segmentation maps of each model for 13B. From the generated maps we can see that the U-TAE model is able to capture the minute intricacies of the area better than the other models. For instance, the SegFormer and U-Net models cannot classify the small grassland regions in the upper left part of the map as well as the U-TAE model. We can also observe that the U-TAE model is better at mapping the built-up class, thus representing the urban regions of the area more accurately. Figure 9 illustrates qualitative prediction results for selected regions of the testing dataset, clearly demonstrating the superior performance of the U-TAE model.

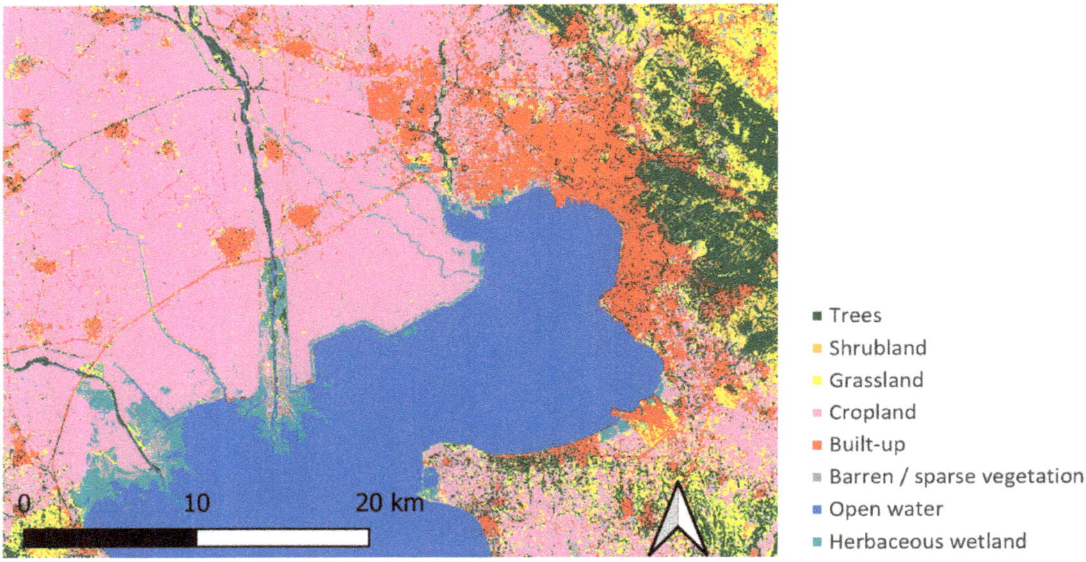

Figure 7. The ground truth segmentation map of the testing set (Thessaloniki).

3.2. Band Combination Ablation Study

We conducted a study on the effects of using different band combinations as input to the modified U-TAE model, which achieved the highest mIoU (57.25%). The purpose of this study is to assess the importance of each band in real-world applications and to determine the band combinations that yield optimal results, avoiding use of unnecessary noisy information.

The combination of the 6B (from S2), S1, and DEM bands appeared to be the most successful input to the U-TAE model (mIoU: 57.58%), surpassing even the previous highest mIoU, which used additionally the IND bands. From Figure 10, it can be inferred that the addition of IND bands generally tends to confuse or hinder the model accuracy. On the contrary, the addition of DEM information seems to significantly improve results, indicating that the model benefits by the knowledge of the geographical terrain.

We also experimented with the 6B combination to examine the relative importance of RGB bands. Specifically, we used only the three thermal infrared S2 bands (NIR, SWIR1, and SWIR2), combined with the S1 and DEM bands. This configuration yields a result that is significantly higher than many other combinations (Figure 10) of information we examined, with a mIoU score that is close to the two highest. Thus, we observe that the infrared information is at least as important as the RGB, if not more, given that the mIoU decreases by only 1.24% when omitting the RGB bands.

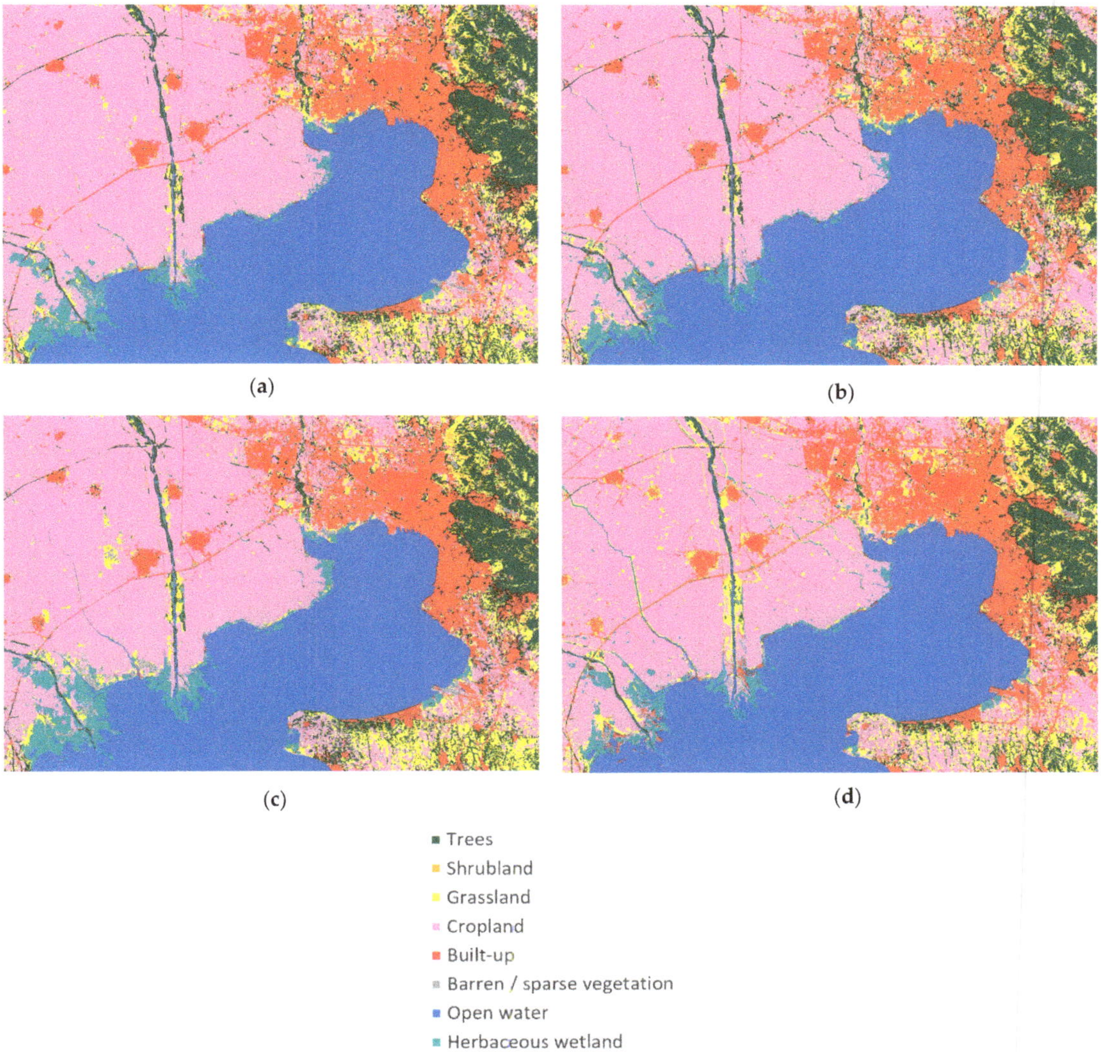

Figure 8. The predicted segmentation maps for each method. From left to right, top to bottom: (**a**) SegFormer B0, (**b**) SegFormer B5, (**c**) U-NET, and (**d**) U-TAE.

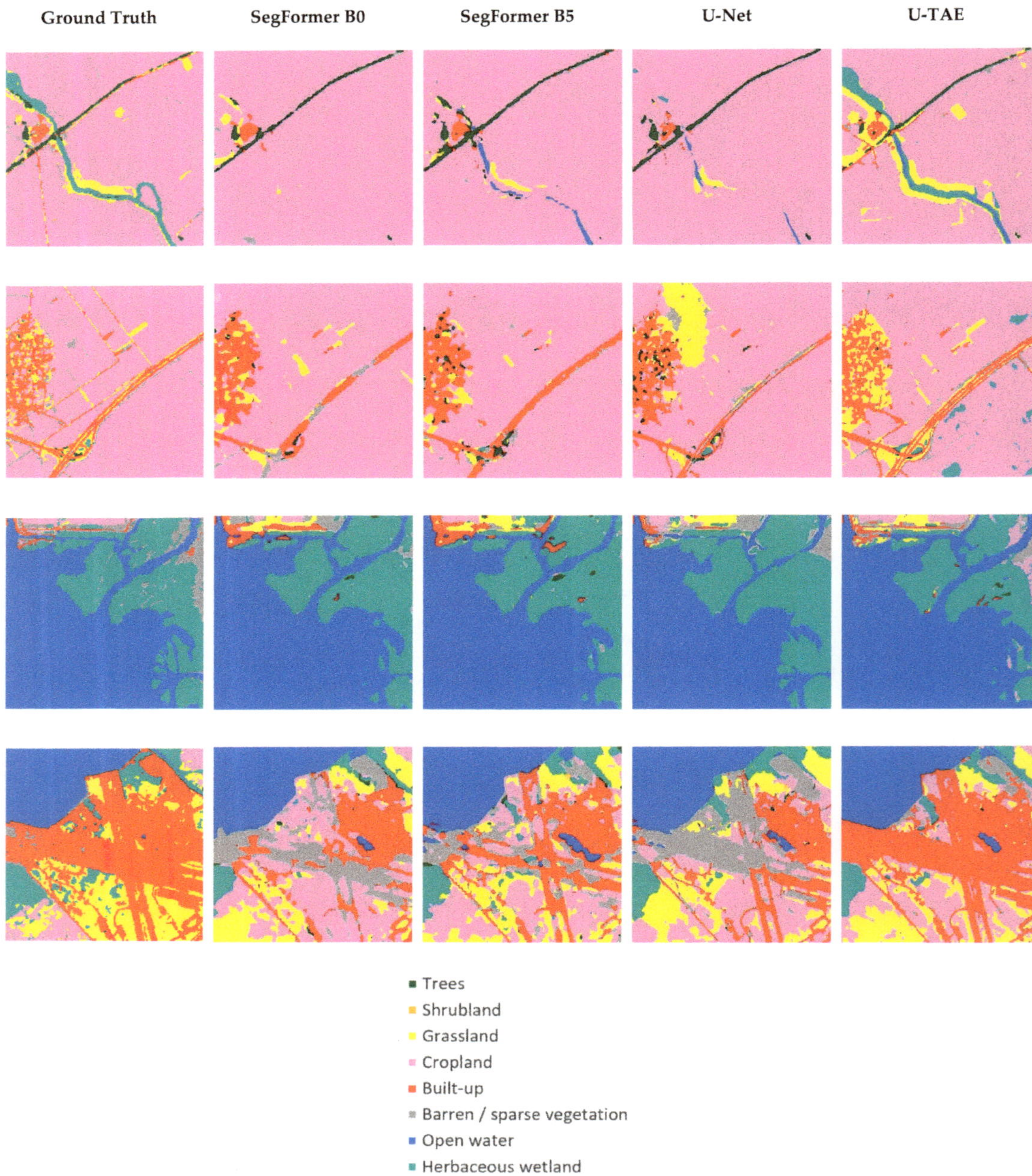

Figure 9. Segmentation map examples for selected regions of the testing dataset. From left to right: ground truth, SegFormer B0, SegFormer B5, U-Net, and U-TAE.

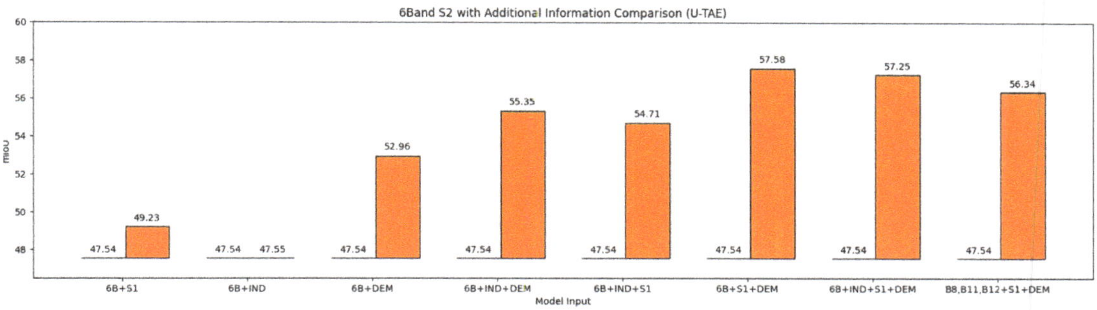

Figure 10. mIoU scores for U-TAE models using different band combinations. At the left of each bar, the mIoU score resulting from the U-TAE model for the 6B band configuration is presented as a baseline.

4. Discussion

Recently, the utilisation of deep learning approaches using input from remote sensing images for land cover and land use classification gained popularity [51], as it can overcome limitations of already available products [52]. In this paper, we utilise a modified version of a robust semantic segmentation approach based on temporal attention, namely U-TAE, for land cover classification. We evaluate its performance by combining different remote sensing data and compare results against two other deep learning approaches, namely U-Net and SegFormer, as well as a traditional ML method, RF.

The use of deep learning for land cover classification by fusing multispectral and SAR data started in early studies [53], where both Sentinel-1 as well as Landsat 8 data are used; Ref. [54] implemented a U-Net network for large-scale land cover classification using only RGB Sentinel-2 images and hand-labeled annotation, achieving promising results. In [55], a dataset containing 27,000 labeled and geo-referenced images is constructed for land use and land cover classification, while a correspondng benchmark based on CNNs is also provided. In [56], a large-scale project is presented, exploiting cloud-based systems and using ML for land cover classification.

While some studies suggested that the NDVI and the NIR band are the most important bands for land cover classification considering vegetation [57], we also recommend incorporating Sentinel-1 data, as they enhance the ability to identify vegetation. This conclusion is supported both by our results as well as by the results of [58], which demonstrate the benefits of fusing Sentinel-2 and Sentinel-1 data. Another study, Ref. [59] highlighted that ML algorithms may outperform conventional DL algorithms when large datasets are not available.

Recently, transformer-based approaches were also widely used for land cover and crop classification with Sentinel 2 imagery [38,60–63], but mainly employ temporal attention. Scheibenreif et al. [64] used large datasets of unlabelled remote sensing data (Sentinel-2 and Sentinel-1 image pairs) for self-supervised pre-training of vision transformers. Such self-supervised transformer-based methods can offer great potential; as the labeling of large remote sensing datasets is a very tedious procedure, such self-supervised approaches have strong potential.

We believe that in the near future, transformer-based architectures, which employ self-attention to differentially weigh parts of the input signal based on their significance, will play a crucial role in advancing remote sensing land cover classification. Transformers already demonstrated impressive results in numerous ML applications, although in certain cases, CNNs may still outperform them. Additionally, transformer models are highly parallelizable, and thus suitable for processing large datasets in a reasonable time frame.

Some limitations of the proposed approach include the fact that temporal attention is actually replaced by band attention, while a combination of both could provide improved results. Furthermore, any errors in the WorldCover product (or other similar products in the future) that are used as ground truth will inevitably be propagated in our proposed approach. Finally, another challenge is the dynamic nature of some land cover classes, such as agriculture, wetlands, etc., which are often changing from time to time. This makes it more difficult to accurately determine land cover, when a longer time frame is used.

5. Conclusions

The results of this paper indicate that deep learning (DL) algorithms demonstrate significantly superior performance against traditional machine learning (ML) algorithms in the land cover classification field. It is also noteworthy that all algorithms demonstrate similar behaviour for specific band combinations, with the best performance obtained by the 13B combination. This suggests that the information contained within these bands is critical for accurate classification. This assumption is justified by the provided ablation study, where we highlighted the importance of S1 (VV and VH bands) as well as DEM (ELEVATION and SLOPE bands), which were seen to significantly affect the algorithm performance. Furthermore, the NIR, SWIR1, and SWIR2 Sentinel-2 bands seem to offer important information, as they result in good performance even without the addition of RGB bands. Furthermore, the modification of the U-TAE algorithm to provide channel (instead of temporal) attention seems to be promising for land cover classification. The U-TAE algorithm's efficiency and simplicity, with the least amount of parameters compared to other DL algorithms, makes it an appealing choice. This is a significant outcome, as it highlights U-TAE's potential to outperform not only conventional DL algorithms, such as U-Net, but also advanced hierarchical transformer-based algorithms, such as SegFormer. Based on the results obtained in this study, future work will involve the classification of 3-month image composites using the U-TAE. Furthermore, we plan to use similar approaches to directly assess land cover changes through time. This approach could provide valuable insights into land use changes and their potential impact on the environment, enabling more effective strategies for land management.

Finally, we need to note that all land cover classification approaches examined in this paper, including U-TAE, are actually pixel-based. Extensions towards object-based (or "panoptic" in ML terminology) segmentation were already proposed [40]. This is an interesting future direction, but also more challenging, as the shape of the land cover "objects" is arbitrary, and is expected to have higher complexity.

Author Contributions: Conceptualization, N.G. and A.T.; methodology, A.T. and N.G.; software, K.M. and A.T.; validation, N.G.; formal analysis, A.T., K.M. and N.G.; investigation, N.G.; resources, N.G.; data curation, A.T.; writing—original draft preparation, A.T.; writing—review and editing, A.T., K.M. and N.G.; visualization, A.T.; supervision, N.G.; project administration, N.G. All authors have read and agreed to the published version of the manuscript.

Funding: This research was funded by European Union and Greece, grant number T7ΔKI-00160 & KMP6-0079153.

Data Availability Statement: The created dataset will be freely provided if the manuscript is accepted.

Acknowledgments: This research was co-financed by the European Regional Development Fund of the European Union and Greek national funds through (a) the Operational Program Competitiveness, Entrepreneurship and Innovation, under the call RESEARCH–CREATE–INNOVATE (project code: T7ΔKI-00160—"ExPert Integrated support system for CoastaL mixed urbAn–industrial—critical infrastructure monitoring using Combined technologies—EPIPELAGIC"), and (b) under the framework of the Action «Investment Plans of Innovation» of the Operational Program «Central Macedonia 2014 2020» as part of the project «INFOROAD—Development of an innovative online tool for mapping and monitoring the forest and rural road network» (Project code: KMP6-0079153).

Conflicts of Interest: The authors declare no conflict of interest.

References

1. Kaplan, G.; Avdan, U. Sentinel-1 and sentinel-2 data fusion for wetlands mapping: Balikdami, turkey. *Int. Arch. Photogramm. Remote Sens. Spat. Inf. Sci.* **2018**, *42*, 729–734. [CrossRef]
2. Solórzano, J.V.; Mas, J.F.; Gao, Y.; Gallardo-Cruz, J.A. Land Use Land Cover Classification with U-Net: Advantages of Combining Sentinel-1 and Sentinel-2 Imagery. *Remote Sens.* **2021**, *13*, 3600. [CrossRef]
3. Drusch, M.; Del Bello, U.; Carlier, S.; Colin, O.; Fernandez, V.; Gascon, F.; Hoersch, B.; Isola, C.; Laberinti, P.; Martimort, P.; et al. Sentinel-2: ESA's optical high-resolution mission for GMES operational services. *Remote Sens. Environ.* **2012**, *120*, 25–36. [CrossRef]
4. Torres, R.; Snoeij, P.; Geudtner, D.; Bibby, D.; Davidson, M.; Attema, E.; Potin, P.; Rommen, B.; Floury, N.; Brown, M.; et al. GMES Sentinel-1 mission. *Remote Sens. Environ.* **2012**, *120*, 9–24. [CrossRef]
5. Roy, D.P.; Wulder, M.A.; Loveland, T.R.; Woodcock, C.E.; Allen, R.G.; Anderson, M.C.; Helder, D.; Irons, J.R.; Johnson, D.M.; Kennedy, R.; et al. Landsat-8: Science and product vision for terrestrial global change research. *Remote Sens. Environ.* **2014**, *145*, 154–172. [CrossRef]
6. Cornelia, A.M. Advantages of Identifying Urban Footprint using Sentinel-1. In Proceedings of the FIG Congress 2018 Embracing Our Smart World Where the Continents Connect: Enhancing the Geospatial Maturity of Societies, Istanbul, Turkey, 6–11 May 2018.
7. Tzouvaras, M.; Kouhartsiouk, D.; Agapiou, A.; Danezis, C.; Hadjimitsis, D.G. The use of Sentinel-1 synthetic aperture radar (SAR) images and open-source software for cultural heritage: An example from Paphos area in Cyprus for mapping landscape changes after a 5.6 magnitude earthquake. *Remote Sens.* **2019**, *11*, 1766. [CrossRef]
8. McCarthy, M.J.; Colna, K.E.; El-Mezayen, M.M.; Laureano-Rosario, A.E.; Méndez-Lázaro, P.; Otis, D.B.; Toro-Farmer, G.; Vega-Rodriguez, M.; Muller-Karger, F.E. Satellite remote sensing for coastal management: A review of successful applications. *Environ. Manag.* **2017**, *60*, 323–339. [CrossRef]
9. Nayak, S. Coastal zone management in India— present status and future needs. *Geo-Spat. Inf. Sci.* **2017**, *20*, 174–183. [CrossRef]
10. Faruque, J.; Vekerdy, Z.; Hasan, Y.; Islam, K.Z.; Young, B.; Ahmed, M.T.; Monir, M.U.; Shovon, S.M.; Kakon, J.F.; Kundu, P. Monitoring of land use and land cover changes by using remote sensing and GIS techniques at human-induced mangrove forests areas in Bangladesh. *Remote Sens. Appl. Soc. Environ.* **2022**, *25*, 100699. [CrossRef]
11. Rakhlin, A.; Davydow, A.; Nikolenko, S. Land cover classification from satellite imagery with u-net and lovász-softmax loss. In Proceedings of the IEEE Conference on Computer Vision and Pattern Recognition Workshops, Salt Lake City, UT, USA, 18–22 June 2018; pp. 262–266.
12. Bossard, M.; Feranec, J.; Otahel, J. *CORINE Land Cover Technical Guide: Addendum*; European Environment Agency: Copenhagen, Denmark, 2000; Volume 40.
13. Zanaga, D.; Van De Kerchove, R.; De Keersmaecker, W.; Souverijns, N.; Brockmann, C.; Quast, R.; Wevers, J.; Grosu, A.; Paccini, A.; Vergnaud, S.; et al. *ESA WorldCover 10 m 2020 v100*; OpenAIRE: Los Angeles, CA, USA, 2021. [CrossRef]
14. Jordi, I.; Marcela, A.; Benjamin, T.; Olivier, H.; Silvia, V.; David, M.; Gerard, D.; Guada-lupe, S.; Sophie, B.; Pierre, D.; et al. Assessment of an operational system for crop type map production using high temporal and spatial resolution satellite optical imagery. *Remote Sens.* **2015**, *7*, 12356–12379.
15. Pelletier, C.; Valero, S.; Inglada, J.; Champion, N.; Dedieu, G. Assessing the robustness of Random Forests to map land cover with high resolution satellite image time series over large areas. *Remote Sens. Environ.* **2016**, *187*, 156–168. [CrossRef]
16. Siachalou, S.; Tsakiri-Strati, M. A hidden Markov models approach for crop classification: Linking crop phenology to time series of multi-sensor remote sensing data. *Remote Sens.* **2015**, *7*, 3633–3650. [CrossRef]
17. Giordano, S.; Bailly, S.; Landrieu, L.; Chehata, N. Improved crop classification with rotation knowledge using Sentinel-1 and -2 time series. *Photogramm. Eng. Remote Sens.* **2020**, *86*, 431–441. [CrossRef]
18. Devadas, R.; Denham, R.J.; Pringle, M. Support vector machine classification of object-based data for crop map-ping, using multi-temporal landsat imagery. International archives of the photogrammetry. *Remote Sens. Spat. Inf. Sci.* **2012**, *39*, 185–190.
19. Hu, Q.; Wu, W.-B.; Song, Q.; Lu, M.; Chen, D.; Yu, Q.-Y.; Tang, H.-J. How do temporal and spectral features matter in crop classification in Heilongjiang Province, China? *J. Integr. Agric.* **2017**, *16*, 324–336. [CrossRef]
20. Nguyen, L.H.; Joshi, D.R.; Clay, D.E.; Henebry, G.M. Characterizing land cover/land use from multiple years of Landsat and MODIS time series: A novel approach using land surface phenology modeling and ran-dom forest classifier. *Remote Sens. Environ.* **2020**, *238*, 111017. [CrossRef]
21. Waldrop, M.M. The chips are down for Moore's law. *Nat. News* **2016**, *530*, 144. [CrossRef] [PubMed]
22. Alem, A.; Kumar, S. Deep learning methods for land cover and land use classification in remote sensing: A review. In Proceedings of the 2020 8th International Conference on Reliability, Infocom Technologies and Optimization (Trends and Future Directions)(ICRITO), Noida, India, 4–5 June 2020; IEEE: Piscataway, NJ, USA, 2020; pp. 903–908.
23. Seydi, S.; Hasanlou, M.; Amani, M. A new end-to-end multi-dimensional CNN framework for land cover/land use change detection in multi-source remote sensing datasets. *Remote Sens.* **2020**, *12*, 2010. [CrossRef]
24. Camalan, S.; Cui, K.; Pauca, V.P.; Alqahtani, S.; Silman, M.; Chan, R.; Plemmons, R.J.; Dethier, E.N.; Fernandez, L.E.; Lutz, D.A. Change detection of amazonian alluvial gold mining using deep learning and sentinel-2 imagery. *Remote Sens.* **2022**, *14*, 1746. [CrossRef]

25. Ronneberger, O.; Fischer, P.; Brox, T. U-net: Convolutional networks for biomedical image segmentation. In Proceedings of the International Conference on Medical Image Computing and Computer-Assisted Intervention, Munich, Germany, 5–9 October 2015; Springer: Cham, Germany, 2015; pp. 234–241.
26. Mohajerani, S.; Saeedi, P. Cloud-Net: An end-to-end cloud detection algorithm for Landsat 8 imagery. In Proceedings of the IGARSS 2019–2019 IEEE International Geoscience and Remote Sensing Symposium, Yokohama, Japan, 28 July–2 August 2019; IEEE: Piscataway, NJ, USA, 2019; pp. 1029–1032.
27. Ye, H.; Liu, S.; Jin, K.; Cheng, H. CT-UNet: An Improved Neural Network Based on U-Net for Building Segmenta-tion in Remote Sensing Images. In Proceedings of the 2020 25th International Conference on Pattern Recognition (ICPR), Milan, Italy, 10–15 January 2021; pp. 166–172. [CrossRef]
28. He, N.; Fang, L.; Plaza, A. Hybrid first and second order attention Unet for building segmentation in re-mote sensing images. *Sci. China Inf. Sci.* **2020**, *63*, 140305. [CrossRef]
29. Jiao, L.; Huo, L.; Hu, C.; Tang, P. Refined UNet: UNet-based refinement network for cloud and shadow precise segmentation. *Remote Sens.* **2020**, *12*, 2001. [CrossRef]
30. Hou, Y.; Liu, Z.; Zhang, T.; Li, Y. C-Unet: Complement UNet for remote sensing road extraction. *Sensors* **2021**, *21*, 2153. [CrossRef]
31. Vaswani, A.; Shazeer, N.; Parmar, N.; Uszkoreit, J.; Jones, L.; Gomez, A.N.; Kaiser, Ł.; Polosukhin, I. Attention Is All You Need. *arXiv e-prints* **2017**, arXiv:1706.03762.
32. Aleissaee, A.A.; Kumar, A.; Anwer, R.M.; Khan, S.; Cholakkal, H.; Xia, G.-S.; Khan, F.S. Transformers in remote sensing: A survey. *Remote Sens.* **2022**, *15*, 1860. [CrossRef]
33. Dosovitskiy, A.; Beyer, L.; Kolesnikov, A.; Weissenborn, D.; Zhai, X.; Unterthiner, T.; Dehghani, M.; Minderer, M.; Heigold, G.; Gelly, S.; et al. An image is worth 16×16 words: Transformers for image recognition at scale. *arXiv* **2020**, arXiv:2010.11929.
34. Zheng, S.; Lu, J.; Zhao, H.; Zhu, X.; Luo, Z.; Wang, Y.; Fu, Y.; Feng, J.; Xiang, T.; Torr, P.H.; et al. Rethinking semantic segmentation from a sequence-to-sequence perspective with transformers. In Proceedings of the IEEE/CVF Conference on Computer Vision and Pattern Recognition, Nashville, TN, USA, 20–25 June 2021; pp. 6881–6890.
35. Wang, W.; Xie, E.; Li, X.; Fan, D.-P.; Song, K.; Liang, D.; Lu, T.; Luo, P.; Shao, L. Pyramid vision transformer: A versatile backbone for dense prediction without convolutions. In Proceedings of the IEEE/CVF International Conference on Computer Vision, Montreal, QC, Canada, 10–17 October 2021; pp. 568–578.
36. Liu, Z.; Lin, Y.; Cao, Y.; Hu, H.; Wei, Y.; Zhang, Z.; Lin, S.; Guo, B. Swin transformer: Hierarchical vision transformer using shifted windows. In Proceedings of the IEEE/CVF International Conference on Computer Vision, Montreal, QC, Canada, 10–17 October 202; pp. 10012–10022.
37. Xie, E.; Wang, W.; Yu, Z.; Anandkumar, A.; Alvarez, J.M.; Luo, P. SegFormer: Simple and efficient design for semantic segmentation with transformers. *Adv. Neural Inf. Process. Syst.* **2021**, *34*, 12077–12090.
38. Garnot, V.S.F.; Landrieu, L.; Giordano, S.; Chehata, N. Satellite image time series classification with pixel-set encoders and temporal self-attention. In Proceedings of the IEEE/CVF Conference on Computer Vision and Pattern Recognition, Seattle, WA, USA, 13–19 June 2020; pp. 12325–12334.
39. Garnot, V.S.F.; Landrieu, L. Lightweight temporal self-attention for classifying satellite images time series. In Proceedings of the Advanced Analytics and Learning on Temporal Data: 5th ECML PKDD Workshop, AALTD 2020, Ghent, Belgium, 18 September 2020; Revised Selected Papers 6. Springer International Publishing: Berlin/Heidelberg, Germany, 2020; pp. 171–181.
40. Garnot, V.S.F.; Landrieu, L. Panoptic segmentation of satellite image time series with convolutional temporal attention networks. In Proceedings of the IEEE/CVF International Conference on Computer Vision, Montreal, QC, Canada, 10–17 October 2021; pp. 4872–4881.
41. Gorelick, N. Google earth engine. In *EGU General Assembly Conference Abstracts*; American Geophysical Union: Vienna, Austria, 2013; Volume 15, p. 11997.
42. Terkenli, T.S. Landscape research in Greece: An overview. *Belgeo. Rev. Belg. Géographie* **2004**, *2–3*, 277–288. [CrossRef]
43. Tzepkenlis, A.; Grammalidis, N.; Kontopoulos, C.; Charalampopoulou, V.; Kitsiou, D.; Pataki, Z.; Patera, A.; Nitis, T. An Integrated Monitoring System for Coastal and Riparian Areas Based on Remote Sensing and Machine Learning. *J. Mar. Sci. Eng.* **2022**, *10*, 1322. [CrossRef]
44. DeFries, R.S.; Townshend, J.R.G. NDVI-derived land cover classifications at a global scale. *Int. J. Remote Sens.* **1994**, *15*, 3567–3586. [CrossRef]
45. Zha, Y.; Gao, J.; Ni, S. Use of normalized difference built-up index in automatically mapping urban areas from TM imagery. *Int. J. Remote Sens.* **2003**, *24*, 583–594. [CrossRef]
46. McFeeters, S.K. The use of the Normalized Difference Water Index (NDWI) in the delineation of open water features. *Int. J. Remote Sens.* **1996**, *17*, 1425–1432. [CrossRef]
47. Long, J.; Shelhamer, E.; Darrell, T. Fully convolutional networks for semantic segmentation. In Proceedings of the IEEE Conference on Computer Vision and Pattern Recognition, Boston, MA, USA, 7–12 June 2015; pp. 3431–3440.
48. L Breiman, L. Random Forests. *Mach. Learn.* **2001**, *45*, 5–32. [CrossRef]
49. Contributors, MMSegmentation. OpenMMLab Semantic Segmentation Toolbox and Benchmark. 2020. Available online: https://github.com/open-mmlab/mmsegmentation (accessed on 7 April 2023).
50. Loshchilov, I.; Hutter, F. Decoupled weight decay regularization. *arXiv* **2017**, arXiv:1711.05101.

51. Digra, M.; Dhir, R.; Sharma, N. Land use land cover classification of remote sensing images based on the deep learning approaches: A statistical analysis and review. *Arab. J. Geosci.* **2022**, *15*, 1003. [CrossRef]
52. Malenovský, Z.; Rott, H.; Cihlar, J.; Schaepman, M.E.; García-Santos, G.; Fernandes, R.; Berger, M. Sentinels for science: Potential of Sentinel-1, -2, and -3 missions for scientific observations of ocean, cryosphere, and land. *Remote Sens. Environ.* **2012**, *120*, 91–101. [CrossRef]
53. Kussul, N.; Lavreniuk, M.; Skakun, S.; Shelestov, A. Deep learning classification of land cover and crop types using remote sensing data. *IEEE Geosci. Remote Sens. Lett.* **2017**, *14*, 778–782. [CrossRef]
54. Karra, K.; Kontgis, C.; Statman-Weil, Z.; Mazzariello, J.C.; Mathis, M.; Brumby, S.P. Global land use/land cover with Sentinel 2 and deep learning. In Proceedings of the 2021 IEEE International Geoscience and Remote Sensing Symposium IGARSS, Brussels, Belgium, 11–16 July 2021; IEEE: Piscataway, NJ, USA, 2021; pp. 4704–4707.
55. Helber, P.; Bischke, B.; Dengel, A.; Borth, D. Eurosat: A novel dataset and deep learning benchmark for land use and land cover classification. *IEEE J. Sel. Top. Appl. Earth Obs. Remote Sens.* **2019**, *12*, 2217–2226. [CrossRef]
56. Verde, N.; Kokkoris, I.P.; Georgiadis, C.; Kaimaris, D.; Dimopoulos, P.; Mitsopoulos, I.; Mallinis, G. National scale land cover classification for ecosystem services mapping and assessment, using multitemporal copernicus EO data and google earth engine. *Remote Sens.* **2020**, *12*, 3303. [CrossRef]
57. Campos-Taberner, M.; García-Haro, F.J.; Martínez, B.; Izquierdo-Verdiguier, E.; Atzberger, C.; Camps-Valls, G.; Gilabert, M.A. Understanding deep learning in land use classification based on Sentinel-2 time series. *Sci. Rep.* **2020**, *10*, 17188. [CrossRef]
58. Ienco, D.; Interdonato, R.; Gaetano, R.; Minh, D.H.T. Combining Sentinel-1 and Sentinel-2 Satellite Image Time Series for land cover mapping via a multi-source deep learning architecture. *ISPRS J. Photogramm. Remote Sens.* **2019**, *158*, 11–22. [CrossRef]
59. Abdi, A.M. Land cover and land use classification performance of machine learning algorithms in a boreal landscape using Sentinel-2 data. *GIScience Remote Sens.* **2020**, *57*, 1–20. [CrossRef]
60. Rußwurm, M.; Körner, M. Self-attention for raw optical satellite time series classification. *ISPRS J. Photogramm. Remote Sens.* **2020**, *169*, 421–435. [CrossRef]
61. Stergioulas, A.; Dimitropoulos, K.; Grammalidis, N. Crop classification from satellite image sequences using a two-stream network with temporal self-attention. In Proceedings of the 2022 IEEE International Conference on Imaging Systems and Techniques (IST), Kaohsiung, Taiwan, 21–23 June 2022; IEEE: Piscataway, NJ, USA, 2022; pp. 1–6.
62. Yuan, Y.; Lin, L. Self-supervised pretraining of transformers for satellite image time series classification. *IEEE J. Sel. Top. Appl. Earth Obs. Remote Sens.* **2020**, *14*, 474–487. [CrossRef]
63. Martini, M.; Mazzia, V.; Khaliq, A.; Chiaberge, M. Domain-adversarial training of self-attention-based networks for land cover classification using multi-temporal Sentinel-2 satellite imagery. *Remote Sens.* **2021**, *13*, 2564. [CrossRef]
64. Scheibenreif, L.; Hanna, J.; Mommert, M.; Borth, D. Self-supervised vision transformers for land-cover segmentation and classification. In Proceedings of the IEEE/CVF Conference on Computer Vision and Pattern Recognition, New Orleans, LA, USA, 18–24 June 2022; pp. 1422–1431.

remote sensing

MDPI

Article

Spatial–Temporal and Driving Factors of Land Use/Cover Change in Mongolia from 1990 to 2021

Junming Hao [1], Qingrun Lin [1], Tonghua Wu [2,*], Jie Chen [2], Wangping Li [1], Xiaodong Wu [2], Guojie Hu [2] and Yune La [2]

[1] School of Civil Engineering, Lanzhou University of Technology, Lanzhou 730050, China
[2] Cryosphere Research Station on the Qinghai-Tibet Plateau, State Key Laboratory of Cryospheric Science, Northwest Institute of Eco-Environment and Resources, Chinese Academy of Sciences, Lanzhou 730000, China
* Correspondence: thuawu@lzb.ac.cn

Abstract: During the past several decades, desertification and land degradation have become more and more serious in Mongolia. The drivers of land use/cover change (LUCC), such as population dynamics and climate change, are increasingly important to local sustainability studies. They can only be properly analyzed at small scales that capture the socio-economic conditions. Several studies have been carried out to examine the pattern of LUCC in Mongolia, but they have been focused on changes in single land types at a local scale. Although some of them were carried out at the national scale, the data interval is more than 10 years. A small-scale and year-by-year dataset of LUCC in Mongolia is thus needed for comprehensive analyses. We obtained year-by-year land use/cover changes in Mongolia from 1990 to 2021 using Landsat TM/OLI data. First, we established a random forest (RF) model. Then, in order to improve the classification accuracy of the misclassification of cropland, grassland, and built and barren areas, the classification and regression trees model (CART) was introduced for post-processing. The results show that 17.6% of the land surface has changed at least once among the six land categories from 1990 to 2021. While the area of barren land has significantly increased, the grassland and forest areas have exhibited a decreasing trend in the past 32 years. The other land types do not show promising changes. To determine the driving factors of LUCC, we applied an RF feature importance ranking to environmental factors, physical factors, socioeconomic factors, and accessibility factors. The mean annual precipitation (MAP), evapotranspiration (ET), mean annual air temperature (MAAT), DEM, GDP, and distance to railway are the main driving factors that have determined the distribution and changes in land types. Interestingly, unlike the global anti-V-shaped pattern, we found that the land use/cover changes show an N-shaped trend in Mongolia. These characteristics of land use/cover change in Mongolia are primarily due to the agricultural policies and rapid urbanization. The results present comprehensive land use/cover change information for Mongolia, and they are of great significance for policy-makers to formulate a scientific sustainable development strategy and to alleviate the desertification of Mongolia.

Keywords: land use/cover; Google Earth engine; random forest model; Mongolia

1. Introduction

Land use/cover change (LUCC) is the result of the comprehensive action of natural objective conditions and human socioeconomic activities, and its formation and evolution process is affected by natural factors and human activities [1]. Over the past 30 years, global warming has enhanced global vegetation greenness and human activities have accelerated land degradation [2]. However, climate change has indeed aggravated desertification and land degradation, especially in the middle and high latitudes of the northern hemisphere [3]. Land use change has affected almost one-third of the global land area in just six decades (1960–2019) and, thus, desertification is around four times greater in its extent than previously estimated from long-term land change assessments [4]. The speed, scale, and spatial scope of human changes to the Earth's surface have reached an unprecedented level [5,6]. At the same time, LUCC can affect the global water and heat balance and biodiversity, thus affecting climate change and ecosystems at a regional and even global scale [7–9]. As an important input parameter of global climate models, LUCC directly affects the accuracy of the simulation and prediction of these models [10]. LUCC and its ecological impact have also become the focus of international research [11,12]. Therefore, understanding the causes, processes, and consequences of land use/cover change is crucial for a better understanding of the climate system, climate effects, and sustainable development [13].

The Mongolian Plateau (MP) lies in an arid and desert belt of the hinterland of Northeast Asia, within a horizontal crisscrossing zone of forest, grassland, and desert. Climate warming and human actions both have negative impacts on the land cover of Mongolia, and are accelerating land degradation [14]. Since 2002–2011, non-desertification and severe desertification areas have increased [15]. By 2016, approximately 72% of the land experienced desertification in Mongolia [16]. This increased to 76.8% in 2017, 24.1% of which experienced slight desertification, 29.8% experienced moderate desertification, 16.8% experienced severe desertification, and 6.1% was heavily degraded [17]. The land degradation of Mongolia has threatened the ecological environment and socioeconomic activities. Thus, it is of great significance to clarify the present situation, and the change processes and driving factors of land use/cover change in Mongolia.

In recent years, the long-term global dataset and various large-scale satellite data have been combined with statistics and machine learning methods to monitor and identify the LUCC of Mongolia and the entire MP. At present, the resolution of the existing global land cover dataset is from 1 km to 10 m, e.g., FROM-GLC (10–30 m, 2015, 2017, China) [18], Esri Land Cover (10 m, 2017–2021, ESA) [19], Globe land 30 (30 m, 2000, 2010, 2020, China) [20], Global Land Survey (30 m, 1975–2012, USGS) [21], Climate Change Initiative Land Cover V2 (300 m, 1992–2020, ESA) [22], and MODIS Land Cover Type/Dynamics (0.5–1 km, 2001–2020, NASA) [23]. These global land cover data can visualize spatial distribution, landscape patterns, and global land use trends. However, due to the limitation of samples and classification models, the results of different models in the same area are different [24,25]. Thus, fine classification and integration of land use/cover type on a regional scale has been mapped in Mongolia. On a local scale, the long-term land use/cover change is analyzed along the transect or economic corridor of the Mongolian plateau using Landsat TM and MODIS data from the 1990s to 2020s [26]. At regional scales, land use changes in Mongolia are mapped in combination with: (1) MODIS data and relevant statistical data in 1992–2005 [27]; (2) ESA 300 m resolution land cover data and TM data in 1970s and 2005 [28,29]; (3) TM data and survey data in 1997, 2007, and 2017 [30]; and TM and ETM+ using the NDVI thresholds method in 1990, 2000, 2010, and 2020 [31–33]. These studies can improve our understanding and provide new knowledge of land use/cover change in Mongolia over long periods. Generally speaking, the research on land use classification methods based on remote sensing is also expanding. Traditional remote sensing image classification methods, such as visual interpretation and statistics-based classification methods, have very simple algorithms [34,35]. However, the classification accuracy has reduced due to the phenomenon of foreign bodies. With the development of artificial intelligence algorithm classification methods, such as neural networks, decision tree classification, and the random forest method, the classification accuracy is higher than that of traditional methods, which is beneficial to a wide range of research results [36,37]. This type of black box operation is prone to overfitting. The classification and regulation tree (CART) method is applicable to a series of binary decisions formed in classification decisions based on human rules [38–40]. The random forest method has low classification error estimation for high-dimensional data [41,42]. Therefore, different methods are combined to improve the accuracy of remote sensing image classification [1,34,35].

The research on land use/cover change in Mongolia mainly focuses on forest decreasing, grassland degradation, urban expansion, and water body and lake change [37–41]. Global warming, logging, grazing, wind erosion, and human activity are considered as the main driving factors [43–46]. Based on the national scale of Mongolia, most research is based on the land use status classification system, cropland, forest, built area, water area, barren land, and grassland [33,47]. They show that the land use in Mongolia is mainly cropland, accounting for 50.5% and increasing gradually, followed by bare land, accounting for 39.9% and gradually decreasing. In Northern Mongolia, due to the influence of climate change, forest area has gradually decreased by 5.4% and degenerated into grassland, and glaciers have turned into water bodies [46,48]. In the past ten years, under the influence of animal husbandry and socio-economic transformation, the area of barren land has decreased by 16.9% and the area of grassland has increased by 17.8% [45,49]. The change of bult area is mainly in the Ulaanbaatar, with an area increase of 140%. The latest national-scale research is more than 10 years old [33,48,50], and there are no long-series data available for year-by-year comparison. However, the large data span and lack of continuous data have seriously affected the scientific understanding of the influencing factors and leading factors of land use/cover change in Mongolia.

Therefore, we mapped the land use/cover of Mongolia year-by-year from 1990 to 2021 based on Landsat TM/OLI and the random forest method, and then analyzed the patterns of land use change and driving factors. To improve the classification accuracy, a CART decision tree algorithm (CART) was developed. The aim of this study was to better measure the patterns and rates of the land use change in Mongolia and understand the influence factors.

2. Materials and Methods

2.1. Study Area

Mongolia is located in north-central Asia and belongs to an arid desert belt (Figure 1). It has an obvious continental climate with four distinct seasons. The average annual temperature was about 1.2 °C and increased by about 0.5 °C during 1998 to 2021 [51]. Annual precipitation rarely exceeds 400 mm and is typically much lower in the south and central desert and steppe regions. In the Gobi Desert, annual rainfall is only 40 mm [52]. The average altitude is 1580 m and the land use/cover types show transitional changes in forest, grassland, and barren land from high latitude to low latitude. The area of Mongolia is 1.56 million. According to the relevant statistics, grassland is the main type of land, accounting for about 59%, followed by bare land, which accounts for 20.8% [53]. At present, due to global warming and increased evaporation, a large number of rivers have dried up, the soil types have obviously changed, the ecological environment is more fragile, and desertification and other disasters are more likely to occur [54].

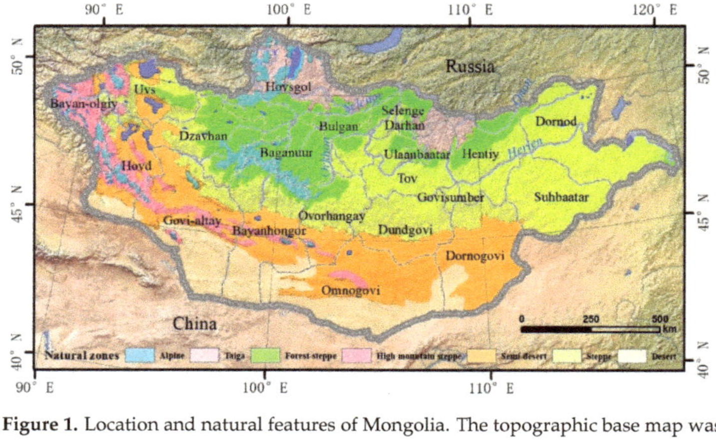

Figure 1. Location and natural features of Mongolia. The topographic base map was from [55].

2.2. Data Source

The types of land use in Mongolia referred to a previous study [56] and the international general LUCC standard (GB/T21010-2017) [57]. We classified the land use types into cropland, forestland, built area, water, barren land, and grassland. The Google Earth Engine (GEE) platform provided the possibility for large-scale remote sensing data processing and factors analysis [58]. The images from June to September were selected in GEE as data sources for 1990–2012 from Landsat 5 TM and for 2013–2021 from Landsat 8 OLI. They were preprocessed by geometric correction and atmospheric correction. Based on the study area, the high-quality remote sensing images were screened in GEE [59]. We found that there were at least 3 images in the same area. Then, median synthesis using a quality assessment band (OA) was used to screen the images with a cloud cover of <20% [60].

The driving factors of LUCC were selected based on a previous study and native environment. A previous study in Mongolia demonstrated that the MAAT, MAP, and ET are the key influencing factors of LUCC [61]. The influence of temperature on plant growth is comprehensive. It is not only an important condition for plant growth and development, but it also affects the growth cycle and greening period of cropland, pasture and forest vegetation, and the migration of the forest line [62–64]. Water is a critical limiting factor in arid and semiarid grassland areas and influences the land use type and regional climate [65,66]. The precipitation increasing by about 100–200 mm can alter the vegetation cover of the desertic Gobi [67]. The ET is related to a number of ecosystem processes, including photosynthesis, soil moisture, and latent heat transfer, and it plays an important role in the process of LUCC [68,69]. Human activities, as an influencing factor, have inhibited forest regrowth by fire or grazing [70]. The socioeconomic factors that are often considered are GDP and demography factors [71–73] due to national development policy and urbanization expansion. At the same time, Mongolia is dominated by herding activities. The increase in human activities and the development of animal husbandry on land degradation mainly influence the changes in grassland and built areas [70,74,75]. Thus, four types of influencing factors were selected: environmental factors, physical factors, socioeconomic factors, and accessibility factors. The factors were MAAT, MAP, ET, DEM, slope, aspect, GDP, population, livestock, distance to road, and distance to railway. (1) The environmental factors, including the mean annual air temperature (MAAT), the mean annual precipitation (MAP), and evapotranspiration (ET), were available from NOVA and PML_V2 at a resolution of 30 m. (2) The physical factors included a digital elevation model (DEM) and the slope and aspect factors. A 30 m DEM was produced using SRTM-V4.1 by NASA. Based on the DEM, the terrain factors were extracted to obtain the corresponding slope and aspect. The aspect value range was 0–360°. (3) The socioeconomic factors comprised population, GDP, and livestock. The socioeconomic data were collected by province from the National Bureau of Statistics of Mongolia. Then, we converted the data into a raster dataset. (4) The accessibility factors, such as the distance to main roads (Dis_road) and distance to railways (Dis_railway), were obtained from the GRIP website. The GRIP dataset consisted of a global vector road dataset and a road density raster dataset at a resolution of 8 km. The accessibility variables were calculated by Euclidean distance in ArcGIS.

Table 1 provides detailed descriptions of these variables with their corresponding references. Figure 2 shows the availability of the Landsat images. Each dataset was generated as a layer in the ArcGIS environment and converted into a 30 × 30 m grid for model-fitting (Figure 3). The coordinate system was from a WGS-84, Baltic Sea level, and UTM projection (an official and unique coordinate system in Mongolia) on 28 January 2009 at a resolution of N28. See ref [76] for the coordinates.

Table 1. Data source description (All data accessed on 1 January 2022).

	Data Type	Time	Data Sources	Data Website Address
Image data	Landsat 5 (TM)SR	1990–2012	USGS	www.usgs.gov
	Landsat 8 (OLI)SR	2013–2021	USGS	www.usgs.gov
	DEM	2000	NASA	https://srtm.csi.cgiar.org
	Land use/cover	2000–2020	GlobeLand30	http://www.globallandcover.com/
Basic geographic data	Livestock, Population	1990–2020	National Bureau of Statistics of Mongolia	http://en.nso.mn
	ET	1990–2020	PML_V2, REA ET	http://poles.tpdc.ac.cn/zh-hans
	GDP	2000–2021	National Bureau of Statistics of Mongolia	http://en.nso.mn
	Road	1990–2021	GRIP global roads database	www.globio.info/download-grip-dataset
	MAAT, MAP	1990–2021	NOVA	www.ncei.noaa.gov
	Land use/cover area	1990–2021	Mongolian Statistical Yearbook, CAS	http://1212.mn

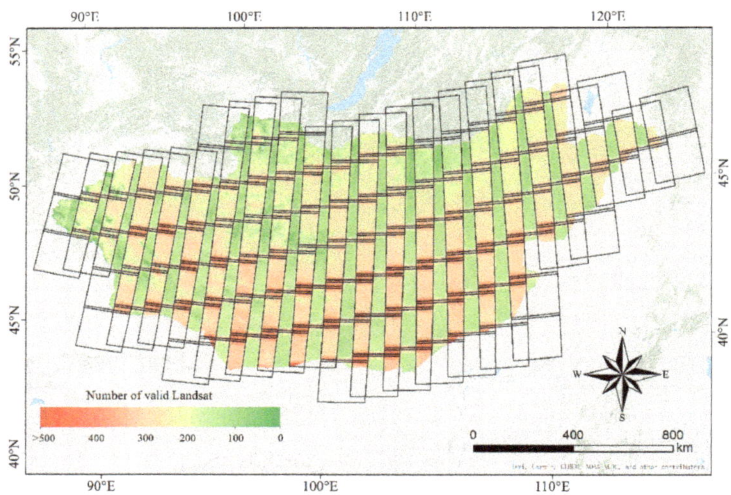

Figure 2. Effective number of Landsat images (cloud cover below 20%) of Mongolia from 1990 to 2021.

2.3. Extraction of Land Use/Cover Type

Our study attempted to develop a long-term land use extraction model for Mongolia. The extraction of land use/cover types with higher accuracy was carried out in the following four steps:

Step I. Data pre-processing

In this study, we selected Landsat bands (1~7) for the classification (Table S1), e.g., blue, green, red, near-infrared, short-wave infrared (SWIR_1/2), and one thermal infrared (TIR) band. Further, surface bio-physical parameters, such as the normalized difference vegetation index (NDVI), the bare soil index (BSI), the normalized difference moisture index (NDWI), and the normalized building index (NDBI) were extracted. Finally, multi-bands imagery with 12 bands was produced for the image classification. The formulas are as follows:

$$NDVI = \frac{NIR - R}{NIR + R} \quad (1)$$

$$BSI = \frac{\rho(SWIR + R) - \rho(NIR + B)}{\rho(SWIR + R) + \rho(NIR + B)} \quad (2)$$

$$NDBI = \frac{SWIR - NIR}{SWIR + NIR} \quad (3)$$

$$MNDWI = \frac{G - MIR}{G + MIR} \quad (4)$$

where *NIR*, *MIR*, *G*, *R*, *B*, and *SWIR* are the reflectance values of the near-infrared band, the mid-infrared band, the green band, the infrared band, the blue band, and the short-wave infrared band, respectively.

Step II. Creation of training and testing samples

Accurate classification is a challenging task, especially with the confusion over the various spectral values such as for barren and built area land. The high-quality training samples and verification samples are needed for feature classification in RF. Therefore, we constructed a year-by-year sample library using visual interpretation in GEE. We used the function (ee.ImageCollection) on the GEE platform to display the Landsat images year by year. Then, the samples of each period in this study area were obtained by visual interpretation. GEE automatically extracted band information of different land use types to generate a sample library. The incorrect samples included in the different land types were deleted by the spectral characteristic curves, which were obtained with the mean band curve (Table 2). Finally, more than 2000 samples were selected from each year to establish a sample library. The sample types included cropland, forest, built area, water, barren land, and grassland. Then, 70% of the sample points were used as training samples for training classifiers and 30% as verification samples for accuracy verification [77,78].

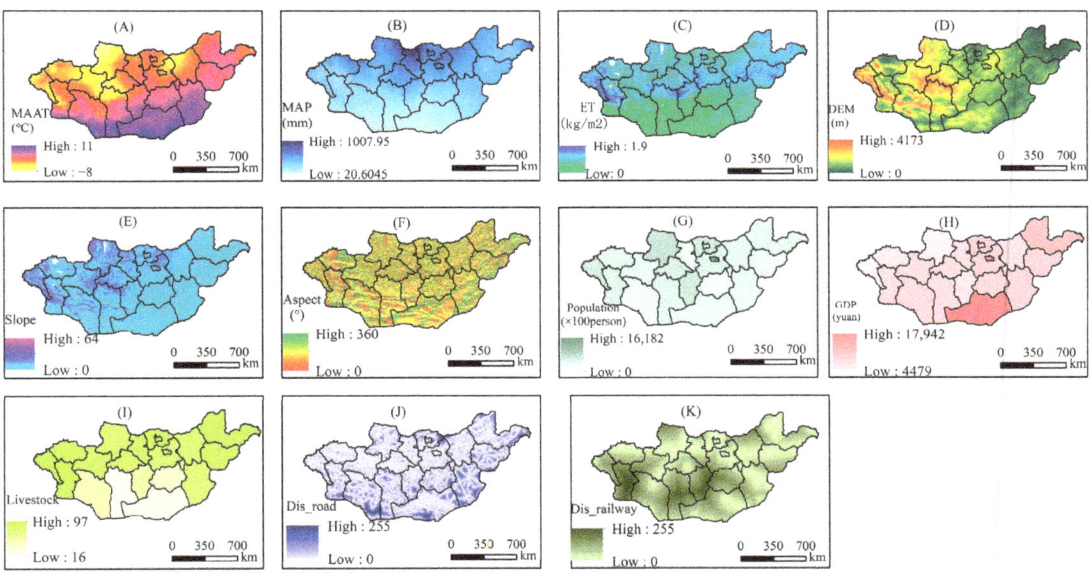

Figure 3. Spatial distribution of driving force variables for land use changes between 1990 and 2021 in Mongolia. (**A**) MAAT, (**B**) MAP, (**C**) ET, (**D**) DEM, (**E**) slope, (**F**) aspect, (**G**) population, (**H**) GDP, (**I**) Livestock, (**J**) distance to road, (**K**) distance to railway.

Table 2. Characteristics of land use types.

ID	Land Type	Characteristic	Image Map	Spectrum Curve
1	Cropland	Land mainly planted with crops, including other economic trees		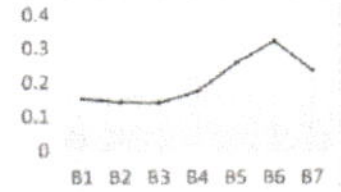
2	Forest	Coverage > 30%, mainly including arbors, bamboos, and other plants		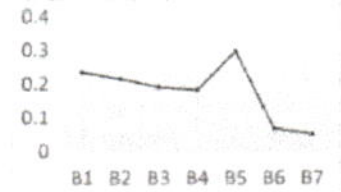
3	Built area	Urban residential land, residential land, transportation facilities, and other buildings	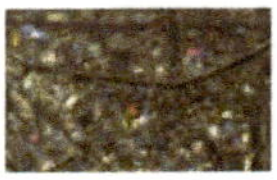	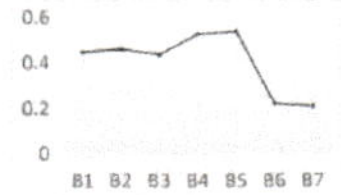
4	Water	Natural lakes, depressions, rivers, and artificial water conservancy		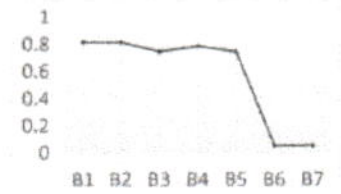
5	Barren	Desert, rock, bare, Gobi, sandy land, and other areas with low vegetation coverage		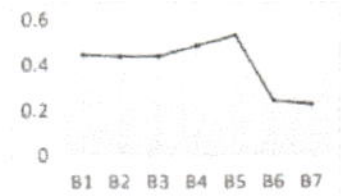
6	Grassland	Coverage > 15%, including desert grassland, meadow grassland, prairie, and other areas		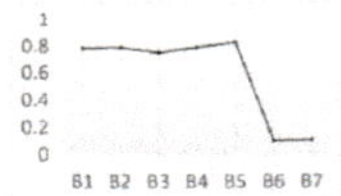

Step III. Random forest

Random forest algorithm was widely used in land use/cover classification. Its basic principle is to construct a set of decision tree classifiers, and each decision tree will give a classification choice. The problem of decision tree over-fitting was improved by the voting mechanism of multiple decision trees, the majority voting mechanism strategy was used to obtain the final output [79], and bootstrap aggregation was used to modify the sample distribution, improving the generalization ability [80]. For the RF, all the training data were placed in a black box and then randomly picked to train the model [81]. Compared with other machine learning methods, the RF classification algorithm has better robustness and can run effectively on large datasets [82]. Moreover, some scholars have achieved excellent research results by using the RF algorithm on GEE to classify land use [83,84].

Land use/cover was performed by using the ee.smileRandomForest function in GEE API, which only needs to determine two parameters: the number of classification trees and the number of characteristic variables input when nodes splitting [83,85]. In order to prevent over-fitting, we tested the number of trees with a threshold of 50 trees. When the number of trees exceeded 100, the model trended to stable. Therefore, 150 trees were finally selected for RF classification, and 6 random variables were selected from the best split when each tree grows.

Step IV. Post-processing of classification

By analyzing the land use/cover classification of the RF, we found that the barren land, built area, grassland, and cropland were misclassified. Therefore, we post-processed the classification results to improve the accuracy of land cover classification. Firstly, we divided the areas that may have been built area or cropland using the CART (Classification and Regression Trees) decision-making method in GEE. Then, the divided areas were further classified by RF through the constructed sample library. The CART method is often applied to land use classification [86]. The CART method usually includes a selection of variables, decision tree generation, and pruning of the decision tree [87]. The accuracy of the analysis can be improved by observing the classification results when modifying the node threshold of the decision tree or optimizing the training data [88]. We found that the cropland was characterized by NDVI ranging from 0.25 to 0.45, while the NDBI values of the built areas were all greater than 0.1, and the BSI value was greater than 0.12. Thus, we constructed a CART decision tree based on the thresholds of the NDVI, NDBI, and BSI. We briefly divided the three areas using the CART. First, the cropland and grassland were divided by the NDVI, in which 0.25 < NDVI < 0.45 was classified as the first partition (including all croplands and part-grasslands). NDBI > 0.1 and BSI > 0.12 were defined as the second partition (including all built areas and part-barren). Other types of land were defined as the third partition. Finally, we used the RF method to reclassify the land use/cover types of the first and second partitions independently, and updated this part of the area using the imagePlough.updateMaskd function in GEE to obtain the final land use classification result.

In addition, we calculated the influence of each variable on the heterogeneity of the observed values at each node of the classification tree by using MeanDecreaseGini on the R platform [89]. The greater the value, the greater the importance of the variable, so we used MeanDecreaseGini to verify the importance of different factors to land use types and rank them. We used the function of Importance() in the R statistical software package to calculate the MeanDecreaseGini value between each driving factor and land use type. Finally, visual representation of the importance of variables was carried out by using the varImpPlot function in the R statistical software package.

2.4. Accuracy Evaluation

A confusion matrix is often used to evaluate the accuracy of each land cover type [90]. For single-date maps, a confusion matrix was built in GEE to evaluate the accuracy of LUCC classification, and then the overall accuracy (OA), producer accuracy (PA), user ac-curacy (UA), and Kappa coefficient of the quantitative indicators were calculated [85]. UA is the probability that a value predicted to be in a certain class really is that class. PA refers to the probability that a value in a given class was classified correctly. For the change of land use type, the accuracy was evaluated with the method of [91]. We selected the samples of the changed areas considering 18 reporting themes in 6 land use/cover types on 2015 and 2021 (Table S2). Then, the change accuracy was analyzed by three indexes: land use gain, loss, and no change. Gain refers to some type of increase, loss refers to some type of decrease, and no change refers to no change during the study period. For example, between 2015 and 2020, cropland loss means a change from cropland to any other class, cropland gain means a change to cropland from any other class, and no change means no change in cropland during this period. Then, the confusion matrix for the accuracy of the change of land use types in two periods was calculated. Moreover, five typical areas in two existing land use products (Google Earth and GlobeLand30) were selected to evaluate the spatial distribution of classification [87]. The related formulas were as follows:

$$UA = \frac{X_{ij}}{X_j} \times 100\% \quad (5)$$

$$PA = \frac{X_{ij}}{X_i} \times 100\% \quad (6)$$

$$OA = \frac{s_d}{n} \times 100\% \quad (7)$$

$$Kappa = \frac{p_o - p_e}{1 - p_e} \quad (8)$$

where X_{ij} is the observed value of the ith row and the jth column in the confusion matrix; X_i, X_j represent the marginal totals of the i-th row and the j-th column, respectively; X_{ii} represents the observed value of the i-th row and the i-th column in the confusion matrix; p_o equals OA; and p_e is the ratio of the product sum of the actual sample number of each class and the predicted sample number of each class to the square of the total sample number n.

2.5. Land Use/Cover Change Dynamic Degree

Based on the statistics of the area changes in different land use/cover change types in each grid, the dynamic degree model and conversion matrix were used to analyze land use change processes. The dynamic degree model developed by [92] was divided into single and comprehensive models to describe the expansion and shrinkage of types of land use. The single land use dynamic degree model (K) can quantity one land use type during a specified period, while the comprehensive land use dynamic degree (LC) was used to characterize the rate and amplitude of land use change in the study area. The formulae are as follows:

$$K = \frac{LU_a - LU_b}{LU_a} \times \frac{1}{T} \times 100\% \tag{9}$$

$$LC = \left[\frac{\sum_{i=1}^{n} \Delta LU_{i-j}}{2 \cdot \sum_{i=1}^{n} \Delta LU_i}\right] \times \frac{1}{T} \times 100\% \tag{10}$$

Here, LU_a and LU_b are the areas of this land use type in the initial year (a) and end year (b) of the study period; LU_i is the area of the i types; ΔLU_{i-j}. is the area of the i-th type that is converted to another type; n is the number of types; and T is the study period, which is five years.

We used ArcGIS spatial analysis to superimpose the land use types in different periods, obtain the structural characteristics and transfer the matrix of land use change, and show the transformation relationship between different land use types in two periods in the form of charts.

In brief, based on the Landsat dataset and the Google Earth Engine platform (GEE), we took the random forest as the main body of the classifier and combined the regression trees (CART), obtained the land types of Mongolia over nearly 30 years, and analyzed the patterns and dynamic and driving forces. The brief details of the methodology applied in this research are presented in Figure 4.

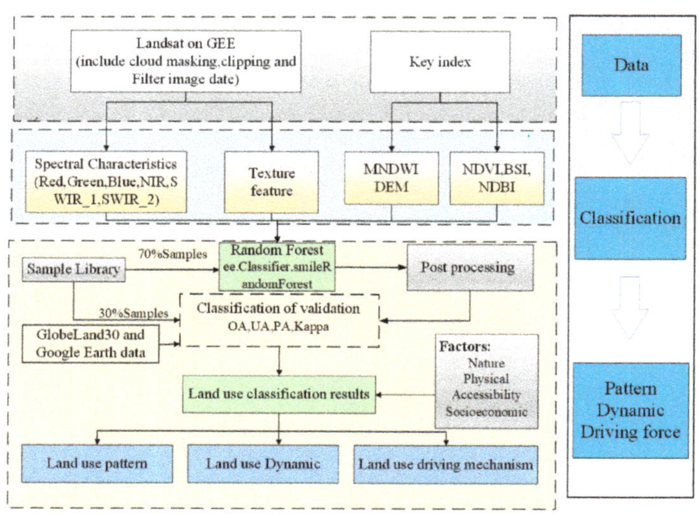

Figure 4. Methodological flowchart for this study.

3. Result

3.1. Land Use/Cover Pattern

Land use/cover in Mongolia is classified into six types: cropland, forest, built area, water, barren land, and grassland. Grassland, barren land, and forest areas are the main land use/cover types in Mongolia.

The spatial distribution of land use/cover in Mongolia from 1990 to 2021 is shown in Figure 5 and Figure S1, and Table 3. Generally speaking, the area of grassland is the largest, accounting for more than 43% of the national area mainly in the central and northern regions of Mongolia. Barren land occupies more than 40%, mainly as arid barren land and desert distributed in southern Mongolia. Sporadic grassland is distributed in the basins and Gobi areas. More than 7% is forest land, mainly distributed in the north of the Kent-Hang'ai Mountain. The water bodies mainly comprise plateau lakes. The cropland and built areas are the smallest, making up about 1% of the whole country. Generally speaking, the grassland, barren land, and forest areas in Mongolia are the most widely distributed types, accounting for more than 95%. The land use/cover type pattern from north to south is a transitional distribution of forest–grassland–barren land.

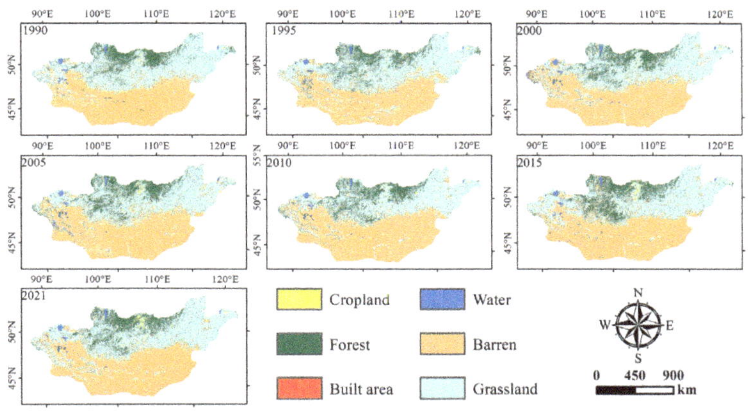

Figure 5. Spatial distribution of land use type in Mongolia from 1990 to 2021.

To intuitively reflect the spatial characteristics of LUCC, we used the grid calculator of ArcGIS to calculate the spatial distribution map of mutual transformation between different land use/cover. The land use change pattern in Mongolia from 1990 to 2021 is shown in Table 3 and Figure 6. We estimate that 27.5×10^4 km^2 of the land surface in Mongolia, accounting for 17.6%, changed at least once among the six land categories from 1990 to 2021. The grassland, water, cropland, and built areas exhibited a growing trend, while the barren land and forest areas decreased. Among all types, the amplitude of land use change was the smallest for forest land and the largest for built areas. The area of barren land decreased by 6.2×10^4 km^2, mainly in central Mongolia. The forest area decreased by 0.5×10^4 km^2, mainly in the Selenge Valley in northern Mongolia. Meanwhile, grassland and water area increased. The grassland increased by 6.6×10^4 km^2, mainly in eastern Mongolia. The water area increased by 1007 km^2, which was mainly related to the changed area of lakes on the plateau. The area with the least amount of change was the built area, but it increased dramatically from 499 km^2 to 1242 km^2, representing approximately twice the growth rate of the other areas. On the time scale, taking 2005 as the boundary, the barren area first showed an increasing and then decreasing trend. In addition, compared to 1990, the increase in barren land mainly occurred in the East Govi and North Donogovi provinces. The change in the grassland area first showed a decreasing and then increasing trend, and the increase in grassland mainly occurred in northern Mongolia, particularly in the Selenge prefecture.

Table 3. The areas of different land use types in Mongolia from 1990 to 2021.

	1990		1995		2000	
	Area (km^2)	**Proportion (%)**	**Area (km^2)**	**Proportion (%)**	**Area (km^2)**	**Proportion (%)**
Cropland	11,985	0.73	9972	0.60	11,881	0.71
Forest	129,645	7.93	141,098	8.56	120,431	7.36
Built area	499	0.03	539	0.03	651	0.04
Water	16,462	1.01	16,989	1.03	17,872	1.09
Barren	723,413	44.25	696,490	42.26	770,983	47.14
Grassland	752,708	46.05	783,298	47.52	710,830	43.47

2005		2010		2015		2021	
Area (km^2)	Proportion (%)	Area (km^2)	Proportion (%)	Area (km^2)	Proportion (%)	Area (km^2)	Proportion (%)
10,901	0.67	10,761	0.66	11,703	0.71	12,720	0.77
116,992	7.15	116,854	7.15	129,406	7.91	124,524	7.61
647	0.04	800	0.05	1037	0.06	1242	0.08
16,649	1.02	16,425	1.01	17,860	1.09	17,469	1.07
777,381	47.50	754,526	46.10	725,936	44.35	661,066	40.40
714,031	43.62	737,004	45.03	751,015	45.88	819,205	50.07

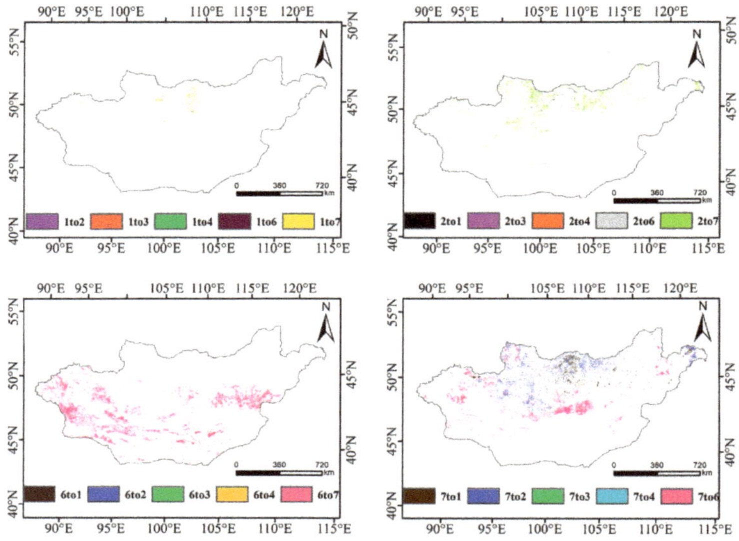

Figure 6. Land use change in Mongolia from 1990 to 2021; 1—cropland, 2—forest, 3—built area, 4—water, 5—barren, and 6—grassland.

3.2. Land Use/Cover Change Dynamic

In order to explain the degree of LUCC, the comprehensive and single land use dynamic degrees were calculated on the classification scale (Figure 7A). Over the past 32 years, the comprehensive land use dynamic degree had an N-shaped trend. Three stages of change were presented. It increased significantly from 0.36% to 0.88% in 1990–2000, then dramatically reduced to 0.09% in 2000–2005. Subsequently, the dynamic degree increased continuously, and the land use/cover change dynamic were 0.24%, 0.29%, and 0.61%. Considering the dynamic degree of single land use, the built areas showed an overall growth tendency. It began to increase substantially after 2005 and reached a peak of 4.94%

in 2010–2015. Cropland decreased in 2005 and then increased. The forest area decreased the most from 1995 to 2000 and gradually increased after 2010. The water area was in a state change, with the biggest change from 1995 to 2000 showing an increasing trend. The barren land increased first and then decreased, with the largest decrease in 2015–2021. The grassland decreased the most in 1995–2000, and the growth rate was the largest in 2015–2021. Generally speaking, the comprehensive land use dynamic degree of Mongolia over the past 32 years showed an upward trend before 1990–2000, and a downward trend from 1995 to 2010. It showed an upward trend from 2005 to 2021. In terms of land types, the increase in built areas began to increase after 2005, reaching a maximum in 2010–2015; cropland also began to increase, especially after 2010.

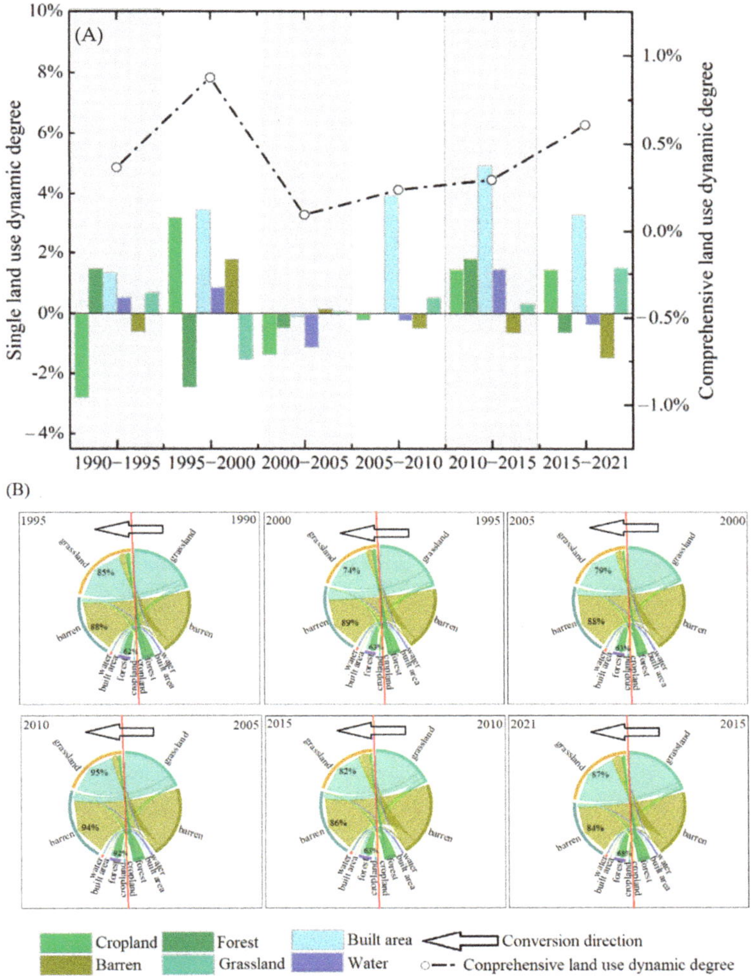

Figure 7. Dynamic degree (**A**) and transfer matrix Sankey diagram (**B**) of land use/cover change in Mongolia from 1990 to 2021. Note: In the figure, the land use change in different years is divided by red lines, the arrow direction represents the transfer direction of land use types, and the percentage indicates how many land use types have not changed during the two periods.

In order to reflect the specific situation of mutual transformation among the different land use/cover change types, we determined the area conversion of each type in each period using a transfer matrix (Figure 7B). It can be seen that, over the past 32 years, area change in Mongolia was mainly a mutual transformation between grassland and barren land. From 1995 to 2000, grassland was mainly transformed into barren land in 2010–2015, and the opposite occurred from 2010 to 2015. The final statistics show that from 1990 to 2021, the area of barren land converted into grassland was 1.1×10^5 km^2. Other land types were also transformed, but this was not obvious because of their small size.

In conclusion, over the past 32 years, the comprehensive dynamic degree in Mongolia showed an N-shaped trend, and 2005 was a turning point. Built area changed dramatically. Particularly from 2010 to 2015, there was obvious mutual transformation between grassland and barren land, and grassland increased and barren land decreased, which indicates that desertification control in Mongolia achieved good results.

3.3. Influencing Factors of LUCC

Based on the R platform, we used the RF algorithm to rank the relative importance of 11 driving factors on land use/cover change (Figure 8). It can be seen that the MAAT, ET, and MAP were the main driving factors in the past 32 years, and the sum of the relative importance of these three categories was more than 50%. As far as environmental factors are concerned, the relative influence of the MAAT, ET, and MAP is high. The environmental factors were the most important factors affecting the land use/cover change types. In terms of the physical factors, the DEM had an obviously high influence, and most of the forests were distributed in 1000–2000 m a.s.l., while the grassland was mainly located above 3000 m a.s.l. in southern Mongolia. Considering the socioeconomic and accessibility factors, the Dis_railway, GDP, and livestock factors had great influence on the LUCC, which is mainly related to policy and lifestyle in Mongolia.

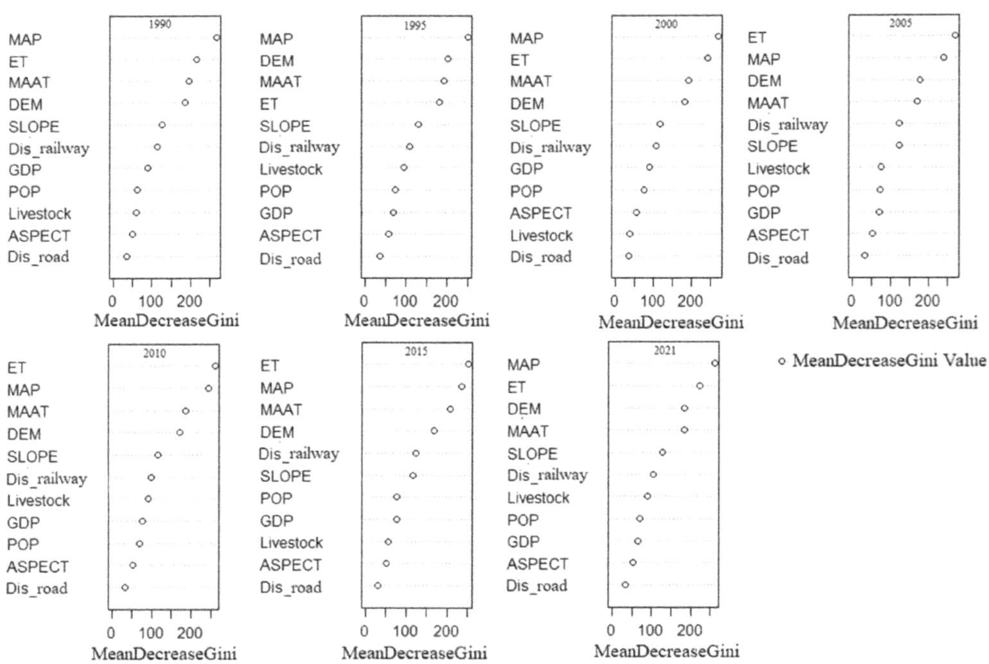

Figure 8. Relative importance (MeanDecreaseGini) of the land use/cover change driving factors.

4. Discussion

4.1. Validation

Based on the RF, we calculated the confusion matrix to evaluate the accuracy of each land type. According to the accuracy of the producers and users, the random forest classification had the highest accuracy for forests, water, and barren land. On the whole, the overall accuracy rate and kappa coefficient were more than 76% and 0.7 (Table 4 and Table S2), which met our research needs. Comparing the Google Earth and GlobeLand30 data [93] (Table 5), it was found that the patterns of different types were similar, while there were some errors between the cropland and water land types, which may have been related to different image dates or image sensors. The difference between the PA and UA can be seen in Table 4. This is because the image features of cropland are similar to those of grassland [94]. Moreover, cropland changed greatly, and its abandonment also led to a greater degree of fragmentation, which interferes with the division of these two land types. Meanwhile, we compared the recent land use areas in Mongolia. Our results are basically consistent with those of [32] and different from those of [48], especially for cropland, grassland, and barren land (Table 6). This may be related to the effect of different approaches, as [48] used the NDVI threshold method to identify the land types, and the threshold value is affected by many uncertain factors, e.g., times, images, and places.

Table 4. Accuracy verification.

LUCC Type	Accuracy Type	1990	1995	2000	2005	2010	2015	2021
Cropland	PA (%)	74	62	63	59	64	59	70
	UA (%)	94	94	92	94	94	93	94
Forest	PA (%)	80	75	78	75	75	82	86
	UA (%)	88	83	89	84	85	90	85
Built area	PA (%)	66	59	77	61	61	60	70
	UA (%)	100	100	95	100	100	100	98
Water	PA (%)	93	93	92	93	94	89	89
	UA (%)	98	97	96	97	95	94	99
Barren	PA (%)	88	83	87	89	89	85	83
	UA (%)	95	89	84	88	89	88	91
Grassland	PA (%)	91	88	83	87	87	87	87
	UA (%)	73	66	68	68	69	67	71
OA (%)	– –	84.63	78.32	80.01	79.94	80.63	79.83	82.42
Kappa	– –	0.7677	0.7127	0.7382	0.7349	0.7445	0.7332	0.7689

Table 5. Verification of classification results.

Data Source	Classification Result Comparison
Google Earth	
Our paper	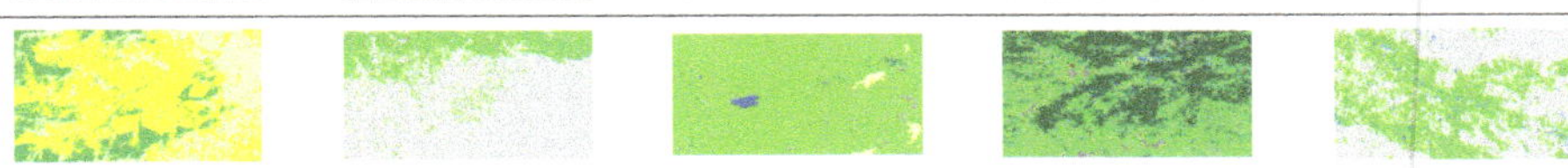
GlobeLand30	

Table 6. Comparison of land use changes in Mongolia and other regions.

Regions	Year	Interval	Land Use Change Rate (%)						Reference
			Cropland	Forest	Grassland	Water	Built	Barren	
Mongolia	1990–2021	1	+8.5	−2.5	+8.5	+7.1	+156.1	−8.6	This paper
Mongolia	1990–2020	10	−27.0	−5.4	+18.4	−8.1	+150.7	−15.2	[48]
Mongolian Plateau	1990–2020	10	+4.9	−1.5	+10.5	−0.6	+47.2	−13.4	[32]
Inner Mongolia	2000–2015	5	−0.2	−1.0	+0.7	+2.3	+22.5	+0.4	[95]
Central Asia	1995–2015	10	+16.1	−0.1	−3.0	−0.2	+223.5	−4.0	[96]
North and West Africa	1985–1995	10	+3.6	−1.5	+64.4	+82.7	+169.4	+1.2	[97]
Global	2001–2012	12	−0.5	−1.2	+2.0	+1.0	−0.1	−6.2	[98]

Note: "+" means an increasing trend, and "−" means a decreasing trend.

At present, the evaluation of error and uncertainty of land use data also included change accuracy [92]. We verified the change accuracy of land cover types, taking 2015 and 2021 as an example (Table S4). It was shown that the no-change-reporting themes had higher agreement and that the change-reporting themes had lower agreement. During the 2015–2021 change periods, forest gain UA was between 80% and 86%. Barren loss PA approached 90.95%. PA for most of the remaining reporting themes ranged from 70% to 80% with agriculture gain and water gain being exceptions, with PA below 65%. The change accuracy may be related to the fragmentation and distribution of land types. The UA of water and grassland was lower; the UA of most of the remaining reporting themes ranged from 70% to 80%. Obviously, the no-change themes had higher consistency. The research showed that UA and PA may be higher in homogeneous regions for the changing region, while the accuracy was lower in heterogeneous regions (such as a single isolated pixel) [86,93].

We selected numbers using parameter optimization. RF can accurately and robustly process high-dimensional and multicollinear data [94]. Over-fitting does not increase according to the number of trees in the RF model, but it tends to be stable with more trees [95]. Therefore, we took 50 trees as increments and finally chose 150 trees to participate in the classification. We improved the accuracy of the image classification using the CART method, but the image quality, natural environment, and dispersion rate may still have affected the accuracy of the classification, especially for the cropland and built area (Table 4 and Table S3). However, the spatial distribution of cropland in Mongolia is discrete. The distribution of cropland and grassland is mixed and led to misclassification. In addition, cropland and grassland have similar spectral characteristics in the growing season, which also affects the classification accuracy. A previous study also found that the classification accuracy of cropland fluctuated greatly compared to other types [96]. For example, the PA of cropland was different between 1990 and 1995, as seen in Table 4. The built land is also very discrete, which is similar to the spectral information of barren land, which also affects the classification accuracy, e.g., the PA and UA.

This method is generally used for large-scale land use classification research. However, the selection of images in different months of the growing season, the influence of cloud cover and the imbalance of sample points will still affect the accuracy. Additionally, solving the problem of land use/cover types with similar spectral characteristics and realize further fine classification is still the focus of current research.

4.2. Reasons for Land Use Change

One of the significant factors leading to vegetation change in Mongolia is climate change [52]. Over the past 32 years, we found that the MAP and ET had the greatest influence on land types, followed by MAAT (Figure 8). We found that cropland in Mongolia only accounts for 1% of the whole country, which is closely related to the typical arid environment of Mongolia. The overall forest in Mongolia still shows a slight decreasing trend, and the proportion of grassland types is gradually increasing. This is consistent with the results of [33,48]. In Mongolia, the MAP in the growing period of plants has decreased by 33 mm since 1961 [99]. The grassland changes in Mongolia are mainly due to global climate change [33]. Over the past 70 years, the MAAT in Mongolia has increased by 2.1 °C, resulting in an increased loss of soil moisture through evapotranspiration [100]. Meanwhile, the increase in temperature and the decrease in precipitation has led to a dry climate, which affects surface evaporation and plant growth. Although the significance of temperature was slightly lower, the change in temperature did have a greater impact on some areas [27]. Meanwhile, the special characteristics of LUCC are consistent with the regions in Mongolia with decreasing precipitation and increasing temperature [101].

Socioeconomic factors also play a key role in land use/cover change in Mongolia. After 1995, grassland in Mongolia showed a trend of degradation, which was mainly related to Mongolia's emphasis on economic development at that time (Figure 7 and Table 3). This is consistent with the results of [102] and may be related to the continuous advancement of privatization and overgrazing and the development of animal husbandry [103]. The concept of "development before governance" has had a profound impact on Mongolia and may have led to increasing built areas and decreasing cropland from 1990 to 2010, showing opposite trends [104,105]. In order to increase its GDP, Mongolia began to exploit mineral resources in large quantities and built railways to facilitate the transportation of resources, which caused great damage to the environment and, thus, had a great impact on land types [106]. Therefore, the significant influences of GDP and Dis_railway on land use/cover types appear to have gradually increased. The implementation of the "Four Modernizations" policy in 2010 led to a sharp increase in the built areas of Ulaanbaatar.

Overall, it holds that the changes in Mongolia are restricted by natural factors such as climate, but they are also driven by policy and economic development.

4.3. Analysis of the LUCC Pattern in Mongolia and Globally

To evaluate the land use/cover change in Mongolia, we summarized the global and regional land changes, such as the surrounding areas of Mongolia and some representative arid areas (Table 6). Globally, barren land and forest land show a decreasing trend. The area of barren land in arid areas such as Mongolia and the Mongolian Plateau has reduced significantly, and the same is true for the global barren land. These changes may be related to global greening [107,108]. However, at present, desertification is still a global phenomenon that cannot be ignored, as it will destroy global biodiversity and aggravate the occurrence of natural disasters [109]. Land use/cover change was the highest for built land and lowest for forest areas during the studied time period in Mongolia, Central Asia, and Africa. Forest land shows a continuous decreasing trend, especially in arid areas such as Mongolia and East Africa. During rapid urbanization in developing countries, the area of built land is greatly increased. According to the research presented here, the natural degradation of forest has exceeded deforestation, which poses a great threat to the natural environment [110]. Although the proportion of grassland has started to increase in Mongolia, this is mainly due to the implementation of a series of new policies and ecological projects by both the Chinese and Mongolian governments [111]. From the perspective of global change, desertification is still an urgent problem to be solved.

Generally speaking, global water and grassland areas show a growth trend, while barren land is decreasing, indicating that global desertification control has achieved initial results. Cropland areas have also begun to increase, but global forest degradation still needs to be solved. According to a regional analysis of East Africa and the Mongolian Plateau, although desertification control in arid and semi-arid areas has achieved results, it cannot be ignored, and forest degradation is still obvious.

The LUCC pattern in Mongolia shows an N-shaped trend and can be divided into three stages—declining in 1995–2005 and increasing in 1990–2000 and 2005–2021. Meanwhile, the dynamics of global land use/cover change show an upside-down V-shaped pattern, accelerating during 1960–2005 and decelerating from 2006 to 2019 [4]. This is contrary to the global land use/cover change processes outlined by [4] from 1995 to 2010. This may be related to policies, rapid urbanization, and climate change factors. The policy and urbanization development in Mongolia have influenced the land use dynamic degree. Mongolia's urbanization started late, and its urbanization rate began to increase significantly after 2005 and 2015 [112]. However, the data show that since 2005, land use/cover change has started to rise, which is due to the global economic crisis; this has led to an increased demand for biofuels and minerals. Unlike other countries that depend on imports (EU, China), Mongolia started to exploit resources to speed up its development, which led to changes in land types to different degrees [113]. Regarding the food demand in Mongolia, global food production changed from agricultural technology intensification to global market and trade growth, especially in 1990 [114]. However, under the influence of the Four Modernizations policy and drought, cropland in Mongolia decreased by 21% from 1994 to 1999 [115]. Grassland restoration has mainly been due to the implementation of the Green Great Wall Project [116]. Due to the implementation of returning farmland to forests and grasslands, grasslands have been expanded from farmland, but from the results, most grassland area in Mongolia still comes from former forests and barren land. The results of this policy are consistent with those of [87]. Generally speaking, the recent implementation of environmental policy has been helpful to realize the sustainable development of grassland. Specifically, the grassland–barren land boundary has gradually recovered since 2010, and the grassland area has started to increase steadily.

Rising temperatures and decreasing precipitation have had a great influence on the land use/cover change in Mongolia [100]. In 1998, the climate of the Mongolian Plateau was identified as being at a turning point from wet to dry, which made Mongolia transition from a wet period to a more severe drought period [117], and the area of barren land increased between 1995 and 2000. After 2005, rainfall began to change from a downward trend to an upward trend [118]. According to the drought grade, the areas with a high frequency of severe drought (6–11%) are mainly located in southern Mongolia and central and western Mongolia. In addition, from 2001 to 2009, land degradation in Mongolia was frequent [119]. During 1990–2000, Mongolia experienced a turning point of rainy and arid years. The forest and grassland areas increased first and then decreased, and the land use/cover changed greatly. From 2000 to 2005, the grassland area remained relatively stable, and the change range of land use/cover in Mongolia decreased. After 2010, the barren area in Mongolia continued to decrease, the grassland area steadily increased, and the desertification process was effectively controlled. The conclusion drawn in this study studies are similar to the conclusions of [48].

5. Conclusions

In our study, we integrated the random forest method and a classification and regression tree model for identifying the land use/cover type in Mongolia between 1990 and 2021. By comparing the land use/cover change patterns and the dynamic and driving factors in different years, we determined the land use change and the significance of influencing factors on land use over the past 32 years. The results are as follows: (1) The grassland, barren land, and forest areas in Mongolia are the most widely distributed types, accounting for more than 95% of the land. (2) Over the past 32 years, 17.6% of the land surface changed at least once among the six land categories, and the land use dynamic degree had an N-shaped trend. (3) Grassland, water, cropland, and built areas exhibited a growing trend, while barren land and forest areas decreased. (4) The MAP, ET, and MAAT are the main environmental factors that determined the distribution and changes of land types, while Dis_railway and GDP are the main socioeconomic and accessibility factors that influenced the change patterns of land use. Meanwhile, the physical environment also plays a key role.

Land use and land cover change are significant for a range of themes and issues central to the study of global environmental change and sustainable development strategy. It is widely acknowledged that a better understanding of land use dynamics over the next 30–50 years is central to the debate on sustainability. We mapped the land use change in Mongolia over the past 32 years at a resolution of 30 m and explained the spatial patterns and change processes, which is valuable for understanding the change information and desertification of Mongolia and the Mongolian Plateau.

Supplementary Materials: The following supporting information can be downloaded at: https://www.mdpi.com/article/10.3390/rs15071813/s1, Figure S1: Spatial distribution of land use type in Mongolia from 1990 to 2021. The lack of land use type in 2012 due to the missing of remote sensing images; Table S1: Landsat5 and Landsat8 band parameters; Table S2: Reporting themes for accuracy results. Table S3: Verification of classification accuracy. Table S4: UA and PA of LUCC.

Author Contributions: J.H. and Q.L. were responsible for the conceptualization, methodology and writing of the original draft, J.C. and Y.L. were responsible for the methodology and data curation; W.L. and G.H. were responsible for data checking—review and editing; T.W. and X.W. helped with the editing and are responsible for the funding acquisition. All authors have read and agreed to the published version of the manuscript.

Funding: This research is supported by the National Natural Science Foundation of China (41961144021 and 32061143032), the CAS "Light of West China" Program (Granted to Tonghua Wu) and Educational the Technology Innovation Project of Gansu Province (2023CXZX-464).

Data Availability Statement: The data has not yet been completed online. Please contact the first or corresponding author if needed.

Acknowledgments: Thanks to the Shuttle Radar Topography Mission (SRTM), the National Bureau of Statistics of Mongolia, the NASA, USGS and NOVA Organization for their data support. We would also like to thank all the staff members of the Google Earth Engine and Google Colaboratory. Special thanks also to the reviewers and editors; your comments and suggestions have greatly improved the paper's content.

Conflicts of Interest: The authors declare that they have no known competing financial interests or personal relationships that could have appeared to influence the work reported in this paper.

References

1. Perminova, T.; Sirina, N.; Laratte, B.; Baranovskaya, N.; Rikhvanov, L. Methods for Land Use Impact Assessment: A Review. *Environ. Impact Assess. Rev.* **2016**, *60*, 64–74. [CrossRef]
2. Affeld, K.; Wiser, S.K.; Payton, I.J.; DeCáceres, M. Using Classification Assignment Rules to Assess Land-Use Change Impacts on Forest Biodiversity at Local-to-National Scales. *For. Ecosyst.* **2018**, *5*, 162–176. [CrossRef]
3. Stoerk, T.; Wagner, G.; Ward, R.E.T. Recommendations for Improving the Treatment of Risk and Uncertainty in Economic Estimates of Climate Impacts in the Sixth Intergovernmental Panel on Climate Change Assessment Report. *Rev. Environ. Econ. Policy* **2018**, *12*, 371–376. [CrossRef]
4. Winkler, K.; Fuchs, R.; Rounsevell, M.; Herold, M. Global Land Use Changes Are Four Times Greater than Previously Estimated. *Nat. Commun.* **2021**, *12*, 2501. [CrossRef]
5. Gilck, F.; Poschlod, P. The History of Human Land Use Activities in the Northern Alps since the Neolithic Age. A Reconstruction of Vegetation and Fire History in the Mangfall Mountains (Bavaria, Germany). *Holocene* **2021**, *31*, 579–591. [CrossRef]
6. McGowan, P.J.K. Conservation: Mapping the Terrestrial Human Footprint. *Nature* **2016**, *537*, 172–173. [CrossRef]
7. Chilukoti, N.; Xue, Y. An Assessment of Potential Climate Impact during 1948–2010 Using Historical Land Use Land Cover Change Maps. *Int. J. Climatol.* **2021**, *41*, 295–315. [CrossRef]
8. He, X.; Liang, J.; Zeng, G.; Yuan, Y.; Li, X. The Effects of Interaction between Climate Change and Land-Use/Cover Change on Biodiversity-Related Ecosystem Services. *Glob. Chall.* **2019**, *3*, 1800095. [CrossRef] [PubMed]
9. Yan, M.; Liu, J.; Wang, Z. Global Climate Responses to Land Use and Land Cover Changes over the Past Two Millennia. *Atmosphere* **2017**, *8*, 64. [CrossRef]
10. Pongratz, J.; Dolman, H.; Don, A.; Erb, K.H.; Fuchs, R.; Herold, M.; Jones, C.; Kuemmerle, T.; Luyssaert, S.; Meyfroidt, P.; et al. Models Meet Data: Challenges and Opportunities in Implementing Land Management in Earth System Models. *Glob. Chang. Biol.* **2018**, *24*, 1470–1487. [CrossRef]
11. Newbold, T.; Hudson, L.N.; Hill, S.L.L.; Contu, S.; Lysenko, I.; Senior, R.A.; Börger, L.; Bennett, D.J.; Choimes, A.; Collen, B.; et al. Global Effects of Land Use on Local Terrestrial Biodiversity. *Nature* **2015**, *520*, 45–50. [CrossRef] [PubMed]
12. Grimm, N.B.; Faeth, S.H.; Golubiewski, N.E.; Redman, C.L.; Wu, J.; Bai, X.; Briggs, J.M. Global Change and the Ecology of Cities. *Science* **2008**, *319*, 756–760. [CrossRef] [PubMed]
13. Findell, K.L.; Berg, A.; Gentine, P.; Krasting, J.P.; Lintner, B.R.; Malyshev, S.; Santanello, J.A.; Shevliakova, E. The Impact of Anthropogenic Land Use and Land Cover Change on Regional Climate Extremes. *Nat. Commun.* **2017**, *8*, 989. [CrossRef] [PubMed]

14. Hessl, A.E.; Anchukaitis, K.J.; Jelsema, C.; Cook, B.; Byambasuren, O.; Leland, C.; Nachin, B.; Pederson, N.; Tian, H.; Hayles, L.A. Past and Future Drought in Mongolia. *Sci. Adv.* **2018**, *4*, 1–8. [CrossRef]
15. Lamchin, M.; Lee, J.; Lee, W.; Lee, E.J.; Kim, M.; Lim, C.; Choi, H.; Kim, S.-R. Assessment of Land Cover Change and Desertification Using Remote Sensing Technology in a Local Region of Mongolia. *Adv. Sp. Res.* **2015**, *57*, 64–77. [CrossRef]
16. Tsalis, T.A.; Malamateniou, K.E.; Koulouriotis, D.; Nikolaou, I.E. New Challenges for Corporate Sustainability Reporting: United Nations' 2030 Agenda for Sustainable Development and the Sustainable Development Goals. *Corp. Soc. Responsib. Environ. Manag.* **2020**, *27*, 1617–1629. [CrossRef]
17. Liang, X.; Li, P.; Wang, J.; Chan, F.K.S.; Togtokh, C.; Ochir, A.; Davaasuren, D. Research Progress of Desertification and Its Prevention in Mongolia. *Sustainability* **2021**, *13*, 6861. [CrossRef]
18. FROM-GLC (10–30 m, 2015, 2017, China). Available online: http//data.ess.tsinghua.edu.cn/ (accessed on 3 February 2022).
19. Esri Land Cover (10 m, 2017–2021, ESA). Available online: https//www.esri.com/en-us/home (accessed on 3 February 2018).
20. Globe Land 30 (30 m, 2000, 2010, 2020, China). Available online: http//data.ess.tsinghua.edu.cn/ (accessed on 3 February 2022).
21. Global Land Survey (30 m, 1975–2012, USGS). Available online: https//www.usgs.gov/ (accessed on 3 February 2022).
22. Climate Change Initiative Land Cover V2 (300 m, 1992–2020, ESA). Available online: https//www.esa-landcover-cci.org/ (accessed on 3 February 2022).
23. MODIS Land Cover Type/Dynamics (0.5–1 Km, 2001–2020, NASA). Available online: https//lpdaac.usgs.gov (accessed on 3 February 2022).
24. Xu, L.; Herold, M.; Tsendbazar, N.-E.; Masiliunas, D.; Li, L.; Lesiv, M.; Fritz, S.; Verbesselt, J. Time Series Analysis for Global Land Cover Change Monitoring: A Comparison across Sensors. *Remote Sens. Environ.* **2022**, *271*, 112905. [CrossRef]
25. Nasiri, V.; Deljouei, A.; Moradi, F.; Sadeghi, S.; Moein, M.; Borz, S.A. Land Use and Land Cover Mapping Using Sentinel-2, Landsat-8 Satellite Images, and Google Earth Engine: A Comparison of Two Composition Methods. *Remote Sens.* **2022**, *14*, 1977. [CrossRef]
26. Choi, M.J.; Enkhbat, U. Distributional Effects of Ger Area Redevelopment in Ulaanbaatar, Mongolia. *Int. J. Urban Sci.* **2020**, *24*, 50–68. [CrossRef]
27. Wei, Y.; Zhen, L.; Liu, X.; Batkhishig, O. Land Use Change and Its Driving Factors in Mongolia from 1992 to 2005. *Chin. J. Appl. Ecol.* **2008**, *19*, 1995–2002. [CrossRef]
28. Vittek, M.; Brink, A.; Donnay, F.; Simonetti, D.; Desclée, B. Land Cover Change Monitoring Using Landsat MSS/TM Satellite Image Data over West Africa between 1975 and 1990. *Remote Sens.* **2013**, *6*, 658–676. [CrossRef]
29. Liu, S.; Liu, X.; Yu, L.; Wang, Y.; Zhang, G.J.; Gong, P.; Huang, W.; Wang, B.; Yang, M.; Cheng, Y. Climate Response to Introduction of the ESA CCI Land Cover Data to the NCAR CESM. *Clim. Dyn.* **2021**, *56*, 4109–4127. [CrossRef]
30. Ersi, C.; Bayaer, T.; Bao, Y.; Bao, Y.; Yong, M.; Zhang, X. Temporal and Spatial Changes in Evapotranspiration and Its Potential Driving Factors in Mongolia over the Past 20 Years. *Remote Sens.* **2022**, *14*, 1856. [CrossRef]
31. Li, G.; Wang, J.; Wang, Y.; Wei, H.; Ochir, A.; Davaasuren, D.; Chonokhuu, S.; Nasanbat, E. Spatial and Temporal Variations in Grassland Production from 2006 to 2015 in Mongolia along the China-Mongolia Railway. *Sustainability* **2019**, *11*, 2177. [CrossRef]
32. Tsalis, T.A.; Malamateniou, K.E.; Koulouriotis, D.; Nikolaou, I.E. Land Cover Change Analysis to Assess Sustainability of Development in the Mongolian Plateau over 30 Years. *Sustainability* **2022**, *14*, 6129. [CrossRef]
33. Wang, J.; Cheng, K.; Liu, Q.; Zhu, J.; Ochir, A.; Davaasuren, D.; Li, G.; Wei, H.; Chonokhuu, S.; Namsrai, O.; et al. Land Cover Patterns in Mongolia and Their Spatiotemporal Changes from 1990 to 2010. *Arab. J. Geosci.* **2019**, *12*, 1–13. [CrossRef]
34. Ma, L.; Li, M.; Ma, X.; Cheng, L.; Du, P.; Liu, Y. A Review of Supervised Object-Based Land-Cover Image Classification. ISPRS J. Photogramm. *Remote Sens.* **2017**, *130*, 277–293. [CrossRef]
35. Xie, T.; Yuan, Z.; Yang, H.; Jiang, M.; Liao, Z.; Xu, C. Review of Land Use/Cover Change Classification Methods Based on Remote Sensing Image. *Front. Earth Sci.* **2020**, *10*, 500–507. [CrossRef]
36. Hassan Khavarian Nehzak, M.A.; Mostafazadeh, R. Hamidreza Rabiei-Dastjerdi Chapter 5 - Assessment of Machine Learning Algorithms in Land Use Classification, Comput. *Earth Environ. Sci.* **2022**, 97–104. [CrossRef]
37. Alshari, E.A.; Gawali, B.W. Development of Classification System for LULC Using Remote Sensing and GIS. *Glob. Transitions Proc.* **2021**, *2*, 8–17. [CrossRef]
38. Ettehadi Osgouei, P.; Kaya, S. Analysis of Land Cover/Use Changes Using Landsat 5 TM Data and Indices. *Environ. Monit. Assess.* **2017**, *189*, 1–11. [CrossRef]
39. Rivera, S.; Lowry, J.H.; Hernandez, A.J.; Ramsey, R.D.; Lezama, R.; Velasquez, M.A. A Comparison between Cluster Busting Technique and a Classification Tree Algorithm of a Moderate Resolution Imaging Spectrometer (MODIS) Land Cover Map of Honduras. *Geocarto Int.* **2012**, *27*, 17–29. [CrossRef]
40. Sang, X.; Guo, Q.; Wu, X.; Fu, Y.; Xie, T.; He, C.; Zang, J. Intensity and Stationarity Analysis of Land Use Change Based on CART Algorithm. *Sci. Rep.* **2019**, *9*, 12279. [CrossRef] [PubMed]
41. Gislason, P.O.; Benediktsson, J.A.; Sveinsson, J.R. Random Forests for Land Cover Classification. *Pattern Recognit. Lett.* **2006**, *27*, 294–300. [CrossRef]
42. Nguyen, L.H.; Joshi, D.R.; Clay, D.E.; Henebry, G.M. Characterizing Land Cover/Land Use from Multiple Years of Landsat and MODIS Time Series: A Novel Approach Using Land Surface Phenology Modeling and Random Forest Classifier. *Remote Sens. Environ.* **2020**, *238*. [CrossRef]

43. Spinoni, J.; Barbosa, P.; Bucchignani, E.; Cassano, J.; Cavazos, T.; Cescatti, A.; Christensen, J.H.; Christensen, O.B.; Coppola, E.; Evans, J.P.; et al. Global Exposure of Population and Land-Use to Meteorological Droughts under Different Warming Levels and SSPs: A CORDEX-Based Study. *Int. J. Climatol.* **2021**, *41*, 6825–6853. [CrossRef]
44. Sakamoto, K.; Tomonari, M.; Ariya, U.; Nakagiri, E.; Matsumoto, T.K.; Akaji, Y.; Otoda, T.; Hirobe, M.; Nachin, B. Effects of Large-Scale Forest Fire Followed by Illegal Logging on the Regeneration of Boreal Forests in Mongolia. *Landsc. Ecol. Eng.* **2021**, *17*, 267–279. [CrossRef]
45. Xu, Y.; Zhang, Y.; Chen, J.; John, R. Livestock Dynamics under Changing Economy and Climate in Mongolia. *Land Use Policy* **2019**, *88*, 104120. [CrossRef]
46. Mandakh, N.; Tsogtbaatar, J.; Dash, D.; Khudulmur, S. Spatial Assessment of Soil Wind Erosion Using WEQ Approach in Mongolia. *J. Geogr. Sci.* **2016**, *26*, 473–483. [CrossRef]
47. Li, Y.; Wang, J.; Zhu, J. Landscape Pattern Analysis of Mongolia Based on the Geographical Partitions. *Arid L. Geogr.* **2016**, *39*, 817–827.
48. Wang, J.; Wei, H.; Cheng, K.; Ochir, A.; Shao, Y.; Yao, J.; Wu, Y.; Han, X.; Davaasuren, D.; Chonokhuu, S.; et al. Updatable Dataset Revealing Decade Changes in Land Cover Types in Mongolia. *Geosci. Data J.* **2022**, *9*, 341–354. [CrossRef]
49. Sanzheev, E.D.; Mikheeva, A.S.; Osodoev, P.V.; Batomunkuev, V.S.; Tulokhonov, A.K. Theoretical Approaches and Practical Assessment of Socio-Economic Effects of Desertification in Mongolia. *Int. J. Environ. Res. Public Health* **2020**, *17*, 4068. [CrossRef] [PubMed]
50. Wang, J.; Wei, H.; Cheng, K.; Ochir, A.; Davaasuren, D.; Li, P.; Shun Chan, F.K.; Nasanbat, E. Spatio-Temporal Pattern of Land Degradation from 1990 to 2015 in Mongolia. *Environ. Dev.* **2020**, *34*, 100497. [CrossRef]
51. The Climate in Mongolia. Available online: https//www.worlddata.info/asia/mongolia/climate.php (accessed on 3 February 2018).
52. Vandandorj, S.; Munkhjargal, E.; Boldgiv, B.; Gantsetseg, B. Changes in Event Number and Duration of Rain Types over Mongolia from 1981 to 2014. *Environ. Earth Sci.* **2017**, *76*, 70. [CrossRef]
53. Dugarsuren, N.; Lin, C.; Tsogt, H. Land Cover Change Detection in Mongolia in Last Decade Using Modis Imagery. *Remote Sens.* **2011**, *4*, 2750–2755.
54. Punsantsogvoo, T. Implications of Rural Settlement Patterns for Development: Case Study in Central and Eastern Economic Region of Mongolia. *Proc. Mong. Acad. Sci.* **2019**, *59*, 14–20. [CrossRef]
55. Map Cruzin Data Research & GIS Project Specialist. Available online: https//mapcruzin.com/ (accessed on 1 May 2022).
56. Fang, X.; Zhao, W.; Zhang, C.; Zhang, D.; Wei, X.; Qiu, W.; Ye, Y. Methodology for Credibility Assessment of Historical Global LUCC Datasets. *Sci. China Earth Sci.* **2020**, *63*, 1013–1025. [CrossRef]
57. Soni, P.K.; Rajpal, N.; Mehta, R.; Mishra, V.K. Urban Land Cover and Land Use Classification Using Multispectral Sentinal-2 Imagery. Multimed. *Tools Appl.* **2022**, *81*, 36853–36867. [CrossRef]
58. Gorelick, N.; Hancher, M.; Dixon, M.; Ilyushchenko, S.; Thau, D.; Moore, R. Google Earth Engine: Planetary-Scale Geospatial Analysis for Everyone. *Remote Sens. Environ.* **2017**, *202*, 18–27. [CrossRef]
59. Mas, J.F.; de Araújo, F.S. Assessing Landsat Images Availability and Its Effects on Phenological Metrics. *Forests* **2021**, *12*, 574. [CrossRef]
60. Cao, R.; Feng, Y.; Chen, J.; Zhou, J. A Supplementary Module to Improve Accuracy of the Quality Assessment Band in Landsat Cloud Images. *Remote Sens.* **2021**, *13*, 4947. [CrossRef]
61. Li, G.; Sun, S.; Fang, C. The Varying Driving Forces of Urban Expansion in China: Insights from a Spatial-Temporal Analysis. Landsc. *Urban Plan.* **2018**, *174*, 63–77. [CrossRef]
62. Wu, X.; Liu, H.; Li, X.; Piao, S.; Ciais, P.; Guo, W.; Yin, Y.; Poulter, B.; Peng, C.; Viovy, N.; et al. Higher Temperature Variability Reduces Temperature Sensitivity of Vegetation Growth in Northern Hemisphere. Geophys. *Res. Lett.* **2017**, *44*, 6173–6181. [CrossRef]
63. Li, X.; Guo, W.; Chen, J.; Ni, X.; Wei, X. Responses of Vegetation Green-up Date to Temperature Variation in Alpine Grassland on the Tibetan Plateau. *Ecol. Indic.* **2019**, *104*, 390–397. [CrossRef]
64. Hansson, A.; Dargusch, P.; Shulmeister, J. A Review of Modern Treeline Migration, the Factors Controlling It and the Implications for Carbon Storage. *J. Mt. Sci.* **2021**, *18*, 291–306. [CrossRef]
65. Bai, Y.; Wu, J.; Xing, Q.; Pan, Q.; Huang, J.; Yang, D.; Han, X. Primary Production and Rain Use Efficiency across a Precipitation Gradient on the Mongolia Plateau. *Ecology* **2008**, *89*, 2140–2153. [CrossRef]
66. Bai, W.; Wan, S.; Niu, S.; Liu, W.; Chen, Q.; Wang, Q.; Zhang, W.; Han, X.; Li, L. Increased Temperature and Precipitation Interact to Affect Root Production, Mortality, and Turnover in a Temperate Steppe: Implications for Ecosystem C Cycling. *Glob. Chang. Biol.* **2010**, *16*, 1306–1316. [CrossRef]
67. Miehe, G.; Schlütz, F.; Miehe, S.; Opgenoorth, L.; Cermak, J.; Samiya, R.; Jäger, E.J.; Wesche, K. Mountain Forest Islands and Holocene Environmental Changes in Central Asia: A Case Study from the Southern Gobi Altay, Mongolia. *Palaeogeogr. Palaeoclimatol. Palaeoecol.* **2007**, *250*, 150–166. [CrossRef]
68. Miao, H.; Chen, S.; Chen, J.; Zhang, W.; Zhang, P.; Wei, L.; Han, X.; Lin, G. Cultivation and Grazing Altered Evapotranspiration and Dynamics in Inner Mongolia Steppes. *Agric. For. Meteorol.* **2009**, *149*, 1810–1819. [CrossRef]
69. Horel, Á.; Zsigmond, T.; Farkas, C.; Gelybó, G.; Tóth, E.; Kern, A.; Bakacsi, Z. Climate Change Alters Soil Water Dynamics under Different Land Use Types. *Sustainability* **2022**, *14*, 3908. [CrossRef]

70. Berihun, M.L.; Tsunekawa, A.; Haregeweyn, N.; Meshesha, D.T.; Adgo, E.; Tsubo, M.; Masunaga, T.; Fenta, A.A.; Sultan, D.; Yibeltal, M. Exploring Land Use/Land Cover Changes, Drivers and Their Implications in Contrasting Agro-Ecological Environments of Ethiopia. *Land Use Policy* **2019**, *87*, 104052. [CrossRef]
71. Wu, F.; Mo, C.; Dai, X. Analysis of the Driving Force of Land Use Change Based on Geographic Detection and Simulation of Future Land Use Scenarios. *Sustainability* **2022**, *14*, 5254. [CrossRef]
72. Ruan, X.; Qiu, F.; Dyck, M. The Effects of Environmental and Socioeconomic Factors on Land-Use Changes: A Study of Alberta, Canada. *Environ. Monit. Assess.* **2016**, *188*, 1–31. [CrossRef]
73. Wang, Y.; Xia, T.; Shataer, R.; Zhang, S.; Li, Z. Analysis of Characteristics and Driving Factors of Land-Use Changes in the Tarim River Basin from 1990 to 2018. *Sustainability* **2021**, *13*, 10263. [CrossRef]
74. Wilkin, S.; Ventresca Miller, A.; Miller, B.K.; Spengler, R.N.; Taylor, W.T.T.; Fernandes, R.; Hagan, R.W.; Bleasdale, M.; Zech, J.; Ulziibayar, S.; et al. Economic Diversification Supported the Growth of Mongolia's Nomadic Empires. *Sci. Rep.* **2020**, *10*, 3916. [CrossRef]
75. Wei, Y.; Zhen, L. The Dynamics of Livestock and Its Influencing Factors on the Mongolian Plateau. *Environ. Dev.* **2020**, *34*, 100518. [CrossRef]
76. Reguzzoni, M.; Carrion, D.; De Gaetani, C.I.; Albertella, A.; Rossi, L.; Sona, G.; Batsukh, K.; Herrera, J.F.T.; Elger, K.; Barzaghi, R.; et al. Open Access to Regional Geoid Models: The International Service for the Geoid. *Earth Syst. Sci. Data* **2021**, *13*, 1653–1666. [CrossRef]
77. Fu, Y.; Zhang, Y. Research on Temporal and Spatial Evolution of Land Use and Landscape Pattern in Anshan City Based on GEE. *Front. Environ. Sci.* **2022**, *10*, 1–17. [CrossRef]
78. Giles, M. Foody Status of Land Cover Classification Accuracy Assessment. *Remote Sens. Environ.* **2002**, *80*, 185–201. [CrossRef]
79. Paul, A.; Mukherjee, D.P.; Member, S.; Das, P.; Gangopadhyay, A. Improved Random Forest for Classification. *IEEE Trans. Image Process.* **2018**, *7149*, 1–13. [CrossRef]
80. Naikoo, M.W.; Talukdar, S.; Ishtiaq, M.; Rahman, A. Modelling Built-up Land Expansion Probability Using the Integrated Fuzzy Logic and Coupling Coordination Degree Model. *J. Environ. Manag.* **2023**, *325*, 116441. [CrossRef] [PubMed]
81. Belgiu, M.; Drăgu, L. Random Forest in Remote Sensing: A Review of Applications and Future Directions. ISPRS J. Photogramm. *Remote Sens.* **2016**, *114*, 24–31. [CrossRef]
82. Ghorbanian, A.; Zaghian, S.; Asiyabi, R.M.; Amani, M.; Mohammadzadeh, A.; Jamali, S. Mangrove Ecosystem Mapping Using Sentinel-1 and Sentinel-2 Satellite Images and Random Forest Algorithm in Google Earth Engine. *Remote Sens.* **2021**, *13*, 2565. [CrossRef]
83. Cui, J.; Zhu, M.; Liang, Y.; Qin, G.; Li, J.; Liu, Y. Land Use/Land Cover Change and Their Driving Factors in the Yellow River Basin of Shandong Province Based on Google Earth Engine from 2000 to 2020. *ISPRS Int. J. Geo-Information* **2022**, *11*, 163. [CrossRef]
84. Zhang, D.D.; Zhang, L. Land Cover Change in the Central Region of the Lower Yangtze River Based on Landsat Imagery and the Google Earth Engine: A Case Study in Nanjing, China. *Sensors* **2020**, *20*, 2091. [CrossRef]
85. Liu, C.; Li, W.; Zhu, G.; Zhou, H.; Yan, H.; Xue, P. Land Use/Land Cover Changes and Their Driving Factors in the Northeastern Tibetan Plateau Based on Geographical Detectors and Google Earth Engine: A Case Study in Gannan Prefecture. *Remote Sens.* **2020**, *12*, 3139. [CrossRef]
86. Fitoka, E.; Tompoulidou, M.; Hatziiordanou, L.; Apostolakis, A.; Höfer, R.; Weise, K.; Ververis, C. Water-Related Ecosystems' Mapping and Assessment Based on Remote Sensing Techniques and Geospatial Analysis: The SWOS National Service Case of the Greek Ramsar Sites and Their Catchments. *Remote Sens. Environ.* **2020**, *245*, 111795. [CrossRef]
87. Zhang, M.; Du, H.; Mao, F.; Zhou, G.; Li, X.; Dong, L.; Zheng, J.; Zhu, D.; Liu, H.; Huang, Z.; et al. Spatiotemporal Evolution of Urban Expansion Using Landsat Time Series Data and Assessment of Its Influences on Forests. *ISPRS Int. J. Geo-Information* **2020**, *9*, 64. [CrossRef]
88. Ghulam, A. Monitoring Tropical Forest Degradation in Betampona Nature Reserve, Madagascar Using Multisource Remote Sensing Data Fusion. IEEE J. Sel. Top. Appl. Earth Obs. *Remote Sens.* **2014**, *7*, 4960–4971. [CrossRef]
89. Zou, D.; Zhao, L.; Liu, G.; Du, E.; Hu, G.; Li, Z.; Wu, T.; Wu, X.; Chen, J. Vegetation Mapping in the Permafrost Region: A Case Study on the Central Qinghai-Tibet Plateau. *Remote Sens.* **2022**, *14*, 232. [CrossRef]
90. Cheng, K.; Wang, J. Forest Type Classification Based on Integrated Spectral-Spatial-Temporal Features and Random Forest Algorithm-A Case Study in the Qinling Mountains. *Forests* **2019**, *10*, 559. [CrossRef]
91. Wickham, J.; Stehman, S.V.; Gass, L.; Dewitz, J.A.; Sorenson, D.G.; Granneman, B.J.; Poss, R.V.; Baer, L.A. The-matic Accuracy Assessment of the 2011 National Land Cover Database (NLCD). *Remote Sens. Environ.* **2017**, *191*, 328–341. [CrossRef] [PubMed]
92. Pontius, R.G.; Huang, J.; Jiang, W.; Khallaghi, S.; Lin, Y.; Liu, J.; Quan, B.; Ye, S. Rules to Write Mathematics to Clarify Metrics Such as the Land Use Dynamic Degrees. *Landsc. Ecol.* **2017**, *32*, 2249–2260. [CrossRef]
93. Jun, C.; Ban, Y.; Li, S. Open Access to Earth Land-Cover Map. *Nature* **2014**, *514*, 434. [CrossRef]
94. Pluto-Kossakowska, J. Review on Multitemporal Classification Methods of Satellite Images for Crop and Arable Land Recognition. *Agriculture* **2021**, *11*, 999. [CrossRef]
95. Li, M.; Li, X.; Liu, S.; Lyu, X.; Dang, D.; Dou, H.; Wang, K. Analysis of the Spatiotemporal Variation of Landscape Patterns and Their Driving Factors in Inner Mongolia from 2000 to 2015. *Land* **2022**, *11*, 1410. [CrossRef]
96. Li, J.; Chen, H.; Zhang, C.; Pan, T. Variations in Ecosystem Service Value in Response to Land Use/Land Cover Changes in Central Asia from 1995-2035. *PeerJ* **2019**, *7*, e7665. [CrossRef]

97. Henchiri, M.; Ali, S.; Essifi, B.; Kalisa, W.; Zhang, S.; Bai, Y. Monitoring Land Cover Change Detection with NOAA-AVHRR and MODIS Remotely Sensed Data in the North and West of Africa from 1982 to 2015. *Environ. Sci. Pollut. Res.* **2020**, *27*, 5873–5889. [CrossRef]
98. De Rosa, M.; Vestergaard Odgaard, M.; Staunstrup, J.K.; Trydeman Knudsen, M.; Hermansen, J.E. Identifying Land Use and Land-Use Changes (LULUC): A Global LULUC Matrix. *Environ. Sci. Technol.* **2017**, *51*, 7954–7962. [CrossRef]
99. Qin, F.; Jia, G.; Yang, J.; Na, Y.; Hou, M.; Narenmandula. Spatiotemporal Variability of Precipita-tion during 1961–2014 across the Mongolian Plateau. *J. Mt. Sci.* **2018**, *15*, 992–1005. [CrossRef]
100. Yang, T.; Li, P.; Wu, X.; Hou, X.; Liu, P.; Yao, G. Assessment of Vulnerability to Climate Change in the Inner Mongolia Steppe at a County Scale from 1980 to 2009. *Rangel. J.* **2014**, *36*, 545–555. [CrossRef]
101. Bliedtner, M.; Strobel, P.; Struck, J.; Salazar, G.; Szidat, S.; Nowaczyk, N.; Bazarradnaa, E.; Lloren, R.; Dubois, N.; Haberzettl, T.; et al. Holocene Temperature Variations in Semi-Arid Central Mongolia—A Chronological and Sedimentological Perspective From a 7400-Year Lake Sediment Record From the Khangai Mountains. *Front. Earth Sci.* **2022**, *10*, 910782. [CrossRef]
102. Ojima, D.S.; Chuluun, T.; Bolortsetseg, B.; Tucker, C.J.; Hicke, J. Eurasian Land Use Impacts on Rangeland Productivity. *Geophys. Monogr. Ser.* **2004**, *153*, 293–301. [CrossRef]
103. Hilker, T.; Natsagdorj, E.; Waring, R.H.; Lyapustin, A.; Wang, Y. Satellite Observed Widespread Decline in Mongolian Grasslands Largely Due to Overgrazing. *Glob. Chang. Biol.* **2014**, *20*, 418–428. [CrossRef]
104. Yao, Z.; Zhang, L.; Tang, S.; Li, X.; Hao, T. The Basic Characteristics and Spatial Patterns of Global Cultivated Land Change since the 1980s. *J. Geogr. Sci.* **2017**, *27*, 771–785. [CrossRef]
105. Buyanduureng. Chelger Agglomeration and Regional Differences of the Mongolia's Agricultural Industry Based on the Space Economics. *Inn. Mong. Soc. Sci.* **2017**, *38*, 184–189. [CrossRef]
106. Zandariya, B. Improving the Policy Framework for Financial Assurance for Mine Closure in Mongolia. *Resour. Policy* **2022**, *77*, 102628. [CrossRef]
107. Wang, Z.; Wang, H.; Wang, T.; Wang, L.; Liu, X.; Zheng, K.; Huang, X. Large Discrepancies of Global Greening: Indication of Multi-Source Remote Sensing Data. *Glob. Ecol. Conserv.* **2022**, *34*, e02016. [CrossRef]
108. Piao, S.; Wang, X.; Park, T.; Chen, C.; Lian, X.; He, Y.; Bjerke, J.W.; Chen, A.; Ciais, P.; Nemani, R.R.; et al. Characteristics, Drivers and Feedbacks of Global Greening Shilong. *Nat. Rev. Earth Environ.* **2019**, *1*, 14–27. [CrossRef]
109. D'Odorico, P.; Bhattachan, A.; Davis, K.F.; Ravi, S.; Runyan, C.W. Global Desertification: Drivers and Feedbacks. *Adv. Water Resour.* **2013**, *51*, 326–344. [CrossRef]
110. Funk, M.; Sugden, A.M. Degradation Exceeds Deforestation. *Science* **2013**, *369*, 1335–1336. [CrossRef]
111. Zhang, Y.; Wang, Q.; Wang, Z.; Yang, Y.; Li, J. Impact of Human Activities and Climate Change on the Grassland Dynamics under Different Regime Policies in the Mongolian Plateau. *Sci. Total Environ.* **2020**, *698*, 134304. [CrossRef]
112. Park, H.; Fan, P.; John, R.; Chen, J. Urbanization on the Mongolian Plateau after Economic Reform: Changes and Causes. *Appl. Geogr.* **2017**, *86*, 118–127. [CrossRef]
113. Dashpurev, B.; Bendix, J.; Lehnert, L.W. Monitoring Oil Exploitation Infrastructure and Dirt Roads with Object-Based Image Analysis and Random Forest in the Eastern Mongolian Steppe. *Remote Sens.* **2020**, *12*, 144. [CrossRef]
114. Krausmann, F.; Langthaler, E. Food Regimes and Their Trade Links: A Socio-Ecological Perspective. *Ecol. Econ.* **2019**, *160*, 87–95. [CrossRef]
115. Yekimovskaya, O.A.; Lopatina, D.N. The Features of Development of Agricultural Land Use in the Republic of Buryatia and Mongolia (the Selenga River Basin). *IOP Conf. Ser. Earth Environ. Sci.* **2019**, *320*, 012007. [CrossRef]
116. Shao, C.; Chen, J.; Li, L.; Dong, G.; Han, J.; Abraha, M.; John, R. Grazing Effects on Surface Energy Fluxes in a Desert Steppe on the Mongolian Plateau. *Ecol. Appl.* **2017**, *27*, 485–502. [CrossRef]
117. Jin, L.; Zhang, J.; Wang, R.; Zhang, M.; Bao, Y.; Guo, E.; Wang, Y. Analysis for Spatio-Temporal Variation Characteristics of Droughts in Different Climatic Regions of the Mongolian Plateau Based on SPEI. *Sustainability* **2019**, *11*, 5767. [CrossRef]
118. Meng, X.; Gao, X.; Li, S.; Li, S.; Lei, J. Monitoring Desertification in Mongolia Based on Landsat Images and Google Earth Engine from 1990 to 2020. *Ecol. Indic.* **2021**, *129*, 107908. [CrossRef]
119. Kimura, R.; Moriyama, M. Use of a Modis Satellite-Based Aridity Index to Monitor Drought Conditions in Mongolia from 2001 to 2013. *Remote Sens.* **2021**, *13*, 2561. [CrossRef]

remote sensing

MDPI

Article

Land Use/Land Cover Mapping Based on GEE for the Monitoring of Changes in Ecosystem Types in the Upper Yellow River Basin over the Tibetan Plateau

Senyao Feng [1], Wenlong Li [1,*], Jing Xu [2], Tiangang Liang [1], Xuanlong Ma [3,4], Wenying Wang [5] and Hongyan Yu [6]

[1] State Key Laboratory of Herbage Improvement and Grassland Agro-Ecosystems, College of Pastoral Agriculture Science and Technology, Lanzhou University, Lanzhou 730000, China
[2] School of Agriculture and Forestry Economic and Management, Lanzhou University of Finance and Economics, Lanzhou 730020, China
[3] College of Earth and Environmental Sciences, Lanzhou University, Lanzhou 730020, China
[4] Institute of Yellow River Basin Green Development, Lanzhou University, Lanzhou 730020, China
[5] Department of Life Sciences, Qinghai Normal University, Xining 810008, China
[6] Service Guarantee Center of Qilian Mountain National Park in Qinghai, Xining 810008, China
* Correspondence: wllee@lzu.edu.cn; Tel.: +86-136-5945-1811

Abstract: The upper Yellow River basin over the Tibetan Plateau (TP) is an important ecological barrier in northwestern China. Effective LULC products that enable the monitoring of changes in regional ecosystem types are of great importance for their environmental protection and macro-control. Here, we combined an 18-class LULC classification scheme based on ecosystem types with Sentinel-2 imagery, the Google Earth Engine (GEE) platform, and the random forest method to present new LULC products with a spatial resolution of 10 m in 2018 and 2020 for the upper Yellow River Basin over the TP and conducted monitoring of changes in ecosystem types. The results indicated that: (1) In 2018 and 2020, the overall accuracy (OA) of LULC maps ranged between 87.45% and 93.02%. (2) Grassland was the main LULC first-degree class in the research area, followed by wetland and water bodies and barren land. For the LULC second-degree class, the main LULC was grassland, followed by broadleaf shrub and marsh. (3) In the first-degree class of changes in ecosystem types, the largest area of progressive succession (positive) was grassland–shrubland (451.13 km^2), whereas the largest area of retrogressive succession (negative) was grassland–barren (395.91 km^2). In the second-degree class, the largest areas of progressive succession (positive) were grassland–broadleaf shrub (344.68 km^2) and desert land–grassland (302.02 km^2), whereas the largest areas of retrogressive succession (negative) were broadleaf shrubland–grassland (309.08 km^2) and grassland–bare rock (193.89 km^2). The northern and southwestern parts of the study area showed a trend towards positive succession, whereas the south-central Huangnan, northeastern Gannan, and central Aba Prefectures showed signs of retrogressive succession in their changes in ecosystem types. The purpose of this study was to provide basis data for basin-scale ecosystem monitoring and analysis with more detailed categories and reliable accuracy.

Keywords: Google Earth Engine; land use/land cover mapping; machine learning; upper Yellow River basin; Sentinel-2; ecosystem types

1. Introduction

Land use/land cover (LULC) provides valid information for demonstrating the complex and continuous interactions between global-to-local ecosystem change and human behavior [1], and it can reveal the main drivers and consequences of global environmental change [2–4]. It is worth noting that although different regions may have radically different LULC expressions [5], their similar processes of change are constantly altering the energy and material cycles at the Earth's surface [5], including the carbon [6], nitrogen [7], and water cycles [8–10]. With the increases in human-led LULC behavior in different parts of the world, such as agricultural expansion [11,12], urban expansion [13], and engineering projects [14], the ecosystem's structure and function [15], biodiversity [16–18], and services [19] are further disturbed, either positively or negatively [20–22]. This creates great uncertainty for regional ecosystems and ecological environments, and this will continue to increase, especially with global climate change [23]. In this context, achieving accurate LULC products at different spatial scales (local, national, and global) is essential for the regular monitoring of existing ecosystems [20], planning and management of natural resources [24], research and assessment of environmental quality [25], and promotion of sustainable regional development [24].

Early LULC product mapping was mainly based on field surveys and manual visual interpretation. This method is inefficient and costly, making it difficult to meet the need for the regular and timely monitoring and mapping of LULC [26]. With the rapid development of remote sensing (RS) technology, the development of LULC products at various temporal and geographical scales has become possible [27]. In the process, satellite data have improved dramatically in four directions: spatial [28,29] and temporal resolution [30], radiation, and spectra [31].For instance, Sentinel-2 has provided unprecedented free images with a spatial resolution of 10 m and a temporal resolution of 5 days to the world since 2015. Meanwhile, with the development of classification methods, machine learning algorithms (e.g., random forest (RF), gradient boosting decision tree (GBDT), and support vector machine (SVM)) have been widely used in the field of LULC mapping [32–35]. These have opened up new possibilities for more detailed LULC mapping [36]. However, they also pose tremendous challenges for the efficiency of RS data collection and processing [37]. To meet these challenges, Google Earth Engine (GEE), a big-data cloud-based network computing resource platform, has emerged, and it has demonstrated strong competitiveness in recent years. It provides massive geospatial datasets with unprecedented online computational and visual analysis capabilities [38,39]. Researchers could use the GEE platform [39] to pre-process or download multi-temporal image data that meet filtering requirements and then use various machine learning algorithms to perform LULC classification and analysis online. As a result, the GEE platform is increasingly being utilized to generate efficient and large-scale LULC products as well as change-monitoring studies [40].

The upper Yellow River basin over the Tibetan Plateau (TP), with a rich variety of ecosystem types, is an important ecological barrier [41] and corridor [42] in the northwest of China and TP. On the one hand, it is important an water supply and connotation area for the entire Yellow River basin and even the interior of northern China [43]. It plays a significant role in soil and water conservation, carbon sequestration, and animal husbandry, in addition to providing water for almost 20% of China's cropland and around 190 million people in the basin [44]. On the other hand, it is one of the most sensitive regions to global climate change [45], with an extremely fragile ecological environment [46]. During recent decades, its LULC and ecosystems have been dramatically altered by both human activities and climate change [47], so there is an urgent need for more detailed and regionally applicable LULC products to monitor and reveal the characteristics (positive or negative) of these changes. Some relevant LULC products are now obtainable for this region, such as FROM-GLC10, the world's first LULC product with a spatial resolution of 10 m based on Sentinel-2 imagery [48], FROM-GLC30 [1], and GlobeLand30 [24], with a spatial resolution of 30 m based on Landsat imagery. However, these existing products have entirely different classification schemes, data sources, methods, and spatial and temporal resolutions, and this has led to significant uncertainty in the classification details they provide. Furthermore, compared to other regions, the study area has more prominent high-altitude landscape heterogeneity, more complex landforms, and richer vegetation communities and ecosystem types, and, as a result, existing products are unable to fully distinguish and describe the LULC variations at the level of ecosystem types in the region. Therefore, it is necessary to take more detailed ecosystem types into account, to develop a new classification scheme, and to produce high-resolution multi-category LULC products for the region to facilitate monitoring at the level of regional ecosystem types.

The study defined a two-level LULC classification scheme based on ecosystem types for the study area and produced LULC maps at 10 m spatial resolution based on the GEE platform, Sentinel-2 image data, and machine learning methods. The main objectives are as follows: (1) to generate products and assess their accuracies and performances; (2) to conduct monitoring of changes in ecosystem types based on LULC products. The results are helpful for the further mapping of LULC products over the whole Yellow River basin and provide new, effective data support for monitoring changes in basin-scale ecosystem types and ecological quality assessment.

2. Materials and Methods

2.1. Study Area

The upper Yellow River basin over TP (32°8′–38°20′N, 95°50′–104°31′E) is located in the westernmost part of the Yellow River basin in northwestern China. It lies on the western edge of the Loess Plateau and the northeastern part of the TP, spanning the Qinghai, Gansu, and Sichuan Provinces (Figure 1). The Yellow River originates here and flows through nine provinces or autonomous regions (Qinghai, Gansu, Sichuan, Ningxia, Inner Mongolia, Shaanxi, Shanxi, Henan, and Shandong) for 5464 km [49] (Figure 1b). The study area is approximately 1.99×10^5 km^2, accounting for about 26.5% of the entire Yellow River basin. It has a complex and diverse landscape, with many high mountains, canyons, basins, and hills sloping from the southwest to the northeast at altitudes of 1650–6254 m [42]. In addition, the study area straddles the plateau's temperate, semi-arid, and subfrigid semi-humid and humid zones and has a typical plateau continental climate, with long sunshine hours, strong solar radiation, long winters, and short summers. It also has an average annual temperature of −4–6 °C, which gradually decreases with increasing altitude, and an average annual precipitation of approximately 420–800 mm [41,43,50,51]. The complex landscapes and climatic types have given rise to a rich variety of LULC types in the study area [52], including mainly grassland, forest, agricultural land, barren land, and wetlands and water bodies (Figure 1d–h), which also makes it an important area for safeguarding ecosystem services in the Yellow River basin. The dynamics of its ecosystem types are closely related to the sustainable development of local socioeconomic development and environment protection.

2.2. Dataset

2.2.1. Satellite Data

In this study, Sentinel-2 multispectral imagery (MSI) from the vegetation growing season (image time range of 1 June to 30 August in 2018 and 2020) was used for composition and classification. Considering the shortage of 2018 Sentinel-2 SR images covering the study area in GEE, the downloading and processing of Sentinel-2 data was divided into two separate parts for 2018 and 2020.

At the ESA Copernicus Open Access Centre (https://scihub.copernicus.eu/dhus/, accessed on 21 March 2022), due to the unique geographical conditions and the phenomenon of high cloud cover in the study area, the threshold for cloud volume was set at 35%. In this context, 92 Sentinel-2A/B Level-1C (L1C) top-of-atmosphere (TOA) reflectance data with <35% cloud cover from the 2018 vegetation growing season were acquired, then batch processed to Level-2A (L2A) surface reflectance data using the radiative transfer model-based Sen2cor tool [53]. All bands were resampled to 10 m by the bilinear method and de-clouded by masking the clouds with a quality scene classification band provided by the L2A images [54]. After median compositing, mosaicking, and cropping the images, the cloud-free Sentinel-2 images from 2018 were finally uploaded to the GEE platform for classification [55].

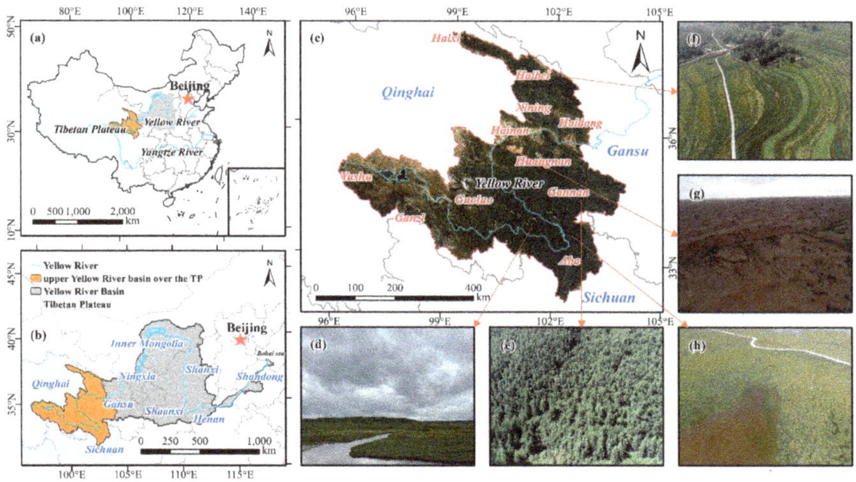

Figure 1. Location of the study area. (**a**) Location of the Yellow River basin in China; (**b**) location of the study area within the Yellow River basin and the Tibetan Plateau; (**c**) examples of true-color images and land use/land cover types for Sentinel-2 from June to August 2020 ((**d**): grassland, (**e**): forest, (**f**): agricultural land, (**g**): barren land, (**h**): wetland and water body).

Sentinel-2 data for 2020 were obtained online through the GEE platform (https://code.earthengine.google.com/, accessed on 6 April 2022). Sentinel-2 L2A images with <35% cloud cover in the study area during the vegetation growing season were retrieved (661 images in total), the cloud was masked through the quality assessment band (QA60) provided by the L2A images, and the subsequent image pre-processing, including resampling, median compositing, mosaicking, and cropping were completed through the GEE platform. Finally, cloud-free Sentinel-2 images from 2020 were obtained [56].

2.2.2. Terrain Data

The Shuttle Radar Topography Mission (SRTM) digital elevation model is available in the GEE platform database [57] with a spatial resolution of 30 m. In this study, the elevation, aspect, hill shade, and slope of the SRTM data were extracted as features for LULC classification using the GEE platform.

2.2.3. Land Use/Land Cover Dataset

The 10 m global-scale land use/land cover dataset FROM-GLC10 provided by Tsinghua University (http://data.ess.tsinghua.edu.cn/, accessed on 25 March 2022) [48], which was generated from Sentinel-2 imagery (spatial resolution of 10 m) from 2017, was used for comparison with the classification results of this study.

2.2.4. Sample Point Data

The field surveys and the collection of field sample points for various LULC types in the study area were conducted between July and August in 2018 and 2020 (Figure 2). Observations were made in areas that were accessible by vehicle, and the Trimble Juno series handheld GPS (https://www.trimble.com, accessed on 10 August 2020) was used to acquire and record information on the type and location of LULCs. Aerial photographs of different land cover types by DJI Phantom 4 and Mavic Pro light unmanned aerial vehicles (UAVs) [58] were used as accurate and reliable field interpretation signs. In 2018 and 2020, we collected a total of 8231 field survey points covering all LULC types in the study area (for 18 second-degree classes). The number of sample points in each class was adjusted with reference to the proportion of land cover area and the actual situation, eliminating the impact of possible sample imbalance on the classification of a small number of rare LULC types [59]. In this case, the effect of sample unevenness on LULC classification was eliminated. In addition, for those remote areas that were difficult to reach by vehicle, such as the western end of the study area (Figure 2), we visually interpreted and collected 15,709 sample points using high-resolution Google Earth Pro and Sentinel-2 RGB true-color composite imagery. Finally, all sample points were checked individually against the high-resolution imagery provided by Google Earth Pro to confirm that the LULC categories for all sample points in 2018 and 2020 were consistent with the actual conditions [60]. In this process, 470 sample points were corrected and added. Then, 80% of the sample points were used as training samples for the classification, and 20% were used to verify the accuracy of the classification through a stratified random sampling method [61]. The set of training and validation points for the years 2018 and 2020 are shown in Table 1.

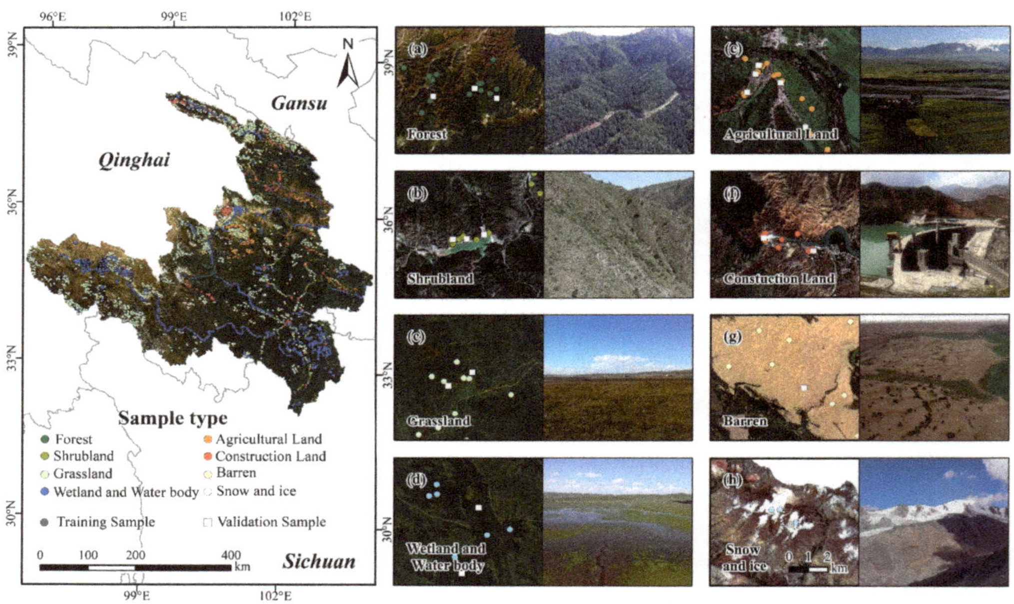

Figure 2. Spatial distribution of LULC first-degree class sample points. (**a**–**h**) Sentinel-2 RGB true-color composite image of LULC first-degree class sample points and aerial photographs of the field.

Table 1. Training and validation sample points in this study.

First-Degree Class	Second-Degree Class	2018		2020	
		Training Samples	Validation Samples	Training Samples	Validation Samples
Forest	Broadleaf Forest	332	83	320	80
	Needleleaf Forest	164	41	196	49
	Mixed Forest	408	102	384	96
Shrubland	Broadleaf Shrub	260	65	196	49
	Needleleaf Shrub	112	28	104	26
Grassland	Grassland	2908	727	2724	681

Table 1. *Cont.*

First-Degree Class	Second-Degree Class	2018 Training Samples	2018 Validation Samples	2020 Training Samples	2020 Validation Samples
Wetland and Water body	Marsh	412	103	448	112
	Lake	380	95	216	54
	River	1304	326	1432	358
Agricultural Land	Cropland	1268	317	1388	347
	Orchard Land	412	103	180	45
Construction Land	Urban	756	189	656	164
	Urban Green Space	92	23	120	30
	Industrial and Transportation Land	272	68	352	88
Barren	Desert Land	244	61	232	58
	Bare Soil	76	19	72	18
	Bare Rock	284	71	248	62
Snow and ice	Snow and ice	384	96	192	48
Total		10068	2517	9460	2365

2.3. Research Methods

The flow chart in Figure 3 shows the mapping and analysis methods applied in this study. Sentinel-2 image pre-processing, feature construction, sample selection, LULC classification, accuracy assessment, and the monitoring of changes in ecosystem types based on LULC are the main steps involved. The following sections describe the classification scheme and several relevant steps in this study in detail.

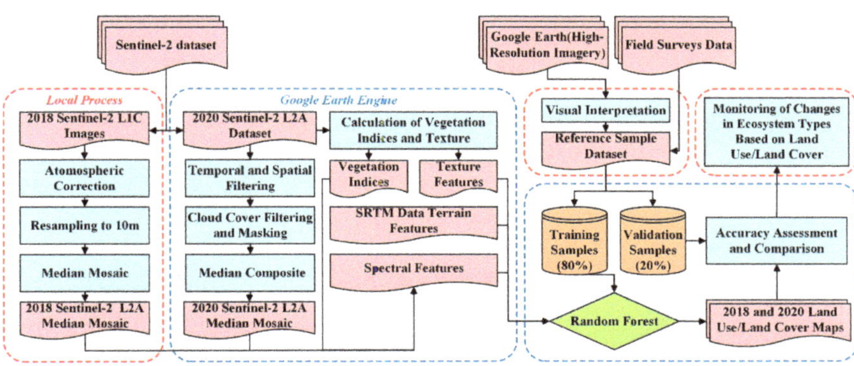

Figure 3. Flow chart of the land use/land cover (LULC) mapping and analysis methodology for this study.

2.3.1. Classification Scheme

Based on ecosystem types [62], taking into account the regional characteristics of the study area and the capability of the satellite data used and referring to the International Geosphere-Biosphere Programme (IGBP) classification system, existing land cover classification products covering it, and LULC classification research from other regions [42,62–65], a new two-level LULC classification scheme was defined (Table 2).

Table 2. The LULC classification scheme of this study.

First-Degree Class	Second-Degree Class	Description
Forest	Broadleaf Forest (BLF)	Dominated by broadleaf trees with a percent cover > 60% and height > 2 m.
	Needleleaf Forest (NLF)	Dominated by needleleaf trees with a percent cover > 60% and height > 2 m.
	Mixed Forest (MF)	Dominated by trees with cover > 60% and height > 2 m. Consists of tree communities with interspersed mixtures or mosaics of the other two forest types. Neither of the forest types exceeded 60%.
Shrubland	Broadleaf Shrub (BLS)	Dominated by broadleaf shrub with a percent cover > 60% and height < 2 m.
	Needleleaf Shrub (NLS)	Dominated by needleleaf shrub with a percent cover > 60% and height < 2 m.
Grassland	Grassland (Grass)	Dominated by herbaceous plants (<2 m) with a percent cover > 60% and height < 2 m.
Wetland and Water body	Marsh (Marsh)	Permanently inundated lands with wet-growing herbaceous or woody plants on the surface.
	Lake (Lake)	Natural, artificially constructed, relatively stationary water surface, including reservoirs, pits, fish ponds, salt ponds.
	River (River)	Naturally flowing linear water surface. Includes canals/channels: artificially constructed, flowing, linear water surfaces > 30 m wide.
Agricultural Land	Cropland (Crop)	Dominated by herbaceous annuals (temporary crops) (<2 m). At least 60% of area is cultivated cropland. Includes post-harvest bare soil period.
	Orchard Land (OL)	Lands with mainly a mosaic of commercial artificial forests or shrublands for cash crops.
Construction Land	Urban (Urban)	At least 60% of area is covered by residential building materials, such as cities, towns, and villages.
	Urban Green Space (UGS)	Artificially cultivated grassland in residential areas, including artificially cultivated recreational and sports grounds in the countryside.
	Industrial and Transportation Land (IAT)	At least 60% of the area is covered by industrial lands, transportation lands, and mining sites.
Barren	Desert Land (DL)	Lands with the ground completely covered by loose sand particles; never has > 10% vegetated cover during any time of the year.
	Bare Soil (BS)	Lands with the surface covered by exposed soil with loose structure; never has > 10% vegetated cover during any time of the year.
	Bare Rock (BR)	Lands with the surface covered by hard rock; gravel-covered surface; never has > 10% vegetated cover during any time of the year.
Snow and ice	Snow and ice (Snow)	At least 80% of the area is covered by snow/ice throughout the year.

Specifically, the ecosystem types of the study area are complex and diverse. On the premise that we were familiar with the basic characteristics of the different types of ecosystems, as well as the meaning of their attributes and their interpretative signs, we divided the LULC into eight primary classes (forest, shrub, grassland, wetland and water body, agricultural land, construction land, barren land, and snow and ice) based on the ecosystem types. On this basis, in order to be able to distinguish and monitor more details within the ecosystem types, the LULC classification scheme was increased to 18 second-degree classes.

For the purpose of LULC classification, we used the same training points for the first- (combining the training points of the second-degree classes into first-degree class for classification) and second-degree classes, then evaluated their accuracy with the same validation points. It is worth noting that to ensure the comparability of LULC area statistics and changes, we only used second-degree class training points to generate all LULC products for the study area (whereby primary first-degree classes were generated by second-degree class products) [60]. In addition, the spatial distribution of sample points for 18 secondary classes and sample selection examples for the Sentinel-2 RGB true color images are shown in Figure 4.

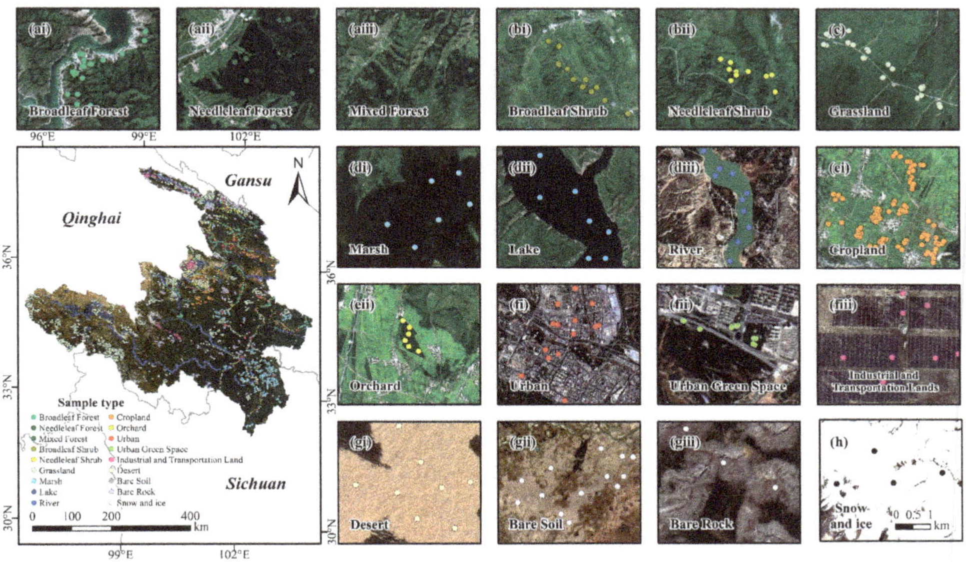

Figure 4. Spatial distribution of LULC second-degree class sample points. (**ai–h**) Sentinel-2 RGB true-color composite image of LULC second-degree class sample points.

2.3.2. Feature Construction

It is important to construct effective feature datasets to participate in LULC classification [66]. For the spectral features, first, based on previous studies and several tests [66–68], the Sentinel-2 L2A multispectral image bands that adequately conveyed the ecosystem types and LULC status of the study area were selected, including B2, B3, B4, B5, B7, B8, and B11. In addition, numerous studies have shown that different vegetation indices (VIs) as input features for classification will effectively improve the accuracy of LULC [26,69–71]. We used the GEE platform to calculate the NDVI (Normalized Difference Vegetation Index [72]), which characterizes the physiological, biochemical, and structural characteristics of the vegetation canopy; the NDWI (Normalized Difference Water Index [73]), which characterizes water bodies; the NDBI (Normalized Difference Building Index [74]), which characterizes manmade land surface features; the SAVI (Soil Adjusted Vegetation Index [75]), which highlights soil characteristics; and the EVI (Enhanced Vegetation Index [76]), which enhances the physiological, biochemical, and structural characteristics of the vegetation canopy. The formulas we used were as follows:

$$\mathrm{NDVI} = \frac{\mathrm{NIR} - \mathrm{Red}}{\mathrm{NIR} + \mathrm{Red}} \tag{1}$$

$$\mathrm{NDWI} = \frac{\mathrm{Green} - \mathrm{NIR}}{\mathrm{Green} + \mathrm{NIR}} \tag{2}$$

$$NDBI = \frac{SWIR - NIR}{SWIR + NIR} \quad (3)$$

$$SAVI = 1.6 \times \frac{NIR - Red}{NIR + Red + 0.6} \quad (4)$$

$$EVI = 2.5 \times \frac{NIR - Red}{NIR + 6 \times Red - 7.5 \times Blue + 1} \quad (5)$$

where NIR, SWIR, Red, Green, and Blue are the reflectance values from the near-infrared (842 nm), short-wave infrared (1610 nm), red (665 nm), green (560 nm), and blue (490 nm) bands of Sentinel-2, respectively.

Texture features are the patterns that result from a specific transformation of a gray-scale image in space. It has been shown that combining texture features with spectral information can improve the accuracy of LULC classification in specific scenarios [77,78]. For example, a recent study has combined GLCM texture features with spectral bands to classify forest tree species and achieved the highest OA (92%) [79], indicating that texture features are important for accurately distinguishing between LULC classes such as forest and shrubland. Currently, Grey-Level Co-occurrence Matrix (GLCM) computation has been widely used in the statistics and analysis of texture features [68,80,81]. As in previous studies [82], here, with the fast computation of the GLCM-based texture feature function *glcmTexture* provided by the GEE platform, and using a gray-level 8-bit image as input, the computation and construction of 18 texture features were completed. In this study, the Sentinel-2 images were first converted to gray-scale as input to the GLCM according to Equation (6), which involves operations between the red, green, and near-infrared bands of Sentinel-2:

$$Gray = (0.3 \times NIR) + (0.59 \times RED) + (0.11 \times GREEN) \quad (6)$$

where NIR, Red, and Green are the reflectance values from the near-infrared (842 nm), red (665 nm), and green (560 nm) bands of Sentinel-2, respectively.

Based on previous studies, we tested the neighborhood of 1, 2, 3, 5, 7, 10, 15, and 30 pixels, respectively, and finally set the GLCM's moving window to 5 × 5 pixels. Compared to other field settings, 5 × 5 pixels provides better recognition of the pixel-level texture in our study area [83]. Then, seven GLCM features including the angular second moment, contrast, correlation, entropy, variance, inverse difference moment, and sum average were calculated.

Topographic features greatly influence the accuracy of LULC classification of forest, shrubland, and agricultural land [50], so we extracted SRTMGL1_003 data provided by the GEE platform [39], including the elevation, slope, slope direction, and mountain shading. These were involved in the construction of the classification features along with texture and spectral features (Table 3).

Table 3. The set of features in this study.

Feature Type	Feature Name
Spectral	B2, B3, B4, B5, B7, B8, B11, NDVI, NDBI, NDWI, SAVI, EVI
Texture	Gray_asm, Gray_contrast, Gray_corr, Gray_ent, Gray_var, Gray_idm, Gray_savg
Terrain	Elevation, Slope, Aspect, Hill shade

2.3.3. GEE-Based Random Forest Classification

The random forest (RF) classifier is a non-parametric integrated learning algorithm based on the composition of multiple independent individual decision trees [84]. During the classification process, the voting process of all trees will determine the final classification result. RF can handle high-dimensional and multicollinear data accurately and robustly [85] and is much less sensitive to overfitting and training data quality [86,87]. At the same time, RF eliminates the need for tree pruning and is computationally fast and less burdensome [88].

These advantages make RF competent for complex remote sensing image classification, and RF is therefore widely used in large-scale, multi-category, and multi-feature LULC mapping [89–91]. This study used a large study area (approximately 200,000 km^2) and large numbers of LULC classes (18 classes in total) and sample points (24,410 samples in total). Following the previous work in [22], the RF classifier set to 500 trees (ntree = 500) was able to be used for LULC classification work after several GEE user computation quota limit tests [86,89]; therefore, this study adopted RF as the classifier for LULC classification. In addition, the importance ranking function based on out-of-bag data error (OOBe) embedded in the RF classifier measures the contribution of individual features to the classification [84,86,92]. During the classification process, the importance scores of all the feature variables involved in the classification using the GEE platform were obtained and plotted [93].

2.3.4. Accuracy Assessment

A random sample of 20% of the sample point data for each of the years 2018 and 2020 was used for the LULC accuracy assessment. This validation set was completely independent of the training dataset and ensured a random distribution of validation points within the study area, with validation points available in each of the 18 second-degree classes.

The confusion matrix is a popular LULC accuracy evaluation method [94,95]. In this study, confusion matrices were constructed in GEE to evaluate LULC classification accuracy, and they then calculated the quantitative metrics' overall accuracy (OA), producer's accuracy (PA), user's accuracy (UA), and Kappa coefficient [96–98]. These indicators reflect the agreement between the LULC classification results and the actual LULC at the sites from different perspectives [99]. Four identical areas were selected for visual evaluation with existing LULC products to compare the differences in the spatial distribution of the classification results [91,100].

2.3.5. The Monitoring of Changes in Ecosystem Types

In this study, the monitoring of changes in ecosystem types was based on the LULC transition matrix and ecological succession theory.

An ecosystem is a comprehensive biological system consisting of all the organisms found and interacting with each other in a particular physical environment [101,102]. It includes multiple types such as the forest ecosystem, shrubland ecosystem, grassland ecosystem, freshwater ecosystem, and wetland ecosystem, among others. Ecosystems are not exactly equivalent to LULC, but the LULC can provide spatially continuous data for monitoring ecosystem change [103], whereby LULC transitions can reflect the change in ecosystem types [104]. Ecological succession is defined as the orderly process of re-assembling vegetation communities or ecosystems following natural or anthropogenic disturbance [105,106]. It can also reflect changes in ecosystem types, including the progressive succession of ecosystem types, for example, the natural growth of grass from abandoned mines or bare soil [107], the artificial restoration of forests [108], and the retrogressive succession of ecosystem types [109], e.g., deforestation by humans or grassland desertification. In addition, some previous studies have shown that LULC transitions can reflect the occurrence of ecological succession [110–112], so it is reasonable to monitor changes in ecosystem types based on ecological succession theory combined with the LULC transition matrix.

For this study, ecological succession relationships between barren land, grassland, freshwater (river and lake), wetland (marsh), shrubland, and forest ecosystem types (including their second-degree classes) were considered. When change in ecosystem types was in the direction of progressive succession (positive), the transfer form was barren/freshwater–grassland/wetland–shrubland–forest, and, conversely, when the change in ecosystem types was in the direction of retrogressive succession (negative), the transfer form was forest–shrubland–grassland/wetland–barren/freshwater. These change relationships are illustrated in Figure 5 and Table S1.

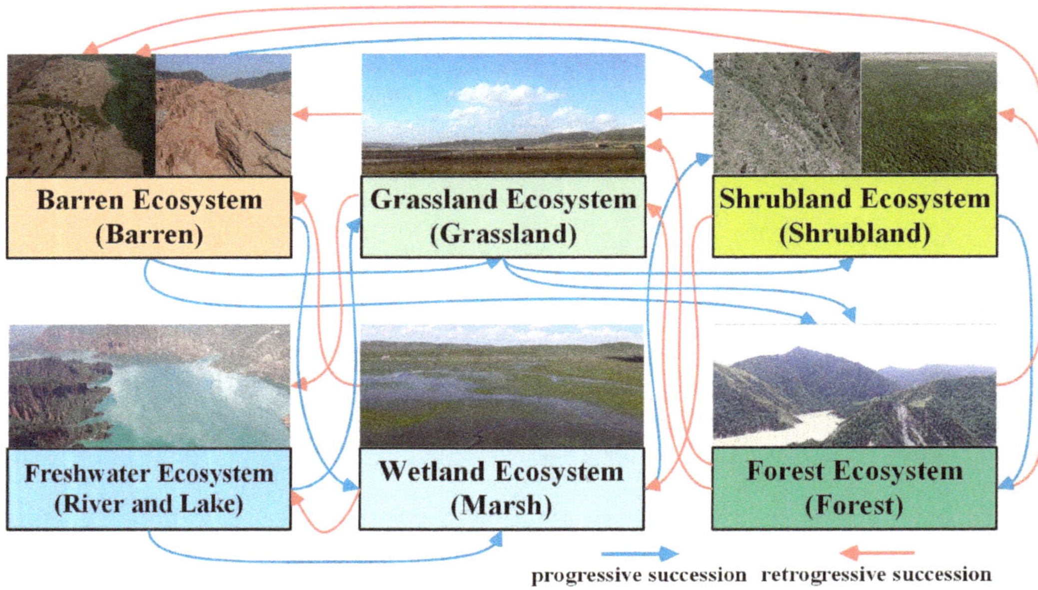

Figure 5. Schematic diagram of the ecological successions related to changes in ecosystem types. The LULC types under each ecosystem type are represented in parentheses.

3. Results

3.1. Feature Importance

Based on the feature importance ranking function, the average importance scores of the 23 feature variables involved in the LULC classification in 2018 and 2020 were plotted (Figure 6). The importance of the spectral features was typically high in all cases; the textural features were low, and, among the topographic features, the elevation (average 2095.82) and slope (average 2100.91) were high, whereas hill shade and aspect were low. Specifically, the B11, B2, B3, and B5 bands of Sentinel-2 were generally high in importance over the two-year period (mean > 1900), with B11 being the most important (average 2228.14). NDBI received the highest importance score (average 2059.84) of the five VIs, and NDWI was also ranked highly (average 1907.12). Of the texture features, Gray_savg had the highest importance (average 1858.31), and several others had the lowest importance (all < 1600). It is noteworthy that although only one of the texture features performed better, we found that the accuracy of LULC decreased when texture features were not included for classification (Table S2); this has been reported in a previous work [60]. Overall, B11, elevation, slope, B2, and NDBI were the most important features, whereas Gray_ent, Gray_asm, hill shade, and aspect contributed less to the LULC classification.

3.2. LULC Accuracy Assessment

The confusion matrix (Tables 4, 5, and S3–S6) showed good classification results for the LULC products generated in both 2018 and 2020. The OA of the first-degree class reached 90.58% and 93.02%, with Kappa coefficients of 0.88 and 0.91, respectively, and the OA of the second-degree class reached 87.45% and 92.14%, with Kappa coefficients of 0.85 and 0.91, respectively. Figure 6 shows that in terms of PA and UA, good accuracies (75.00% to 97.88%) were achieved for most of the LULC first-degree class, such as wetland

and water body, barren land, grassland, agricultural land, and forest. For both the 2018 and 2020 class I categories, water performed best in both PA (95.99% and 97.67%) and UA (97.33% and 97.88%), probably due to the better applicability of NDWI for identifying wetland and the more pronounced spectral response of water bodies compared to other LULC types [100]; the accuracies for shrubland were generally lower, and there were some misclassifications with forests and grasslands (Figure 7a,c and Tables S3 and S4). The addition of more subclasses led to an overall decreasing trend in the classification performance of the LULC second-degree classes, especially for the orchard and urban green space (Tables S5 and S6); these two showed more mixed and misclassified scores, with both PA and UA fluctuating around 80%. There were some misclassifications between urban land, cropland, and orchard, as all of these are classes that are relatively heavily influenced by human activities [42] (Table S5). Further, there was a small amount of misclassification between cropland, urban land, and grassland (Tables S5 and S6) due to the similar spectral characteristics between degraded grassland and post-harvest cropland and urban land. In contrast, broadleaf, needleleaf, and mixed forest showed good classification performance, with PA and UA generally above 80%. Of all the second-degree classes, the river, lake, marsh, farmland, grassland, bare rock, and snow were the best classified, with both PA and UA above 83% (Figure 7b,d and Tables S5 and S6). Considering the large number of LULC types involved in this work, it achieved more satisfactory results for all indicators of accuracy assessment.

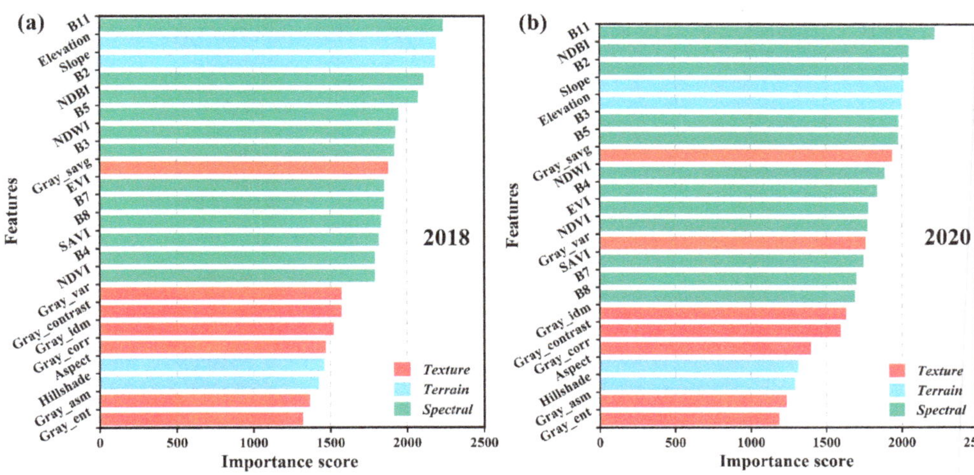

Figure 6. Average importance of features. (**a**) 2018; (**b**) 2020.

3.3. LULC Visual Assessment

For a more comprehensive assessment of LULC results, Figure 8 shows the high-resolution image provided by Google Earth Pro, the Sentinel-2 true-color image, the FROM-GLC10 product, and our LULC map, respectively. Four areas covering multiple LULC types for comparison were selected. It was found that our LULC results had a high fit with the field conditions of the high-resolution and Sentinel-2 true-color images, and there were some differences to the FROM-GLC10 product. For example, the extraction of town, road, and river information in our LULC results was more complete compared to FROM-GLC10 (Figure 8a). In the identification of desert (Figure 8b) and marsh (Figure 8c), our spatial distribution was similar to FROM-GLC10, but we had relatively less noise and provided more LULC detail, especially in the more accurate spatial distribution of marsh that we produced. However, although we did better in extracting the types of marsh, urban land, cropland, and roads, there was still some localized mixing and misclassification of forest, shrubland, and grassland (Figure 8d).

Table 4. The confusion matrix of LULC first-degree class accuracy (PA: producer's accuracy; UA: user's accuracy; OA: overall accuracy).

LULC First-Degree Class	2018		2020	
	PA (%)	UA (%)	PA (%)	UA (%)
Forest	91.59	87.34	95.56	92.67
Shrubland	75.27	81.39	85.33	86.49
Grassland	92.02	90.53	92.66	92.66
Wetland and Water body	95.99	97.67	97.33	97.88
Agricultural Land	90.00	88.94	93.11	91.94
Construction Land	81.79	86.09	86.88	90.74
Barren	88.74	88.16	97.10	88.74
Snow and ice	93.75	92.78	75.00	92.31
OA (%)	90.58		93.02	
Kappa Coefficient	0.88		0.91	

Table 5. The confusion matrix of LULC second-degree class accuracy (PA: producer's accuracy; UA: user's accuracy; OA: overall accuracy).

LULC Second-Degree Class	2018		2020	
	PA (%)	UA (%)	PA (%)	UA (%)
Broadleaf Forest	89.16	80.43	90.00	83.72
Needleleaf Forest	80.49	80.49	89.80	91.67
Mixed Forest	79.41	81.00	87.50	91.30
Broadleaf Shrub	80.00	83.87	81.63	80.00
Needleleaf Shrub	89.29	89.29	80.77	87.50
Grassland	90.10	91.61	94.57	93.33
Marsh	90.29	94.90	90.18	87.07
Lake	86.32	94.25	94.44	94.44
River	93.25	87.61	97.49	98.59
Cropland	87.38	83.94	94.52	94.80
Orchard Land	72.82	75.00	82.22	80.43
Urban	80.95	88.44	88.41	90.63
Urban Green Space	82.61	86.36	76.67	76.67
Industrial and Transportation Land	85.29	81.69	85.23	87.21
Desert Land	96.72	81.94	91.38	94.64
Bare Soil	84.21	88.89	94.44	85.00
Bare Rock	87.32	96.88	87.10	87.10
Snow and ice	86.46	85.57	85.42	91.11
OA (%)	87.45		92.14	
Kappa Coefficient	0.85		0.91	

Furthermore, when comparing the proportion of the study area occupied by the different LULC types in our product with FROM-GLC10 (Figure S1), it was found that the proportions of grassland, construction land, agricultural land, snow, and ice were not much different from FROM-GLC10. However, there were obvious differences in the proportions of the forest, shrubland, wetland, and waterbody areas. Further visual assessment using the 2018 validation points of marsh, shrub, and forest and the UAV aerial photographs was conducted (Figure S2). In these areas, we found that our products performed relatively better for marsh (wetland), shrubland, and forest compared to FROM-GLC10 (Figure S2), which may be due to differences in the definitions of LULC types in different classification systems and classification methods. Overall, the LULC products for the upper Yellow River basin over TP were reasonably consistent with the FROM-GLC10 products and have high accuracy, and they can be used in subsequent analysis.

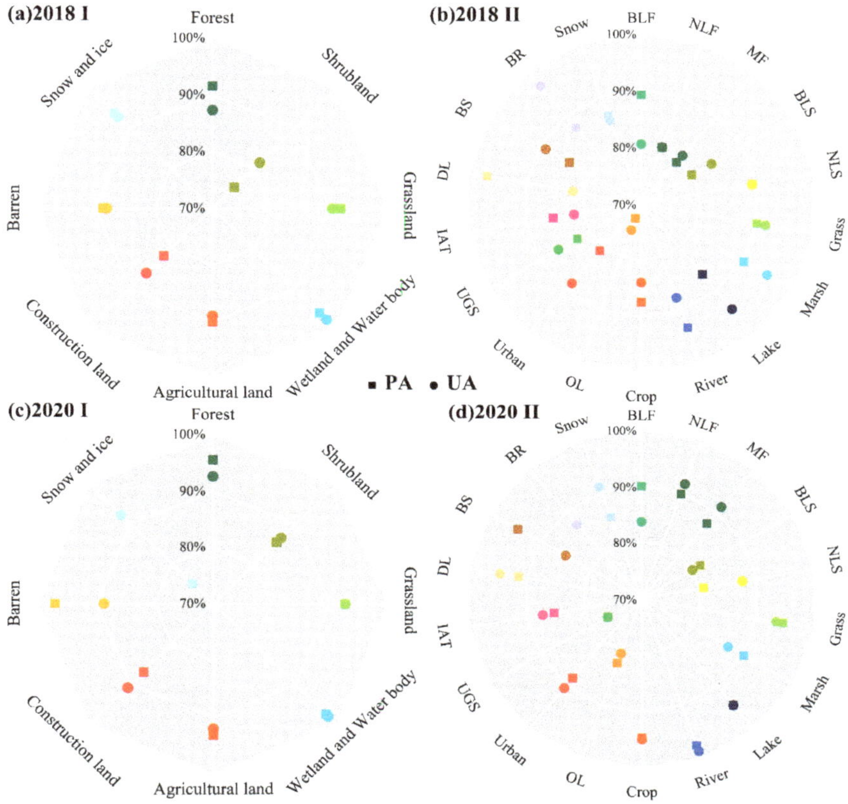

Figure 7. Producer's accuracy (PA) and user's accuracy (UA) for all LULC types: (**a**) first-degree class in 2018; (**b**) second-degree class 2018; (**c**) first-degree class 2020; (**d**) second-degree class 2018 in 2020.

3.4. LULC Classification Results

The results of the LULC classification and area statistics for the study area are shown in Figure 9 and Figure S3. In both 2018 and 2020, the most dominant LULC first-degree class in the study area was grassland (75.43% in 2018 and 75.44% in 2020), mainly in the western part of the Haibei, Hainan, Huangnan, Gannan, Guoluo, and Yushu Prefectures (Figure 8a,b). This was followed by the wetland and water body (6.09% in 2018, 5.89% in 2020), which was mainly concentrated in the western Guoluo, northern Hainan, southern Huangnan, southwestern Gannan and Aba Prefectures. Barren land (3.99% in 2018, 3.98% in 2020) was widely distributed at higher altitudes in the Yushu, Guoluo, central Hainan, and northern Haibei Prefectures. Shrubland (3.31% in 2018 and 3.22% in 2020) and agricultural land (3.13% in 2018 and 3.10% in 2020) were concentrated in areas with lower elevations, more abundant water resources, and suitable temperatures, such as Xining, Haidong, and the western Haibei and eastern Gannan Prefectures. In addition to grassland, the main LULC second-degree classes in the study area were broadleaf shrub (2.42% in 2018 and 2.31% in 2020), marsh (3.22% in 2018 and 3.07% in 2020), cropland (3.02% in 2018 and 2.99% in 2020), bare rock (2.78% in 2018 and 2020), and urban land (2018 1.51% and

1.66% in 2020); the urban areas were mainly located in the northern and central parts of the study area.

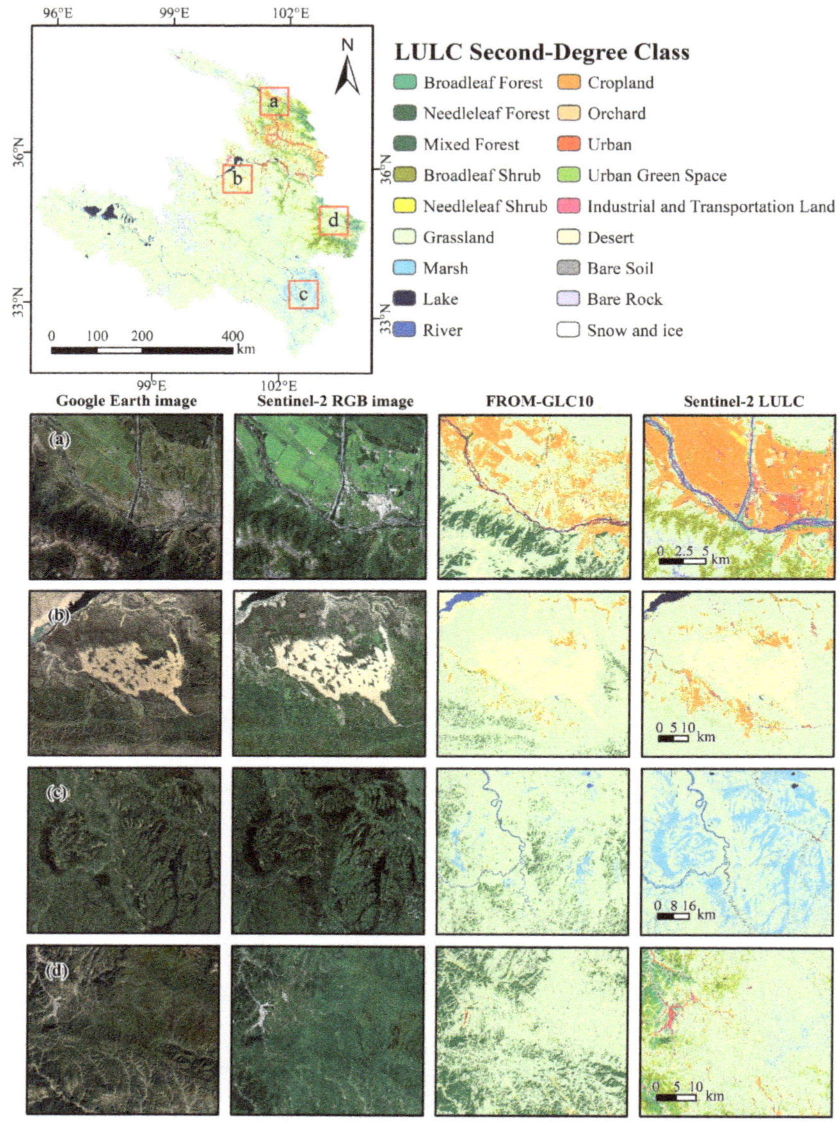

Figure 8. Visual comparison of this study with Google Earth, Sentinel-2 RGB, and FROM-GLC10 products. (**a**–**d**) Four typical areas were selected for close-ups.

3.5. Monitoring Changes in of Ecosystem Types Based on LULC Products

Based on the LULC transition matrices (Figure 10a,b) and ecological succession theory (Figure 5, Table S1), the changes in ecosystem types (Figure 10c,d) and their spatial distribution (Figure 10e) were obtained.

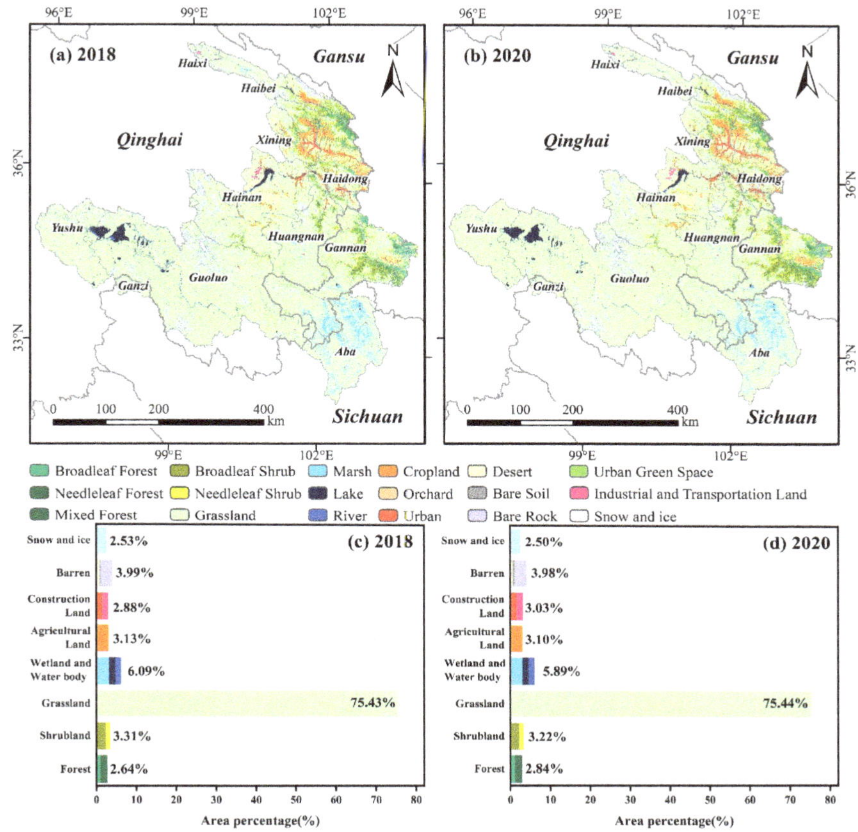

Figure 9. Spatial distribution of LULC second-degree classes in the study area in 2018 (**a**) and 2020 (**b**); area share of different LULC types in 2018 (**c**) and 2020 (**d**).

In the first-degree class of changes in ecosystem types (Figure 10c), the largest areas of progressive succession (positive) were grassland–shrubland, barren–grassland, and grassland–forest, with areas of 451.13 km^2, 425.93 km^2 and 405.95 km^2, respectively, whereas the smallest areas of progressive succession (positive) were freshwater–wetland, barren–wetland and freshwater–grassland, with areas of 34.75 km^2, 28.66 km^2, and 20.51 km^2, respectively. The largest areas of retrogressive succession (negative) were grassland–barren, shrubland–grassland, and shrubland–barren, with areas of 395.91 km^2, 326.84 km^2, and 240.38 km^2, respectively, whereas the smallest areas of retrogressive succession (negative) were wetland–barren, shrubland–wetland, and wetland–freshwater, with areas of 107.63 km^2, 68.13 km^2 and 39.94 km^2, respectively.

In the second-degree class of changes in ecosystem types (Figure 10d), the largest areas of progressive succession (positive) were grassland–broadleaf shrub, desert land–grassland, grassland–broadleaf forest, and desert land–broadleaf shrub, with areas of 344.68 km^2, 302.02 km^2, 267.91 km^2, and 144.75 km^2, respectively, whereas the smallest areas of progressive succession (positive) were river–grassland, desert land–mixed forest, desert land–marsh, and needleleaf shrub–broadleaf forest, with areas of 7.71 km^2, 9.11 km^2, 9.32 km^2, and 10.03 km^2, respectively. The largest areas of retrogressive succession (negative) were broadleaf shrub–grassland, grassland–bare rock, grassland–desert land, and broadleaf shrub–desert land, with areas of 309.08 km^2, 193.89 km^2, 191.16 km^2, and 77.04 km^2, respectively, whereas the smallest areas of retrogressive succession (negative) were needleleaf forest–bare soil, mixed forest–desert land, needleleaf forest–desert land, and broadleaf forest–needleleaf shrub, with areas of 3.55 km^2, 5.27 km^2, 6.30 km^2, and 10.57 km^2, respectively.

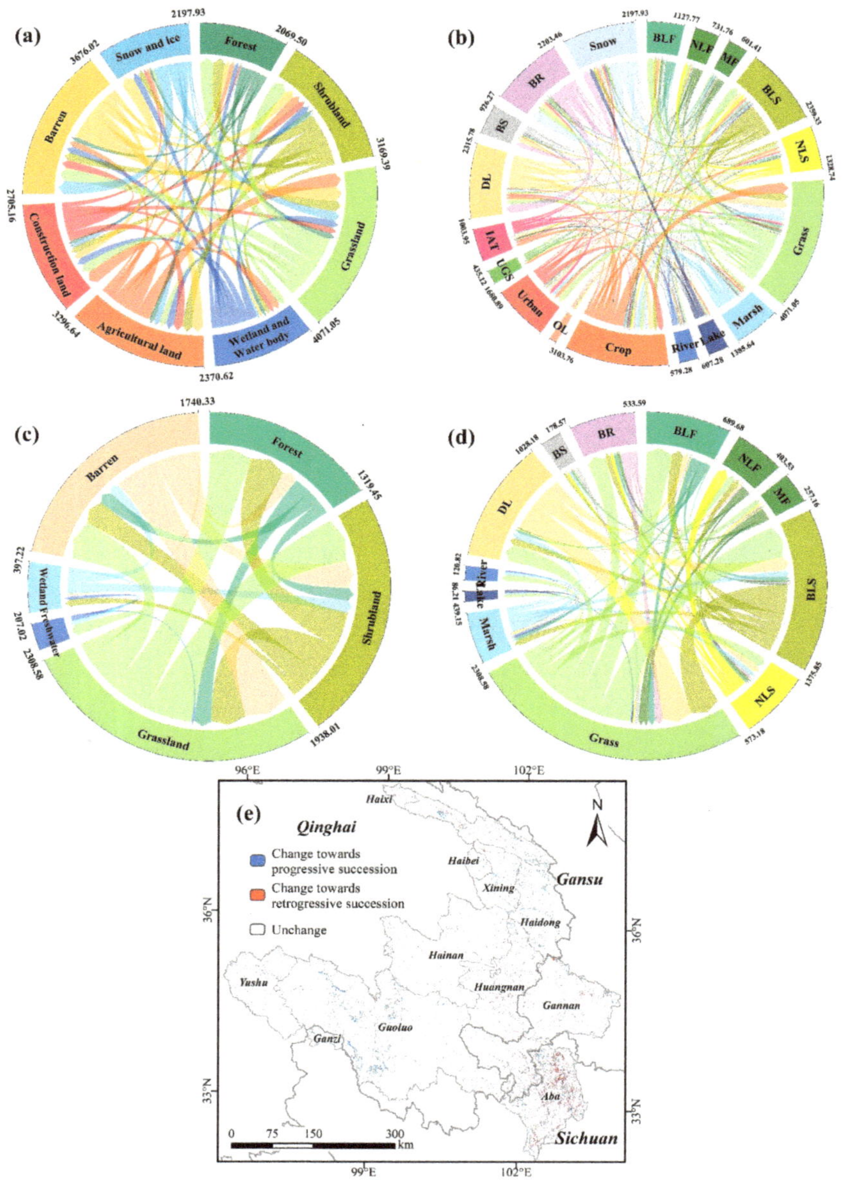

Figure 10. 2018–2020 LULC transition matrix and changes in ecosystem types; the width of the chord represents the area of the transfer. (**a**) First-degree class transition matrix chord diagram; (**b**) second-degree class transition matrix chord diagram; (**c**) first-degree class chord diagram of changes in ecosystem types; (**d**) second-degree class chord diagram of changes in ecosystem types; (**e**) distribution of changes in ecosystem types based on ecological succession and LULC.

The distribution of changes in ecosystem types based on LULC types for the source and the upper Yellow River basin from 2018 to 2020 was quantified and produced (Figure 10e). As can be seen from the figure, the changes in ecosystem types showed a trend towards positive succession (positive) in the northern and southwestern parts of the study area, whereas in the south-central Huangnan, northeastern Gannan, and central Aba Prefectures, it showed signs of retrogressive succession (negative) in ecosystem change.

4. Discussion

4.1. Comparison of Classification Methods

To further verify and assess the performance of the RF method for high-resolution LULC mapping in the study area, a comparison of the classification accuracy with three machine learning methods widely used in GEE were obtained, including classification and regression trees (CART), the gradient boosting decision tree (GBDT), and the support vector machine (SVM) (Table 6). The results showed that RF performed best among all the classifiers (87.45–92.14%), followed by GBDT [113] (86.65–90.60%) and CART [114] (78.76–86.99%), whereas the SVM [115] (74.64–84.30%) classifier had the lowest performance. The difference in effectiveness between the two integrated learning algorithms, GBDT and RF, was not significant, with RF slightly more effective than GBDT in this study; however, GBDT has shown better classification performance in some studies [116]. It is worth noting that although SVM performed far worse than RF and GBDT here, it has achieved good results in some classification studies [117–119], particularly those with relatively small training samples and LULC types [120], in which it can outperform RF. Although some studies have shown little difference in classification performance between SVM and RF [121,122], the classification performance of SVM is clearly limited when working with large-scale LULC mapping with a large number of training samples as in this study; RF and GBDT perform better in this respect.

Table 6. Comparison of the accuracy of several machine learning methods.

Machine Learning Methods	2018		2020	
	OA (%)	Kappa Coefficient	OA (%)	Kappa Coefficient
Random Forest	87.45	0.85	92.14	0.91
CART	78.76	0.75	86.99	0.85
Gradient Boosting Decision Tree	86.65	0.84	90.60	0.89
Support Vector Machine	74.64	0.70	84.30	0.81

4.2. Comparison of LULC Products Accuracy

In a visual comparison with the FROM-GLC10 product, our product provided more detail and had better performance in the extraction of cropland, urban land, rivers, and desert, but forest, shrubland, and grasslands showed some misclassification. In the first-degree class (Table 4), satisfactory accuracy (PA and UA > 80%) was achieved in almost all LULC types, except for shrubland (PA = 75.27%) and snow and ice. This result is similar to the GEE-based classification of the Qilian Mountains by Yang et al. [50]; the PA of the shrubland only reached 66.67% in their work. There was some visual similarity between both shrubland and degraded grassland and sparse forest, and that led to significant misclassification in transitional areas. The misclassification of snow and ice was mainly attributed to the persistence of clouds in the southern part of the study area that were hard to remove.

In the second-degree class, a large proportion of the LULC classes (15 out of 18) showed good accuracy (PA and UA > 80%), with reduced misclassification of shrubland and lower accuracy for the orchard (PA = 72.82%), urban green space, and mixed forest (Table 5). This result is similar to the LULC of the Yellow River basin by Ji et al. [42]; the PA of the orchard and terrace reached 0.74 in their work. The fact that orchard and urban green spaces have relatively similar vegetation types in composition makes them extremely easy to confuse both spectrally and texturally [120]. Similarly, the only difference between mixed, needleleaf, and broadleaf forest classes is the ratio of needleleaf/broadleaf forest. The same problem affects the classification of second-degree classes such as construction land, shrubland, and barren land; in other words, even if we could accurately identify these secondary categories from field-collected UAV data or Google Earth Pro high-resolution imagery, there would still be a high degree of uncertainty in the classification, and it would still be challenging to capture these LULC types accurately [68]. Although we achieved a high degree of accuracy in our study, there are still some confusion problems in these secondary classes.

4.3. Changes in Ecosystem Types Based on Ecological Succession and LULC

Human activities and natural environmental changes, through their impact on LULC, largely influence the process of ecological succession and ultimately lead to the occurrence of changes in ecosystem types [112]. Traditionally, change in ecosystem types is relatively slow [123], especially for areas with low anthropogenic disturbance such as the upper Yellow River basin over TP; however, our study shows that approximately 3955 km^2 of the study area underwent changes in ecosystem types between 2018 and 2020. This implies that changes in ecosystem types are accelerating under the predominant role of anthropogenic disturbance.

The monitoring of changes in ecosystem types revealed both positive and negative effects of anthropogenic disturbances on ecosystem changes during this process. For the retrogressive succession, artificial disturbances often play a negative role: the change in ecosystem types of the grassland–desert land (191.16 km^2) may be associated with desertification of grasslands due to overgrazing; the change in ecosystem types of the broadleaf forest–grassland (70.49 km^2) is possibly associated with deforestation.

However, for the progressive succession, anthropogenic disturbances often play a positive role: the changes in ecosystem types of the grassland–broadleaf shrub (344.68 km^2) and desert land–broadleaf shrub (144.75 km^2) may be related to the local ecological barrier project in recent years; the change in ecosystem types of the desert land–grassland (302.02 km^2) confirms the overall increase in ecosystem services in the Sanjiangyuan area of the TP, as noted in recent reports [124], and that the ongoing efforts to combat desertification and related ecological restoration have been effective [47]. The change in ecosystem types of the grassland–broadleaf forest (267.91 km^2) may be related to a combination of ecological restoration projects such as reforestation and afforestation [22,47,125]. These findings illustrate the important role of LULC products that can be used for monitoring of regional ecosystem types and further confirm the reliability of our LULC products for monitoring of changes in ecosystem types.

4.4. Limitations and Prospects

Relying on the free GEE platform, we were able to carry out large-scale high-resolution LULC mapping work [36,38,126]. Unfortunately, Sentinel-2 cannot currently be used to monitor LULC changes over long time series. In addition, our research area is located at the eastern edge of the TP, where high altitudes and frequent variable weather rendered it permanently covered by clouds. Although multi-temporal compositing and cloud masking with GEE can compensate for this to some extent, monitoring still suffered in some areas, especially in the densely cloudy upper Yellow River basin [22,127]. Moreover, for the study area, GEE has not yet provided Sentinel-2 L2A(SR) data for 2018 and earlier; only L1C (TOA) data are available. Even though the relevant studies concluded that the effect of TOA data on classification is minimal [128], we still chose to use SR data for classification to better map the complex LULC distribution. This meant we spent more time on image pre-processing and reduced the efficiency of mapping LULC.

Despite these limitations, based on the processing framework and the new classification scheme of this study, one or more (2022, 2024, and so on) LULC maps that can be used for monitoring of changes in ecosystem types will be easily obtained. At the same time, with the enrichment of classification algorithms' multi-source RS data and cloud computing platform, larger scale, higher resolution, and higher accuracy products can be provided when monitoring at the level of ecosystem types. In future research, we will generate more detailed high-resolution LULC products to promote regional ecological surveys and conservation.

5. Conclusions

The study combined an 18-class LULC classification scheme based on ecosystem types with Sentinel-2 multi-temporal data, the GEE platform, and the RF machine learning method to present new high-resolution LULC products with a spatial resolution of 10 m for the upper Yellow River basin over the TP and conducted monitoring of changes in ecosystem types. The main findings were as follows: (1) Combining Sentinel-2 imagery and RF classification methods, the products of the upper Yellow River basin over TP that were obtained based on the GEE platform have high classification accuracies and are suitable for monitoring changes in ecosystem types in the study area. (2) The LULC products based on a two-level classification system show that in the LULC first-degree class, the grassland, wetland and water body, barren land, shrubland, and agricultural land are the main LULC types. In the LULC second-degree class, grassland, broadleaf shrub, marsh, cropland, bare rock, and urban land are the main LULC types. (3) In 2018–2020, in the first-degree class of changes in ecosystem types, the largest areas of progressive succession (positive) were grassland–shrubland and barren–grassland, whereas the smallest areas were freshwater–wetland and barren–wetland. The largest areas of retrogressive succession (negative) were grassland–barren and shrubland–grassland, whereas the smallest areas were wetland–barren and shrubland–wetland. In the second-degree class of changes in ecosystem types, the largest areas of progressive succession (positive) were grassland–broadleaf and shrubland desert land–grassland, whereas the smallest areas of progressive succession were river–grass and desert land–mixed forest. The largest areas of retrogressive succession (negative) were broadleaf shrub–grassland and grassland–bare rock, whereas the smallest areas were needleleaf forest–bare soil and mixed forest–desert land. The northern and southwestern parts of the study area showed a trend towards positive succession, whereas the south-central Huangnan, northeastern Gannan, and central Aba Prefectures showed signs of retrogressive succession in changes in ecosystem types. This study can provide effective data support for the monitoring of regional scale ecosystem types, quality assessments and protection, natural resource management, and sustainable development of the upper Yellow River basin over the TP.

Supplementary Materials: The following supporting information can be downloaded at: https://www.mdpi.com/article/10.3390/rs14215361/s1, Figure S1: Ratios (%) of different LULC types areas between this study and FROM-GLC10; Figure S2: Visual comparison and assessment of this study with Sentinel-2 RGB, FROM-GLC10 products and UAV aerial photographs; Figure S3: Spatial distribution of LULC first-degree classes in the study area in 2018 and 2020; Figure S4: The map of changes in ecosystem types based on LULC types; Table S1: Relationships of changes in ecosystem types; Table S2: The accuracies with different feature combinations in 2018 and 2020; Table S3: Confusion matrix of LULC first-degree classes accuracy assessment in 2018; Table S4: Confusion matrix of LULC first-degree classes accuracy assessment in 2020; Table S5: Confusion matrix of LULC second-degree classes accuracy assessment in 2018; Table S6: Confusion matrix of LULC second-degree classes accuracy assessment in 2020; Table S7: Transition matrix of LULC first-degree classes in 2018-2020; Table S8: Transition matrix of LULC second-degree classes in 2018-2020; Table S9: Accuracy comparison (%) of the LULC classification result obtained in this study with the FROM-GLC10 (2017) land cover products in 2018.

Author Contributions: Conceptualization, W.L.; Formal analysis, S.F.; Funding acquisition, W.W. and H.Y.; Investigation, S.F.; Methodology, T.L.; Project administration, J.X.; Resources, X.M.; Supervision, T.L.; Validation, W.L.; Writing—original draft, S.F.; Writing—review and editing, S.F. and W.L. All authors have read and agreed to the published version of the manuscript.

Funding: This research was funded by the National Natural Science Foundation of China (grant No. 31170430, grant No. 41471450); the earmarked fund for the China Agriculture Research System (grant No. CARS-34); the Fundamental Research Funds for the Central Universities (grant No. lzujbky-2021-ct11); the funds for investigation and monitoring of typical forest, shrub, and meadow ecosystems in the vertical distribution zone of the Haibei region of the National Park (grant No. QHXH-2021-017); the Protection and Restoration of Muli Alpine Wetland in Tianjun County (grant No. QHTM-2021-003); the Key Entrusted Program of Lanzhou University of Finance and Economics (grant No. Lzufe2021W-003); and the Science and Technology Program of Gansu Province, China (grant No. 20CX9ZA060).

Institutional Review Board Statement: Not applicable.

Informed Consent Statement: Not applicable.

Data Availability Statement: The data presented in this study are available in the Supplementary Materials, or please contact the first author of this article.

Acknowledgments: Thanks are given to Pengfei Xue, Chenli Liu, Jing Ge, Mengjing Hou, Yuanchun Li, Jie Liu, Ruijing Wang, Xinyue Meng, Weiye Huang, Bin Liu, and Xuejing Shang from Lanzhou University, China, for their help in the field survey data collection. Specific thanks are given to the editors and four anonymous reviewers for their valuable suggestions and comments in the improvement of this work and manuscript.

Conflicts of Interest: The authors declare no conflict of interest.

References

1. Gong, P.; Wang, J.; Yu, L.; Zhao, Y.; Zhao, Y.; Liang, L.; Niu, Z.; Huang, X.; Fu, H.; Liu, S. Finer resolution observation and monitoring of global land cover: First mapping results with Landsat TM and ETM+ data. *Int. J. Remote Sens.* **2013**, *34*, 2607–2654. [CrossRef]
2. Lambin, E.F.; Meyfroidt, P. Global land use change, economic globalization, and the looming land scarcity. *Proc. Natl. Acad. Sci. USA* **2011**, *108*, 3465–3472. [CrossRef] [PubMed]
3. Turner, B.L.; Lambin, E.F.; Reenberg, A. The emergence of land change science for global environmental change and sustainability. *Proc. Natl. Acad. Sci. USA* **2007**, *104*, 20666–20671. [CrossRef]
4. Song, X.-P.; Hansen, M.C.; Stehman, S.V.; Potapov, P.V.; Tyukavina, A.; Vermote, E.F.; Townshend, J.R. Global land change from 1982 to 2016. *Nature* **2018**, *560*, 639–643. [CrossRef] [PubMed]
5. Foley, J.A.; DeFries, R.; Asner, G.P.; Barford, C.; Bonan, G.; Carpenter, S.R.; Chapin, F.S.; Coe, M.T.; Daily, G.C.; Gibbs, H.K. Global consequences of land use. *Science* **2005**, *309*, 570–574. [CrossRef]
6. Le Quéré, C.; Andrew, R.M.; Canadell, J.G.; Sitch, S.; Korsbakken, J.I.; Peters, G.P.; Manning, A.C.; Boden, T.A.; Tans, P.P.; Houghton, R.A. Global carbon budget 2016. *Earth Syst. Sci. Data* **2016**, *8*, 605–649. [CrossRef]
7. Gruber, N.; Galloway, J.N. An Earth-system perspective of the global nitrogen cycle. *Nature* **2008**, *451*, 293–296. [CrossRef]
8. Grimm, N.B.; Faeth, S.H.; Golubiewski, N.E.; Redman, C.L.; Wu, J.; Bai, X.; Briggs, J.M. Global change and the ecology of cities. *Science* **2008**, *319*, 756–760. [CrossRef]
9. Sterling, S.M.; Ducharne, A.; Polcher, J. The impact of global land-cover change on the terrestrial water cycle. *Nat. Clim. Change* **2013**, *3*, 385–390. [CrossRef]
10. Liu, X.; Liu, Y.; Wang, Y.; Liu, Z. Evaluating potential impacts of land use changes on water supply-demand under multiple development scenarios in dryland region. *J. Hydrol.* **2022**, *610*, 127811. [CrossRef]
11. Macedo, M.N.; DeFries, R.S.; Morton, D.C.; Stickler, C.M.; Galford, G.L.; Shimabukuro, Y.E. Decoupling of deforestation and soy production in the southern Amazon during the late 2000s. *Proc. Natl. Acad. Sci. USA* **2012**, *109*, 1341–1346. [CrossRef] [PubMed]
12. Morton, D.C.; DeFries, R.S.; Shimabukuro, Y.E.; Anderson, L.O.; Arai, E.; del Bon Espirito-Santo, F.; Freitas, R.; Morisette, J. Cropland expansion changes deforestation dynamics in the southern Brazilian Amazon. *Proc. Natl. Acad. Sci. USA* **2006**, *103*, 14637–14641. [CrossRef] [PubMed]
13. Xu, X.; Xie, Y.; Qi, K.; Luo, Z.; Wang, X. Detecting the response of bird communities and biodiversity to habitat loss and fragmentation due to urbanization. *Sci. Total Environ.* **2018**, *624*, 1561–1576. [CrossRef] [PubMed]
14. Llerena-Montoya, S.; Velastegui-Montoya, A.; Zhirzhan-Azanza, B.; Herrera-Matamoros, V.; Adami, M.; de Lima, A.; Moscoso-Silva, F.; Encalada, L. Multitemporal analysis of land use and land cover within an oil block in the Ecuadorian Amazon. *ISPRS Int. J. Geo-Inf.* **2021**, *10*, 191. [CrossRef]
15. Simmons, J.A.; Currie, W.S.; Eshleman, K.N.; Kuers, K.; Monteleone, S.; Negley, T.L.; Pohlad, B.R.; Thomas, C.L. Forest to reclaimed mine land use change leads to altered ecosystem structure and function. *Ecol. Appl.* **2008**, *18*, 104–118. [CrossRef] [PubMed]
16. Newbold, T.; Hudson, L.N.; Hill, S.L.; Contu, S.; Lysenko, I.; Senior, R.A.; Börger, L.; Bennett, D.J.; Choimes, A.; Collen, B. Global effects of land use on local terrestrial biodiversity. *Nature* **2015**, *520*, 45–50. [CrossRef] [PubMed]

17. Betts, M.G.; Wolf, C.; Ripple, W.J.; Phalan, B.; Millers, K.A.; Duarte, A.; Butchart, S.H.; Levi, T. Global forest loss disproportionately erodes biodiversity in intact landscapes. *Nature* **2017**, *547*, 441–444. [CrossRef] [PubMed]
18. Hou, M.; Bao, X.; Ge, J.; Liang, T. Land cover pattern and habitat suitability on the global largest breeding sites for Black-necked Cranes. *J. Clean. Prod.* **2021**, *322*, 128968. [CrossRef]
19. Nelson, E.; Sander, H.; Hawthorne, P.; Conte, M.; Ennaanay, D.; Wolny, S.; Manson, S.; Polasky, S. Projecting global land-use change and its effect on ecosystem service provision and biodiversity with simple models. *PLoS ONE* **2010**, *5*, e14327. [CrossRef] [PubMed]
20. Jones, K.R.; Venter, O.; Fuller, R.A.; Allan, J.R.; Maxwell, S.L.; Negret, P.J.; Watson, J.E. One-third of global protected land is under intense human pressure. *Science* **2018**, *360*, 788–791. [CrossRef]
21. Tassi, A.; Gigante, D.; Modica, G.; Di Martino, L.; Vizzari, M. Pixel-vs. Object-based landsat 8 data classification in google earth engine using random forest: The case study of maiella national park. *Remote Sens.* **2021**, *13*, 2299. [CrossRef]
22. Liu, C.; Li, W.; Zhu, G.; Zhou, H.; Yan, H.; Xue, P. Land use/land cover changes and their driving factors in the Northeastern Tibetan Plateau based on Geographical Detectors and Google Earth Engine: A case study in Gannan Prefecture. *Remote Sens.* **2020**, *12*, 3139. [CrossRef]
23. Xu, P.; Herold, M.; Tsendbazar, N.-E.; Clevers, J.G. Towards a comprehensive and consistent global aquatic land cover characterization framework addressing multiple user needs. *Remote Sens. Environ.* **2020**, *250*, 112034. [CrossRef]
24. Chen, J.; Chen, J.; Liao, A.; Cao, X.; Chen, L.; Chen, X.; He, C.; Han, G.; Peng, S.; Lu, M. Global land cover mapping at 30 m resolution: A POK-based operational approach. *ISPRS J. Photogramm. Remote Sens.* **2015**, *103*, 7–27. [CrossRef]
25. González-González, A.; Clerici, N.; Quesada, B. A 30 m-resolution land use-land cover product for the Colombian Andes and Amazon using cloud-computing. *Int. J. Appl. Earth Obs. Geoinf.* **2022**, *107*, 102688. [CrossRef]
26. Meng, B.; Zhang, Y.; Yang, Z.; Lv, Y.; Chen, J.; Li, M.; Sun, Y.; Zhang, H.; Yu, H.; Zhang, J. Mapping Grassland Classes Using Unmanned Aerial Vehicle and MODIS NDVI Data for Temperate Grassland in Inner Mongolia, China. *Remote Sens.* **2022**, *14*, 2094. [CrossRef]
27. Yao, J.; Wu, J.; Xiao, C.; Zhang, Z.; Li, J. The Classification Method Study of Crops Remote Sensing with Deep Learning, Machine Learning, and Google Earth Engine. *Remote Sens.* **2022**, *14*, 2758. [CrossRef]
28. Drusch, M.; Del Bello, U.; Carlier, S.; Colin, O.; Fernandez, V.; Gascon, F.; Hoersch, B.; Isola, C.; Laberinti, P.; Martimort, P. Sentinel-2: ESA's optical high-resolution mission for GMES operational services. *Remote Sens. Environ.* **2012**, *120*, 25–36. [CrossRef]
29. Huang, B.; Zhao, B.; Song, Y. Urban land-use mapping using a deep convolutional neural network with high spatial resolution multispectral remote sensing imagery. *Remote Sens. Environ.* **2018**, *214*, 73–86. [CrossRef]
30. Roy, D.P.; Wulder, M.A.; Loveland, T.R.; Woodcock, C.E.; Allen, R.G.; Anderson, M.C.; Helder, D.; Irons, J.R.; Johnson, D.M.; Kennedy, R. Landsat-8: Science and product vision for terrestrial global change research. *Remote Sens. Environ.* **2014**, *145*, 154–172. [CrossRef]
31. Chen, Y.; Lin, Z.; Zhao, X.; Wang, G.; Gu, Y. Deep learning-based classification of hyperspectral data. *IEEE J. Sel. Top. Appl. Earth Obs. Remote Sens.* **2014**, *7*, 2094–2107. [CrossRef]
32. Liu, Z.J.; Liu, Y.S.; Dong, J.W.; Baig, M.H.A.; Chi, W.F.; Peng, L.Y.; Wang, J.Y. Patterns and causes of winter wheat and summer maize rotation area change over the North China Plain. *Environ. Res. Lett.* **2022**, *17*, 044056. [CrossRef]
33. You, N.; Dong, J.; Huang, J.; Du, G.; Zhang, G.; He, Y.; Yang, T.; Di, Y.; Xiao, X. The 10-m crop type maps in Northeast China during 2017–2019. *Sci. Data* **2021**, *8*, 1–11. [CrossRef]
34. Wang, M.; Liu, Z.; Baig, M.H.A.; Wang, Y.; Li, Y.; Chen, Y. Mapping sugarcane in complex landscapes by integrating multi-temporal Sentinel-2 images and machine learning algorithms. *Land Use Policy* **2019**, *88*, 104190. [CrossRef]
35. Forkuor, G.; Dimobe, K.; Serme, I.; Tondoh, J.E. Landsat-8 vs. Sentinel-2: Examining the added value of sentinel-2's red-edge bands to land-use and land-cover mapping in Burkina Faso. *Giscience Remote Sens.* **2018**, *55*, 331–354. [CrossRef]
36. Jin, Z.; Azzari, G.; You, C.; Di Tommaso, S.; Aston, S.; Burke, M.; Lobell, D.B. Smallholder maize area and yield mapping at national scales with Google Earth Engine. *Remote Sens. Environ.* **2019**, *228*, 115–128. [CrossRef]
37. Ma, Y.; Wu, H.; Wang, L.; Huang, B.; Ranjan, R.; Zomaya, A.; Jie, W. Remote sensing big data computing: Challenges and opportunities. *Future Gener. Comput. Syst.* **2015**, *51*, 47–60. [CrossRef]
38. Tamiminia, H.; Salehi, B.; Mahdianpari, M.; Quackenbush, L.; Adeli, S.; Brisco, B. Google Earth Engine for geo-big data applications: A meta-analysis and systematic review. *ISPRS J. Photogramm. Remote Sens.* **2020**, *164*, 152–170. [CrossRef]
39. Gorelick, N.; Hancher, M.; Dixon, M.; Ilyushchenko, S.; Thau, D.; Moore, R. Google Earth Engine: Planetary-scale geospatial analysis for everyone. *Remote Sens. Environ.* **2017**, *202*, 18–27. [CrossRef]
40. Xu, H.; Qi, S.; Li, X.; Gao, C.; Wei, Y.; Liu, C. Monitoring three-decade dynamics of citrus planting in Southeastern China using dense Landsat records. *Int. J. Appl. Earth Obs. Geoinf.* **2021**, *103*, 102518. [CrossRef]
41. Ge, J.; Meng, B.; Liang, T.; Feng, Q.; Gao, J.; Yang, S.; Huang, X.; Xie, H. Modeling alpine grassland cover based on MODIS data and support vector machine regression in the headwater region of the Huanghe River, China. *Remote Sens. Environ.* **2018**, *218*, 162–173. [CrossRef]
42. Ji, Q.; Liang, W.; Fu, B.; Zhang, W.; Yan, J.; Lü, Y.; Yue, C.; Jin, Z.; Lan, Z.; Li, S. Mapping land use/cover dynamics of the Yellow River Basin from 1986 to 2018 supported by Google Earth Engine. *Remote Sens.* **2021**, *13*, 1299. [CrossRef]

43. Hou, M.; Ge, J.; Gao, J.; Meng, B.; Li, Y.; Yin, J.; Liu, J.; Feng, Q.; Liang, T. Ecological risk assessment and impact factor analysis of alpine wetland ecosystem based on LUCC and boosted regression tree on the Zoige Plateau, China. *Remote Sens.* **2020**, *12*, 368. [CrossRef]
44. Wohlfart, C.; Liu, G.; Huang, C.; Kuenzer, C. A River Basin over the course of time: Multi-temporal analyses of land surface dynamics in the Yellow River Basin (China) based on medium resolution remote sensing data. *Remote Sens.* **2016**, *8*, 186. [CrossRef]
45. Wang, Y.; Huang, X.; Wang, J.; Zhou, M.; Liang, T. AMSR2 snow depth downscaling algorithm based on a multifactor approach over the Tibetan Plateau, China. *Remote Sens. Environ.* **2019**, *231*, 111268. [CrossRef]
46. Ye, C.; Sun, J.; Liu, M.; Xiong, J.; Zong, N.; Hu, J.; Huang, Y.; Duan, X.; Tsunekawa, A. Concurrent and lagged effects of extreme drought induce net reduction in vegetation carbon uptake on Tibetan Plateau. *Remote Sens.* **2020**, *12*, 2347. [CrossRef]
47. Bryan, B.A.; Gao, L.; Ye, Y.; Sun, X.; Connor, J.D.; Crossman, N.D.; Stafford-Smith, M.; Wu, J.; He, C.; Yu, D. China's response to a national land-system sustainability emergency. *Nature* **2018**, *559*, 193–204. [CrossRef]
48. Gong, P.; Liu, H.; Zhang, M.N.; Li, C.C.; Wang, J.; Huang, H.B.; Clinton, N.; Ji, L.Y.; Li, W.Y.; Bai, Y.Q.; et al. Stable classification with limited sample: Transferring a 30-m resolution sample set collected in 2015 to mapping 10-m resolution global land cover in 2017. *Sci. Bull.* **2019**, *64*, 370–373. [CrossRef]
49. Chen, Y.; Syvitski, J.P.; Gao, S.; Overeem, I.; Kettner, A.J. Socio-economic impacts on flooding: A 4000-year history of the Yellow River, China. *Ambio* **2012**, *41*, 682–698. [CrossRef]
50. Yang, Y.; Yang, D.; Wang, X.; Zhang, Z.; Nawaz, Z. Testing accuracy of land cover classification algorithms in the qilian mountains based on gee cloud platform. *Remote Sens.* **2021**, *13*, 5064. [CrossRef]
51. Gao, J.; Liang, T.; Liu, J.; Yin, J.; Ge, J.; Hou, M.; Feng, Q.; Wu, C.; Xie, H. Potential of hyperspectral data and machine learning algorithms to estimate the forage carbon-nitrogen ratio in an alpine grassland ecosystem of the Tibetan Plateau. *ISPRS J. Photogramm. Remote Sens.* **2020**, *163*, 362–374. [CrossRef]
52. Ma, Y.; Huang, X.; Feng, Q.; Liang, T. Alpine Grassland Reviving Response to Seasonal Snow Cover on the Tibetan Plateau. *Remote Sens.* **2022**, *14*, 2499. [CrossRef]
53. Chen, C.; Ma, Y.; Ren, G.; Wang, J. Aboveground biomass of salt-marsh vegetation in coastal wetlands: Sample expansion of in situ hyperspectral and Sentinel-2 data using a generative adversarial network. *Remote Sens. Environ.* **2022**, *270*, 112885. [CrossRef]
54. Zekoll, V.; Main-Knorn, M.; Alonso, K.; Louis, J.; Frantz, D.; Richter, R.; Pflug, B. Comparison of masking algorithms for sentinel-2 imagery. *Remote Sens.* **2021**, *13*, 137. [CrossRef]
55. Wang, Z.; Wu, B.; Zhang, M.; Zeng, H.; Yang, L.; Tian, F.; Ma, Z.; Wu, H. Indices enhance biological soil crust mapping in sandy and desert lands. *Remote Sens. Environ.* **2022**, *278*, 113078. [CrossRef]
56. Xi, Z.; Xu, H.; Xing, Y.; Gong, W.; Chen, G.; Yang, S. Forest Canopy Height Mapping by Synergizing ICESat-2, Sentinel-1, Sentinel-2 and Topographic Information Based on Machine Learning Methods. *Remote Sens.* **2022**, *14*, 364. [CrossRef]
57. Farr, T.G.; Rosen, P.A.; Caro, E.; Crippen, R.; Duren, R.; Hensley, S.; Kobrick, M.; Paller, M.; Rodriguez, E.; Roth, L. The shuttle radar topography mission. *Rev. Geophys.* **2007**, *45*. [CrossRef]
58. Meng, B.; Yang, Z.; Yu, H.; Qin, Y.; Sun, Y.; Zhang, J.; Chen, J.; Wang, Z.; Zhang, W.; Li, M.; et al. Mapping of Kobresia pygmaea Community Based on Umanned Aerial Vehicle Technology and Gaofen Remote Sensing Data in Alpine Meadow Grassland: A Case Study in Eastern of Qinghai–Tibetan Plateau. *Remote Sens.* **2021**, *13*, 2483. [CrossRef]
59. Huang, H.; Chen, Y.; Clinton, N.; Wang, J.; Wang, X.; Liu, C.; Gong, P.; Yang, J.; Bai, Y.; Zheng, Y. Mapping major land cover dynamics in Beijing using all Landsat images in Google Earth Engine. *Remote Sens. Environ.* **2017**, *202*, 166–176. [CrossRef]
60. Dubertret, F.; Le Tourneau, F.-M.; Villarreal, M.L.; Norman, L.M. Monitoring Annual Land Use/Land Cover Change in the Tucson Metropolitan Area with Google Earth Engine (1986–2020). *Remote Sens.* **2022**, *14*, 2127. [CrossRef]
61. Waśniewski, A.; Hościło, A.; Chmielewska, M. Can a Hierarchical Classification of Sentinel-2 Data Improve Land Cover Mapping? *Remote Sens.* **2022**, *14*, 989. [CrossRef]
62. Ouyang, Z.; Zhang, L.; Wu, B.; Li, X.; Xu, W.; Xiao, Y.; Zheng, H. An ecosystem classification system based on remote sensor information in China. *Acta Ecol. Sin.* **2015**, *35*, 219–226.
63. Loveland, T.R.; Reed, B.C.; Brown, J.F.; Ohlen, D.O.; Zhu, Z.; Yang, L.; Merchant, J.W. Development of a global land cover characteristics database and IGBP DISCover from 1 km AVHRR data. *Int. J. Remote Sens.* **2000**, *21*, 1303–1330. [CrossRef]
64. Congalton, R.G.; Gu, J.; Yadav, K.; Thenkabail, P.; Ozdogan, M. Global land cover mapping: A review and uncertainty analysis. *Remote Sens.* **2014**, *6*, 12070–12093. [CrossRef]
65. Roy, P.S.; Roy, A.; Joshi, P.K.; Kale, M.P.; Srivastava, V.K.; Srivastava, S.K.; Dwevidi, R.S.; Joshi, C.; Behera, M.D.; Meiyappan, P.; et al. Development of Decadal (1985–1995–2005) Land Use and Land Cover Database for India. *Remote Sens.* **2015**, *7*, 2401–2430. [CrossRef]
66. Vizzari, M. PlanetScope, Sentinel-2, and Sentinel-1 Data Integration for Object-Based Land Cover Classification in Google Earth Engine. *Remote Sens.* **2022**, *14*, 2628. [CrossRef]
67. Pott, L.P.; Amado, T.J.C.; Schwalbert, R.A.; Corassa, G.M.; Ciampitti, I.A. Satellite-based data fusion crop type classification and mapping in Rio Grande do Sul, Brazil. *ISPRS J. Photogramm. Remote Sens.* **2021**, *176*, 196–210. [CrossRef]
68. Verde, N.; Kokkoris, I.P.; Georgiadis, C.; Kaimaris, D.; Dimopoulos, P.; Mitsopoulos, I.; Mallinis, G. National scale land cover classification for ecosystem services mapping and assessment, using multitemporal copernicus EO data and google earth engine. *Remote Sens.* **2020**, *12*, 3303. [CrossRef]

69. Tuvdendorj, B.; Zeng, H.; Wu, B.; Elnashar, A.; Zhang, M.; Tian, F.; Nabil, M.; Nanzad, L.; Bulkhbai, A.; Natsagdorj, N. Performance and the Optimal Integration of Sentinel-1/2 Time-Series Features for Crop Classification in Northern Mongolia. *Remote Sens.* **2022**, *14*, 1830. [CrossRef]
70. Fremout, T.; Cobián-De Vinatea, J.; Thomas, E.; Huaman-Zambrano, W.; Salazar-Villegas, M.; Limache-de la Fuente, D.; Bernardino, P.N.; Atkinson, R.; Csaplovics, E.; Muys, B. Site-specific scaling of remote sensing-based estimates of woody cover and aboveground biomass for mapping long-term tropical dry forest degradation status. *Remote Sens. Environ.* **2022**, *276*, 113040. [CrossRef]
71. Amini, S.; Saber, M.; Rabiei-Dastjerdi, H.; Homayouni, S. Urban Land Use and Land Cover Change Analysis Using Random Forest Classification of Landsat Time Series. *Remote Sens.* **2022**, *14*, 2654. [CrossRef]
72. Tucker, C.J. Red and photographic infrared linear combinations for monitoring vegetation. *Remote Sens. Environ.* **1979**, *8*, 127–150. [CrossRef]
73. Gao, B.-C. NDWI—A normalized difference water index for remote sensing of vegetation liquid water from space. *Remote Sens. Environ.* **1996**, *58*, 257–266. [CrossRef]
74. Zha, Y.; Gao, J.; Ni, S. Use of normalized difference built-up index in automatically mapping urban areas from TM imagery. *Int. J. Remote Sens.* **2003**, *24*, 583–594. [CrossRef]
75. Huete, A.R. A soil-adjusted vegetation index (SAVI). *Remote Sens. Environ.* **1988**, *25*, 295–309. [CrossRef]
76. Huete, A.; Didan, K.; Miura, T.; Rodriguez, E.P.; Gao, X.; Ferreira, L.G. Overview of the radiometric and biophysical performance of the MODIS vegetation indices. *Remote Sens. Environ.* **2002**, *83*, 195–213. [CrossRef]
77. Johansen, K.; Phinn, S. Mapping structural parameters and species composition of riparian vegetation using IKONOS and Landsat ETM+ data in Australian tropical savannahs. *Photogramm. Eng. Remote Sens.* **2006**, *72*, 71–80. [CrossRef]
78. Fassnacht, F.E.; Latifi, H.; Stereńczak, K.; Modzelewska, A.; Lefsky, M.; Waser, L.T.; Straub, C.; Ghosh, A. Review of studies on tree species classification from remotely sensed data. *Remote Sens. Environ.* **2016**, *186*, 64–87. [CrossRef]
79. Mohammadpour, P.; Viegas, D.X.; Viegas, C. Vegetation Mapping with Random Forest Using Sentinel 2 and GLCM Texture Feature—A Case Study for Lousã Region, Portugal. *Remote Sens.* **2022**, *14*, 4585. [CrossRef]
80. Haralick, R.M.; Shanmugam, K.; Dinstein, I.H. Textural features for image classification. *IEEE Trans. Syst. Man Cybern.* **1973**, *SMC-3*, 610–621. [CrossRef]
81. Kupidura, P. The comparison of different methods of texture analysis for their efficacy for land use classification in satellite imagery. *Remote Sens.* **2019**, *11*, 1233. [CrossRef]
82. Tassi, A.; Vizzari, M. Object-oriented lulc classification in google earth engine combining snic, glcm, and machine learning algorithms. *Remote Sens.* **2020**, *12*, 3776. [CrossRef]
83. Zhang, X.; Zeraatpisheh, M.; Rahman, M.M.; Wang, S.; Xu, M. Texture is important in improving the accuracy of mapping photovoltaic power plants: A case study of Ningxia Autonomous Region, China. *Remote Sens.* **2021**, *13*, 3909. [CrossRef]
84. Breiman, L. Random forests. *Mach. Learn.* **2001**, *45*, 5–32. [CrossRef]
85. Pelletier, C.; Valero, S.; Inglada, J.; Champion, N.; Dedieu, G. Assessing the robustness of Random Forests to map land cover with high resolution satellite image time series over large areas. *Remote Sens. Environ.* **2016**, *187*, 156–168. [CrossRef]
86. Belgiu, M.; Drăguţ, L. Random forest in remote sensing: A review of applications and future directions. *ISPRS J. Photogramm. Remote Sens.* **2016**, *114*, 24–31. [CrossRef]
87. Hemmerling, J.; Pflugmacher, D.; Hostert, P. Mapping temperate forest tree species using dense Sentinel-2 time series. *Remote Sens. Environ.* **2021**, *267*, 112743. [CrossRef]
88. Wei, P.; Zhu, W.; Zhao, Y.; Fang, P.; Zhang, X.; Yan, N.; Zhao, H. Extraction of Kenyan Grassland Information Using PROBA-V Based on RFE-RF Algorithm. *Remote Sens.* **2021**, *13*, 4762. [CrossRef]
89. Rodriguez-Galiano, V.F.; Ghimire, B.; Rogan, J.; Chica-Olmo, M.; Rigol-Sanchez, J.P. An assessment of the effectiveness of a random forest classifier for land-cover classification. *ISPRS J. Photogramm. Remote Sens.* **2012**, *67*, 93–104. [CrossRef]
90. Talukdar, S.; Singha, P.; Mahato, S.; Pal, S.; Liou, Y.-A.; Rahman, A. Land-use land-cover classification by machine learning classifiers for satellite observations—A review. *Remote Sens.* **2020**, *12*, 1135. [CrossRef]
91. Li, W.; Dong, R.; Fu, H.; Wang, J.; Yu, L.; Gong, P. Integrating Google Earth imagery with Landsat data to improve 30-m resolution land cover mapping. *Remote Sens. Environ.* **2020**, *237*, 111563. [CrossRef]
92. Huang, W.; Li, W.; Xu, J.; Ma, X.; Li, C.; Liu, C. Hyperspectral Monitoring Driven by Machine Learning Methods for Grassland Above-Ground Biomass. *Remote Sens.* **2022**, *14*, 2086. [CrossRef]
93. Qi, S.; Song, B.; Liu, C.; Gong, P.; Luo, J.; Zhang, M.; Xiong, T. Bamboo Forest Mapping in China Using the Dense Landsat 8 Image Archive and Google Earth Engine. *Remote Sens.* **2022**, *14*, 762. [CrossRef]
94. Congalton, R.G. A review of assessing the accuracy of classifications of remotely sensed data. *Remote Sens. Environ.* **1991**, *37*, 35–46. [CrossRef]
95. Foody, G.M. Status of land cover classification accuracy assessment. *Remote Sens. Environ.* **2002**, *80*, 185–201. [CrossRef]
96. Rosenfield, G.H.; Fitzpatrick-Lins, K. A coefficient of agreement as a measure of thematic classification accuracy. *Photogramm. Eng. Remote Sens.* **1986**, *52*, 223–227.
97. Stehman, S.V. Selecting and interpreting measures of thematic classification accuracy. *Remote Sens. Environ.* **1997**, *62*, 77–89. [CrossRef]

98. Mostafa, E.; Li, X.; Sadek, M.; Dossou, J.F. Monitoring and Forecasting of Urban Expansion Using Machine Learning-Based Techniques and Remotely Sensed Data: A Case Study of Gharbia Governorate, Egypt. *Remote Sens.* **2021**, *13*, 4498. [CrossRef]
99. Olofsson, P.; Foody, G.M.; Herold, M.; Stehman, S.V.; Woodcock, C.E.; Wulder, M.A. Good practices for estimating area and assessing accuracy of land change. *Remote Sens. Environ.* **2014**, *148*, 42–57. [CrossRef]
100. Naboureh, A.; Li, A.; Ebrahimy, H.; Bian, J.; Azadbakht, M.; Amani, M.; Lei, G.; Nan, X. Assessing the effects of irrigated agricultural expansions on Lake Urmia using multi-decadal Landsat imagery and a sample migration technique within Google Earth Engine. *Int. J. Appl. Earth Obs. Geoinf.* **2021**, *105*, 102607. [CrossRef]
101. Tsujimoto, M.; Kajikawa, Y.; Tomita, J.; Matsumoto, Y. A review of the ecosystem concept—Towards coherent ecosystem design. *Technol. Forecast. Soc. Change* **2018**, *136*, 49–58. [CrossRef]
102. Tansley, A.G. The use and abuse of vegetational concepts and terms. *Ecology* **1935**, *16*, 284–307. [CrossRef]
103. Sparrow, B.D.; Edwards, W.; Munroe, S.E.; Wardle, G.M.; Guerin, G.R.; Bastin, J.F.; Morris, B.; Christensen, R.; Phinn, S.; Lowe, A.J. Effective ecosystem monitoring requires a multi-scaled approach. *Biol. Rev.* **2020**, *95*, 1706–1719. [CrossRef] [PubMed]
104. Vogelmann, J.E.; Xian, G.; Homer, C.; Tolk, B. Monitoring gradual ecosystem change using Landsat time series analyses: Case studies in selected forest and rangeland ecosystems. *Remote Sens. Environ.* **2012**, *122*, 92–105. [CrossRef]
105. Odum, E.P.; Barrett, G.W. *Fundamentals of Ecology*; Saunders: Philadelphia, PA, USA, 1971; Volume 3.
106. Luken, J.O. *Directing Ecological Succession*; Springer Science & Business Media: Berlin/Heidelberg, Germany, 1990.
107. Coradini, K.; Krejčová, J.; Frouz, J. Potential of vegetation and woodland cover recovery during primary and secondary succession, a global quantitative review. *Land Degrad. Dev.* **2022**, *33*, 512–526. [CrossRef]
108. Chang, C.C.; Turner, B.L. Ecological succession in a changing world. *J. Ecol.* **2019**, *107*, 503–509. [CrossRef]
109. Peltzer, D.A.; Wardle, D.A.; Allison, V.J.; Baisden, W.T.; Bardgett, R.D.; Chadwick, O.A.; Condron, L.M.; Parfitt, R.L.; Porder, S.; Richardson, S.J. Understanding ecosystem retrogression. *Ecol. Monogr.* **2010**, *80*, 509–529. [CrossRef]
110. Yan, J.; Zhang, Y.; Bai, W.; Liu, Y.; Bao, W.; Liu, L.; Zheng, D. Land cover changes based on plant successions: Deforestation, rehabilitation and degeneration of forest in the upper Dadu River watershed. *Sci. China Ser. D Earth Sci.* **2005**, *48*, 2214–2230. [CrossRef]
111. Bergen, K.M.; Dronova, I. Observing succession on aspen-dominated landscapes using a remote sensing-ecosystem approach. *Landsc. Ecol.* **2007**, *22*, 1395–1410. [CrossRef]
112. Temesgen, H.; Nyssen, J.; Zenebe, A.; Haregeweyn, N.; Kindu, M.; Lemenih, M.; Haile, M. Ecological succession and land use changes in a lake retreat area (Main Ethiopian Rift Valley). *J. Arid. Environ.* **2013**, *91*, 53–60. [CrossRef]
113. Friedman, J.H. Greedy function approximation: A gradient boosting machine. *Ann. Stat.* **2001**, *29*, 1189–1232. [CrossRef]
114. De'ath, G.; Fabricius, K.E. Classification and regression trees: A powerful yet simple technique for ecological data analysis. *Ecology* **2000**, *81*, 3178–3192. [CrossRef]
115. Cortes, C.; Vapnik, V. Support-vector networks. *Mach. Learn.* **1995**, *20*, 273–297. [CrossRef]
116. Chen, Z.; Kang, Y.; Sun, Z.; Wu, F.; Zhang, Q. Extraction of Photovoltaic Plants Using Machine Learning Methods: A Case Study of the Pilot Energy City of Golmud, China. *Remote Sens.* **2022**, *14*, 2697. [CrossRef]
117. Burai, P.; Deák, B.; Valkó, O.; Tomor, T. Classification of Herbaceous Vegetation Using Airborne Hyperspectral Imagery. *Remote Sens.* **2015**, *7*, 2046–2066. [CrossRef]
118. Ghayour, L.; Neshat, A.; Paryani, S.; Shahabi, H.; Shirzadi, A.; Chen, W.; Al-Ansari, N.; Geertsema, M.; Pourmehdi Amiri, M.; Gholamnia, M.; et al. Performance Evaluation of Sentinel-2 and Landsat 8 OLI Data for Land Cover/Use Classification Using a Comparison between Machine Learning Algorithms. *Remote Sens.* **2021**, *13*, 1349. [CrossRef]
119. Saboori, M.; Homayouni, S.; Shah-Hosseini, R.; Zhang, Y. Optimum Feature and Classifier Selection for Accurate Urban Land Use/Cover Mapping from Very High Resolution Satellite Imagery. *Remote Sens.* **2022**, *14*, 2097. [CrossRef]
120. Rapinel, S.; Mony, C.; Lecoq, L.; Clement, B.; Thomas, A.; Hubert-Moy, L. Evaluation of Sentinel-2 time-series for mapping floodplain grassland plant communities. *Remote Sens. Environ.* **2019**, *223*, 115–129. [CrossRef]
121. Xiong, K.; Adhikari, B.R.; Stamatopoulos, C.A.; Zhan, Y.; Wu, S.; Dong, Z.; Di, B. Comparison of Different Machine Learning Methods for Debris Flow Susceptibility Mapping: A Case Study in the Sichuan Province, China. *Remote Sens.* **2020**, *12*, 295. [CrossRef]
122. Liu, S.; Qi, Z.; Li, X.; Yeh, A.G.-O. Integration of convolutional neural networks and object-based post-classification refinement for land use and land cover mapping with optical and SAR data. *Remote Sens.* **2019**, *11*, 690. [CrossRef]
123. Holl, K.D.; Aide, T.M. When and where to actively restore ecosystems? *For. Ecol. Manag.* **2011**, *261*, 1558–1563. [CrossRef]
124. Ouyang, Z.; Zheng, H.; Xiao, Y.; Polasky, S.; Liu, J.; Xu, W.; Wang, Q.; Zhang, L.; Xiao, Y.; Rao, E. Improvements in ecosystem services from investments in natural capital. *Science* **2016**, *352*, 1455–1459. [CrossRef] [PubMed]
125. Li, W.; Wang, W.; Chen, J.; Zhang, Z. Assessing effects of the Returning Farmland to Forest Program on vegetation cover changes at multiple spatial scales: The case of northwest Yunnan, China. *J. Environ. Manag.* **2022**, *304*, 114303. [CrossRef] [PubMed]
126. Calderón-Loor, M.; Hadjikakou, M.; Bryan, B.A. High-resolution wall-to-wall land-cover mapping and land change assessment for Australia from 1985 to 2015. *Remote Sens. Environ.* **2021**, *252*, 112148. [CrossRef]
127. Li, W.; Xue, P.; Liu, C.; Yan, H.; Zhu, G.; Cao, Y. Monitoring and Landscape Dynamic Analysis of Alpine Wetland Area Based on Multiple Algorithms: A Case Study of Zoige Plateau. *Sensors* **2020**, *20*, 7315. [CrossRef]
128. You, N.; Dong, J. Examining earliest identifiable timing of crops using all available Sentinel 1/2 imagery and Google Earth Engine. *ISPRS J. Photogramm. Remote Sens.* **2020**, *161*, 109–123. [CrossRef]

MDPI AG
Grosspeteranlage 5
4052 Basel
Switzerland
Tel.: +41 61 683 77 34

Remote Sensing Editorial Office
E-mail: remotesensing@mdpi.com
www.mdpi.com/journal/remotesensing

www.ingramcontent.com/pod-product-compliance
Lightning Source LLC
Chambersburg PA
CBHW040828250926
51850CB00007BA/1286
* 9 7 8 3 7 2 5 8 7 1 7 0 4 *